Techniques
in Photomorphogenesis

Biological Techniques Series

J. E. TREHERNE
Department of Zoology
University of Cambridge
England

P. H. RUBERY
Department of Biochemistry
University of Cambridge
England

Ion-sensitive Intracellular Microelectrodes, *R. C. Thomas*, 1978
Time-lapse Cinemicroscopy, *P. N. Riddle*, 1979
Immunochemical Methods in the Biological Sciences: Enzymes and Proteins,
    R. J. Mayer and *J. H. Walker*, 1980
Microclimate Measurement for Ecologists, *D. M. Unwin*, 1980
Whole-body Autoradiography, *C. G. Curtis, S. A. M. Cross,*
    R. J. McCulloch and *G. M. Powell*, 1981
Microelectrode Methods for Intracellular Recording and Ionophoresis,
    R. D. Purves, 1981
Red Cell Membranes—A Methodological Approach, *J. C. Ellory* and
    J. D. Young, 1982
Techniques of Flavonoid Identification, *K. R. Markham*, 1982
Techniques of Calcium Research, *M. V. Thomas*, 1982
Isolation of Membranes and Organelles from Plant Cells, *J. L. Hall* and
    A. L. Moore, 1983
Intracellular Staining of Mammalian Neurones, *A. G. Brown and*
    R. E. W. Fyffe, 1984

Techniques in Photomorphogenesis

Edited by

Harry Smith

Department of Botany
University of Leicester
Leicester, England

M. G. Holmes

Smithsonian Environmental Research Center
Rockville, Maryland

1984

ACADEMIC PRESS

(Harcourt Brace Jovanovich, Publishers)

London   Orlando   San Diego   New York
Toronto   Montreal   Sydney   Tokyo

ACADEMIC PRESS, INC. (LONDON) LTD.
24-28 Oval Road,
London NW1 7DX

United States Edition published by
ACADEMIC PRESS, INC.
Orlando, Florida 32887

British Library Cataloguing in Publication Data

Main entry under title:

Techniques in photomorphogenesis.

  Includes index.
  1. Plants--Photomorphogenesis--Laboratory manuals.
I. Smith, Harry   II. Holmes, Martin Geoffrey.
QK757.T43   1984   581.3   84-11190
ISBN 0-12-652990-6 (alk. paper)

PRINTED IN THE UNITED STATES OF AMERICA

84 85 86 87   9 8 7 6 5 4 3 2 1

Contents

1
Introduction
Harry Smith

2
Criteria for Photoreceptor Involvement
H. Mohr

3
Light Sources
M. G. Holmes

11
Phytochrome in Membranes
Stanley J. Roux

12
Blue-Light Photoreceptor
Dieter Dörnemann and Horst Senger

Contributors

Numbers in parentheses indicate the pages on which the authors' contributions begin.

DIETER DÖRNEMANN (279), *Fachbereich Biologie/Botanik, Philipps-Universität Marburg, 3550 Marburg, West Germany*

LEONID FUKSHANSKY (109, 131), *Institute for Biology II, University of Freiburg, Schänzlestrasse 1, D-7800 Freiburg, West Germany*

J. GROSS (131), *Institute for Biology II, University of Freiburg, Schänzlestrasse 1, D-7800 Freiburg, West Germany*

P. M. HAYWARD (159), *Department of Botany, University of Leicester, Leicester LE1 7RH, England*

M. G. HOLMES[1] (43, 81), *Smithsonian Environmental Research Center, Rockville, Maryland 20852*

H. MOHR (13), *Institute for Biology II, University of Freiburg, Schänzlestrasse 1, D-7800 Freiburg, West Germany*

LEE H. PRATT (175, 201), *Department of Botany, University of Georgia, Athens, Georgia 30602*

STANLEY J. ROUX (257), *Department of Botany, The University of Texas at Austin, Austin, Texas 78712*

EBERHARD SCHÄFER (109, 131), *Institute for Biology II, University of Freiburg, Schänzlestrasse 1, D-7800 Freiburg, West Germany*

HUGO SCHEER (227), *Botanisches Institut der Universitaet, D-8000 Muenchen 19, West Germany*

HORST SENGER (279), *Fachbereich Biologie/Botanik, Philipps-Universität Marburg, 3550 Marburg, West Germany*

M. SEYFRIED (131), *Institute for Biology II, University of Freiburg, Schänzlestrasse 1, D-7800 Freiburg, West Germany*

HARRY SMITH (1), *Department of Botany, University of Leicester, Leicester LE1 7RH, England*

[1]Present address: 49 Markby Way, Lower Earley, Reading, Berks, England

ix

Preface

Photomorphogenesis, the control of growth and development by light, is studied for many different purposes. For some investigators, photomorphogenesis provides an elegant experimental approach to the elucidation of the control mechanisms that underly development itself. By varying the quantity, spectral quality, direction, and timing of the actinic radiation, profound changes in the pattern of plant development may be induced. Such precise and effective developmental control, using an environmental stimulus that may be applied and removed without residuum or damage, offers a unique opportunity to construct unambiguous tests of models and hypotheses, unconstrained by the doubts that always accompany alternative approaches, such as those involving chemical or surgical treatments. To others, it is the special photochemistry of the photoreceptors themselves that is the attraction; thousands of biological molecules absorb light, but only a handful have been selected through evolution to act as sensors of the radiation environment. Photomorphogenesis in plants offers us two photosensors for study, one that appears to be present in all classes of living organisms except viruses, the other unique to green plants and possessed of a remarkable degree of functional photochromicity. Yet other researchers of photomorphogenesis are concerned with broad ecological issues of the relationship between plants and their environment. Light in the natural environment is obviously of major importance in the life of the plant, with unpredictable climatic and vegetational fluctuations imposed upon a background of regular, predictable variations in timing, quantity, spectral quality, and direction. The information content of the light environment is considerable, and photomorphogenesis is seen by some as merely the means whereby the information is perceived and translated into physiological responses which, in turn, determine the plant's ecological relationships.

Photomorphogenesis also has much of applied interest, particularly for those interested in topics such as the efficiency of light interception, the effects of plant density in crop stands, and the design of controlled environment facilities. Even certain of those who construct instruments for the measurement of light have recently begun to realise that light is not only an energy source for plants and that wavelengths outside the 400–700-nm range ought to be considered.

Finally, there are those—like the editors of this book and many, if not all, of its contributors—for whom the study of photomorphogenesis has become a *raison d'etre,* a discrete topic of investigation, of consuming interest and continuing challenge, worthy in its own right of the investment of one's total experimental effort.

Although put together by the members of this latter group, who might perhaps consider themselves immodestly as cognoscenti of an esoteric field, the book is definitely not intended to be used only by photomorphogeneticists. On the contrary, the principal purpose of the book is evangelical, to spread the methodology and technology of photomorphogenesis widely and to make it possible for those intending to use light in their experiments to do so in proper accordance with current practice. It is profoundly depressing to reflect on the amount of otherwise valuable research that has been rendered vague and equivocal by lack of attention to the nature of the light treatments employed. Similarly, an apparently deep-seated reluctance to use standardized nomenclature and internationally recognized units and symbols has contributed immeasurably to the confused perceptions which sometimes seem to abound among those to whom light is just something to switch on and off. It is hoped that this book will provide a coherent and authoritative guide for all those intending, for whatever purpose, to carry out experiments on the control of plant development by light. It is intended to be a laboratory manual, with practical advice taking precedence over theory, although in some cases an exposition of the theory underlying the techniques has been regarded as essential.

The book began with one of the editors (H.S.) attempting to write it himself; he quickly realised that photomorphogenesis is so broad a topic that no one person could deal authoritatively with all the techniques being used. Once the decision was made to move to an edited book, he soon realised that for each topic there was someone else more suited to write it than he was himself! The contributors have been particularly patient during the 3 or 4 years it has taken to compile the book. Because of editorial inertia, solved only by the recruitment of the second editor (M.G.H.), many of the authors found themselves needing to rewrite their chapters extensively. We are, therefore, grateful to all of the contributors for their patience, fortitude, and continuing cooperation.

Harry Smith
M. Geoffrey Holmes

1
Introduction
Harry Smith

I. DEFINITION AND SCOPE OF PHOTOMORPHOGENESIS

Photomorphogenesis is not an easy term to define. The traditional approach has been to consider "photomorphogenesis" to mean "the effects of light on plant development." Thus Mohr, in "Lectures in Photomorphogenesis," published in 1972, defined photomorphogenesis as "control . . . (by light of) . . . growth and differentiation (and therewith development) of a plant independently of photosynthesis." Similarly, Smith, in "Phytochrome and Photomorphogenesis," published in 1975, using the same approach in a rather more restricted manner, defined photomorphogenesis as "developmental responses to nondirectional, nonperiodic, light stimuli." The two definitions differ only to the extent that the latter sought to exclude phototropism and photoperiodism from being considered part of photomorphogenesis. In both cases, light is seen as an environmental "effector," eliciting diversions from a basic developmental groundplan, which is, by implication, the pattern of development exhibited by plants growing in darkness. It seems somehow ridiculous to consider the developmental pattern of an etiolated plant as the fundamental groundplan against which the developmental effects of light should be contrasted, and in recent years, our view of photomorphogenesis has become more sophisticated. It is now general practice to use the term photomorphogenesis in an umbrellalike manner to denote all those manifold ways in which plants sense and react developmentally to the fluctuating light environment. For example, Mohr now distinguishes between two strategies of development, "photomorphogenesis" and "skotomorphogenesis" (Mohr and Shropshire, 1983). Photomorphogenesis is the strategy of development if, and as long as, light is available, whereas skotomorphogenesis is the strategy of development in darkness (i.e., etiolation). On this basis, all development in the light, even aspects which are not affected directly by light, is considered to fall into the

TECHNIQUES IN PHOTOMORPHOGENESIS
0-12-652990-6

category of photomorphogenesis. A different approach, but based on an analogous principle is, again, that of Smith (1982) who emphasizes the information content of the natural light environment; in his terms, photomorphogenesis is seen as being essentially those processes through which the plant acquires and interprets environmental information. A crucially important point is that light is only an "elective factor" (Mohr, 1972; Mohr and Shropshire, 1983). Even though the light environment to which a plant is exposed may contain a great deal of information, that information does not carry any specificity with respect to morphogenesis and must be interpreted in accordance with the genetic and epigenetic constraints extant within the reactive cells. In other words, the developmental reaction of a given plant to a particular light stimulus is a function of the expression of genetic information, and different plants — or even a single plant at different stages of developmental competence — may "elect" different patterns of gene expression as reactions to an identical stimulus.

It is as difficult to describe collectively the activities of those who would regard themselves as investigators into photomorphogenesis as it is to produce a generally acceptable definition of the term itself. Photomorphogeneticists include: organic chemists, photochemists, biophysicists, biological mathematicians, molecular biologists, biochemists, geneticists, physiologists, and ecologists, plus agriculturalists and horticulturalists on the applied side. Photomorphogenesis, as a topic of study, is as multidisciplinary as it is possible to be. The common factor amongst this diverse range of expertise is an interest in, and often a fascination for, the remarkable products of chemical evolution which we know as "photoreceptors." The first law of photochemistry (the Grotthus–Draper law) states that only absorbed light can cause photochemical reactions (see Chapter 6). Consequently, since each photobiological process is necessarily the biological outcome of a photochemical reaction, the pivotal agents in photomorphogenesis are those pigments which absorb light, become photochemically active, and thereby initiate the developmental changes. Chapter 2 presents the criteria by which at least two, and perhaps three, distinct photoreceptors operative in photomorphogenesis may be recognized, whilst the second half of the book (i.e., Chapters 8–12) detail the principal techniques used to study the specific molecular properties of these photoreceptors.

When considered against the background of the many thousands of kinds of light-absorbing molecules (pigments) in the biosphere, it may seem at first sight surprising that only a small number, perhaps ten at the most, serve as photoreceptors. There are good reasons, however, why the potential to act as a biological photoreceptor is rare (see Presti and Delbruck, 1978). Table I gives a list of the classes of photoreceptors which have been identified; even in this restricted list, it can be seen that some of the examples are chemically

Table I. Known biological photoreceptors.[a]

UV-reactivating enzyme
Provitamins D
Rhodopsin
Bacteriorhodopsin
Chlorophyll
Bacteriochlorophyll
Protochlorophyll
Haem pigments
Flavoproteins
Carotenoids
Phytochrome

[a] From Presti and Delbruck (1978).

closely related, emphasizing the conservative nature of photoreceptor evolution. It is also instructive to note that most of the listed photoreceptors are operative in plants — perhaps an indication of the importance to plants of possessing mechanisms for sensing the light environment. The common aim of those investigating photomorphogenesis is to elucidate the role and function of the individual photoreceptors in the information – transduction chain between the environmental signals and the selective regulation of the expression of genetic information and to place the understanding so gained in a proper ecological and evolutionary context.

The scope of this book is therefore very wide indeed, ranging from methods of analyzing the natural light environment to techniques for the chemical analysis of photoreceptor chromophores. Many techniques have been borrowed or adapted from other disciplines, and none, except perhaps for the basic red – far-red photoreversibility test for phytochrome (see Chapter 2), can be considered to be exclusively applicable to photomorphogenesis. Some techniques, such as those involved in the purification and immunological characterization of photoreceptors (Chapters 8 and 9), are developed from standard protein biochemistry procedures, whilst others, such as action spectroscopy (Chapter 6) and *in vivo* spectroscopy (Chapter 7), are applicable within the whole field of photobiology. Remaining chapters provide comprehensive information on the generation (Chapter 5) and the measurement (Chapters 3 and 4) of light, on membrane biochemistry as applied to phytochrome (Chapter 11), on phytochrome chromophore analysis (Chapter 10), and on the special problems of investigating the ubiquitous but elusive blue-light-absorbing photoreceptor (Chapter 12).

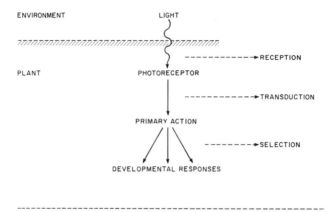

Fig. 1. The partial processes of photomorphogenesis.

II. THE PARTIAL PROCESSES OF PHOTOMORPHOGENESIS

Thinking about photomorphorgenesis in terms of the acquisition and inter-pretation of environmental information allows the identification of three principal partial processes (see Fig. 1). First, the information contained within the incident light is captured by the process known as "photorecep-tion." This occurs when specific photoreceptor molecules absorb photons and become photochemically active. Photoreception, therefore, represents the essential communication link across the interface between the environ-ment and the plant. The information acquired by photoreception is then transformed into a form interpretable within the cell. This partial process is generally termed "transduction" and is commonly equated with the primary actions of the photoreceptor, the details of which are in each case still elu-sive. The final, interpretative, partial process is the "selection" of appropri-ate pathways of metabolism and development. In certain situations, such as the photoperiodic induction of flowering, the selection of gene expression at the genomic level would seem to be a necessary step in the overall photo-morphogenetic process; in other cases, however, such as the very rapid control of stem growth by blue light (Cosgrove, 1981), any genome-level regulation is likely to be a consequence, rather than a cause, of the develop-mental change.

III. EXPERIMENTAL APPROACHES TO THE STUDY OF PHOTOMORPHOGENESIS

From the scheme in Fig. 1, three major categories of experimental approach to the study of photomorphogenesis may be recognized: (a) analysis of the light signals received by the photoreceptors; (b) investigations into the structure, properties, and molecular mechanisms of action of the photoreceptors; and (c) elaboration of the cascade of cellular events which presumably lie between photoreceptor action and the ultimate observed developmental responses.

A. Analysis and simulation of the light environment

It is clearly of prime importance to define the light environment to which the photoreceptors are exposed. To be effective, such investigations should include radiometric and spectroradiometric analysis and should be concerned with the light environment within, as well as outside, the plant tissues. Physiological experiments have shown that plants can acquire at least five different categories of information from the light environment, i.e., quantity, quality, direction, timing (periodicity), and state of polarization (see Table II). In the natural environment, both light quantity (i.e., fluence rate) and light quality (i.e., the spectral distribution of radiation) vary daily, seasonally, climatically, geographically, and topographically (e.g., as a function of the spatial arrangement or orientation of receptive surfaces). Variations in the direction of actinic radiation are experienced not only as a function of the "daily march of the sun," but also, for example, because of inhomogeneities of canopy-cover or partial burial under the soil. The periodicity of the natural light environment is perhaps the most predictable of all the parameters, varying only with latitude and season. The perception of light quantity, light quality, direction, and periodicity is of great importance in the plant's continual acclimation to its environment. Plants can also detect and react to polarized light, and the application of experimental procedures based on polarized light has been extremely important in elucidating some aspects of the biology of photoreceptor action (see Haupt, 1982); it seems likely, however, that such responses do not have a functional role in the natural environment and are no more than a manifestation of the intracellular localization of the photoreceptors.

Table II. Information that may be derived from the light environment.[a]

Information	Perception mechanism
Light quantity	Photon-counting
Light quality	Photon ratios
Direction	Photon gradients
Duration	Timing of light–dark transitions
Polarization	Dichroic photoreceptor arrangement

[a] From Smith (1983).

Partial processes	Techniques
Reception	Radiometry
	Spectroradiometry
	Simulation
Transduction	Action Spectroscopy
	In Vivo Spectroscopy
	Physiological Kinetic Analysis
	Photoreceptor Isolation
	Chromophore Chemistry
	Photochemistry
	Analytical Spectroscopy
	Flash Photolytic Spectroscopy
	Photoreceptor Molecular Biology
	Photoreceptor Localization
	Light-Induced Absorption Changes
Selection	*In Vivo* Dichroism Studies
	Biophysical Analysis
	Biochemical Analysis
	Growth Measurements

Fig. 2. Application of the range of techniques used in photomorphogenesis.

It cannot be stressed too strongly that photomorphogenesis is concerned with understanding the behaviour of plants in relation to the natural light environment. Most of the "classical" experiments in photomorphogenesis are based on techniques which have little relevance to the natural light environment, even though they have often proved to be very powerful investigative tools. To comprehend the perceptive function of a photoreceptor it is necessary to match its physical and biological properties with those of the natural environment. The role of phytochrome in the perception of canopy-shade is a relevant example of this approach. Vegetation canopies both reduce the fluence rate and change the spectrum of the light passing through them. The chlorophyll in the vegetation strongly attenuates the 400–700-nm wavelength bands, but allows most of the 700–800-nm radiation to pass through, thereby markedly depressing the ratio of red to far-red radiation. Comparison of the environmental changes with the spectral properties of phytochrome, in association with simulation experiments to test the developmental responses of plants to the natural range of red:far-red ratio, has led to the hypothesis that the fundamental function of phytochrome is the perception of environmental fluctuations in light quality (see Smith, 1982, for review). The main techniques employed in this general approach to photomorphogenesis are outlined in Fig. 2.

B. Photoreceptor structure and properties

Perhaps the most conspicuously successful investigations in photomorphogenesis have been those concerned with the structure and properties of
phytochrome; as yet, unfortunately, the blue-absorbing receptor has not
been found equally amenable to study. Even conceptually simple physiological experiments can provide information of great predictive value, as
exemplified by the brilliant action spectrum and photoreversibility studies of
S. B. Hendricks and H. A. Borthwick in the 1940s and 1950s. As detailed by
Briggs (1976), these studies allowed Hendricks and Borthwick not only to
propose the photochromic nature of phytochrome, but also to construct the
pigment's probable absorption spectrum and to predict the tetrapyrrole
nature of the chromophore. Hendricks and Borthwick also suggested, principally on the basis of the massive amplification observed in the responses to
small amounts of red light, that the photoreceptor was a protein. That all
these predictions have turned out to be accurate is a testament to the power
of the physiological approach. At the present time, the physiological approach to photoreceptor properties tends to be centered around action spectroscopy and kinetic analysis, together with *in vivo* spectroscopy (see Chapters 6 and 7).

Particularly in the case of phytochrome, direct physicochemical analysis
of the purified photoreceptor has in recent years proved very productive.
Much detailed information has been accumulated concerning the chemistry
of the chromophore and its photoconversions, the molecular weight, amino
acid composition, isoelectric point and immunochemical characterstics of
the apoprotein, and the spectral properties and transformation kinetics of
the whole molecule, both *in vivo* and *in vitro*. Rapid progress has been made
in investigations into the molecular biology of phytochrome, and it is likely
that soon the gene (or genes) for phytochrome will have been cloned. Such
investigations have as their motivation the entirely reasonable principle that
knowledge of the molecular properties of the photoreceptors will ultimately
lead to an understanding of how photon absorption is transduced into biological signals. Although a full theoretical treatment would be out of place
here, it is important to remember that perception of all four informational
parameters of the natural light environment, i.e., quantity, quality, direction
and periodicity, must in each case be a function of an essentially simple
initial event—the absorption of photons. Thus in light quantity perception, mechanisms must exist for registering the rate of photon absorption,
implying the counting of photons on a time basis. Similarly, the perception
of light quality must involve the comparison of the rates of photon absorption in at least two distinct wavebands, implying the involvement of either
two separate photoreceptors or two different forms of a single sensor. The

perception of the direction of the actinic beam clearly requires comparison of the rates of photon absorption in different parts of a nonuniformly illuminated cell or organ.   Finally, photoperiodic perception, at the simplest level, requires measuring both photon absorption rate and elapsed time.   A central problem in photomorphogenesis, therefore, is to understand the molecular mechanism(s) whereby the rate of photon absorption is transduced into a quantitative cellular signal.   In spite of the accumulated data on the physico–chemical properties of phytochrome, this problem has, as yet, only been approached by formal, and largely nonquantitative, modeling techniques, and explanations at the molecular level are singularly lacking.

C. Cellular and developmental responses

From a historical point of view, the majority of investigations into photomorphogenesis have been descriptive observations of responses of plants, or parts of plants, to defined — or, more often, ill-defined — light treatments. Although research of this nature is valuable in establishing the "phenomenology" of photomorphogenesis, much of it is destined to lie forgotten within the pages of learned journals, partly because of the prevailing lack of control over growth and treatment conditions, but also because experiments have generally been designed only to collect and amass the details of responses, rather than to analyze those responses in relation to some hypothesis of photoreceptor action.   There are, of course, notable exceptions to this sweeping generalization, but the record of research into the cellular and developmental aspects of photomorphogenesis is very limited in terms of contributions of lasting value.   The principal objective should be to elucidate the successive steps in the cascade of events between the primary actions of the photoreceptors and the developmental responses.   The obvious approach is kinetic, i.e., to search for those cellular events which occur in rapid temporal succession after the onset of any light treatment.   The application of this approach to studies of phytochrome action is detailed admirably by Quail (1983), who nevertheless shows that even though events within a few seconds of photon absorption have been documented, it is still not possible even to make the relatively broad conclusion that phytochrome action involves modification of membrane properties.

There are major questions of principle — and considerable controversy — at this level of research into photomorphogenesis.   For example, there is not even agreement over whether there is a single primary action of phytochrome, the alternative view being that the different response kinetics observed can only be explained on the basis of multiple "primary" actions (see Quail, 1983, for discussion).   This contentious issue spills over into disagreement over the relative importance of genetic and epigenetic control mechanisms in photomorphogenesis. Whilst it is obvious that developmental changes of the magnitude of the induction of flowering necessarily involve the selective regulation of gene expression at the genome level, the development of plants is so plastic that many other overtly developmental changes — such as, perhaps, the photocontrol of stem extension — may be mediated through controls at a considerably "lower" level of organization. The application of the modern techniques of molecular biology, which have already revealed the transcriptional control of the synthesis of a small number of enzymes in photomorphogenesis, will help to resolve some of these currently intransigent problems.

IV. THE PHOTOMORPHOGENESIS LABORATORY

To most of those working in the field of photomorphogenesis it has always seemed somewhat unjust that those who elect to study the responses of plants to light have to spend most of their working life in the dark! The darkroom, however, is the first essential for research into photomorphogenesis. It need not be elaborate; in fact, it is probably true that most of the best work has been done in simple, but well-designed, facilities. The dark laboratory must normally serve two purposes: to provide a temperature-controlled environment in which plant materials may be grown in the absence of light, and to allow experiments to be performed on that material using defined light sources. The first of these functions requires stray light to be minimized, whilst the second clearly produces stray light! Although good work can be done in a single, dual-purpose laboratory, great care is needed to prevent light from experimental treatment sources reaching plants being grown up for experimentation. Consequently, the more convenient approach is to divide the available space into at least two separate rooms, one of which is temperature-controlled and used solely as a dark growth room, whilst the other can be used for experimental treatments without stray light causing any problems. More elaborate designs are clearly possible if space and funds permit; for example, there are definite advantages in an interconnecting, three-room arrangement, with one room for plant growth, one for the experimental light sources, and a further room for the manipulation of plant materials, including spectrophotometry. Going further, for those interested in photoreceptor isolation, or in the biochemical effects of light treatment, it is extremely valuable to have a dark cold-room with direct access from the dark laboratory suite. Access to the dark laboratory should be so constructed that persons carrying trays or pushing trolleys may enter without allowing the simultaneous entry of stray light. The most effective light trap is a baffle consisting of a separate antechamber with an inner and outer door, the chamber being large enough to accommodate a laboratory trolley. If space is limited, however, a black curtain, hung on a semicircular track enclosing an area outside the access door sufficient for a person plus a trolley, will often be found to be satisfactory. Rarely are photomorphogenesis laboratories constructed *de novo,* most having been established through modification of existing facilities. If suitable rooms for modification are not available, a simple solution is to purchase a free-standing controlled-environment chamber of the walk-in type, which can be erected within an existing laboratory. For most purposes, a chamber having an internal floor area of ~ 4 m², if efficiently shelved, would provide ample space for plant growth.

Within the dark laboratory the major problem, as mentioned above, is stray light. It is advisable to have all surfaces, with the exception of working surfaces, painted matt-black. Chapter 4 deals with the subject of safelights in detail. At this point, however, it should be stressed that no light is truly "safe" in the photomorphogenetic sense; consequently, all safelights should be on short delay switches. Experience has shown that most operations in the dark growth room — e.g., watering — can be accomplished within a minute or two, and thus it is not necessary to expose plants for longer periods; it is particularly important not to leave safelights on continuously. Many investigators now grow plants up in total darkness; if this is intended, it is generally advisable to use light-tight containers for the plants, rather than to rely on all personnel remembering not to turn on the safelight. Some seedlings, however, grow poorly in closed boxes and it is necessary to maintain the whole room in total darkness, even though some protection can be given by hanging black curtains over the shelving on which the plants are grown. When operating without safelights, it is clearly important to make sure there are no hazards in the darkroom and to ensure that all personnel are very familiar with the layout of the room — it is surprising how easy it is to become disoriented in total darkness. Because of their very nature, darkrooms rapidly become dirty and begin to harbour fungal infections; these can be prevented only by a regular routine of cleaning and decontamination. It is useful to have the rooms equipped with white-light sources for these operations, but the switches should be located in such a position that they are only activated intentionally. All service, maintenance, and security personnel should be made aware of the consequences of switching on lights or using hand torches in the dark laboratory.

REFERENCES

Briggs, W. R. (1976). *In* "Light and Plant Development" (H. Smith), pp. 1–6. Butterworth, London.

Cosgrove, D. J. (1981). *Plant Physiol.* **67**, 584–590.

Haupt, W. (1982). *Annu. Rev. Plant Physol.* **33**, 205–233.

Mohr, H. (1972). "Lectures on Photomorphogenesis." Springer-Verlag, Berlin and New York.

Mohr, H. and Shropshire, W., Jr. (1983). *Encycl. Plant Physiol. New Ser.* **16A**, 24–38.

Presti, D. and Delbruck, M. (1978). *Plant Cell Environ.* **1**, 81–100.

Quail, P. H. (1983). *Encycl. Plant Physiol., New. Ser.* **16A**, 178–212 (see Mohr and Shropshire, 1983).

Smith, H. (1975). "Phytochrome and Photomorphogenesis." (McGraw-Hill, London.)

Smith, H. (1982). *Annu. Rev. Plant Physiol.* **33**, 481–518.

Smith, H. (1983). *Symp. Soc. Exp. Biol* **36**, 1–18.

2
Criteria for Photoreceptor Involvement
H. Mohr

I. SENSOR PIGMENTS

The perception of specific photomorphogenetic light stimuli implies the operation of specific sensor pigments. Two of these "photoreceptors" are known to exist in plants: phytochrome and the blue/UV-A photoreceptor ("cryptochrome"). A third specific sensor pigment may be involved in the absorption of quanta in the UV-B range (UV-B photoreceptor).

A. The Blue/UV-A photoreceptor (cryptochrome)

Even though this photoreceptor is ubiquitous, being active in algae, mosses, liverworts, ferns, fungi, and higher plants, as well as animals, its molecular identity is not established unambiguously. While strong evidence points to its being a yellow flavoprotein (Senger, 1980; Drumm-Herrel and Mohr, 1982b; see also Chapter 11) it is still defined by its action spectrum, which exhibits three peaks (or shoulders) in the blue spectral range (400–500 nm) and another peak at approximately 370 nm in the UV-A waveband (320–400 nm) (Fig. 1). It is almost certain that both maxima of action (in the blue as well as in the UV-A range) are due to light absorption by the same molecule, very probably a flavoprotein. Action spectra very similar to the one in Fig. 1 have been determined for many responses throughout the plant Kingdom (see Senger, 1980).

B. The UV-B photoreceptor

At present, the UV-B photoreceptor, which was originally discovered by Wellmann (1971, 1974) in biochemical studies with cell suspension cultures

TECHNIQUES IN PHOTOMORPHOGENESIS
0-12-652990-6

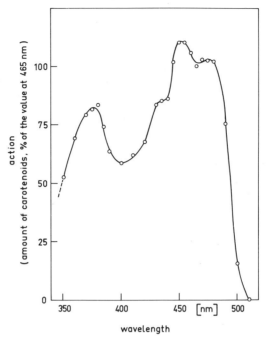

Fig. 1. An action spectrum for light-dependent carotenoid synthesis in the mycelium of the fungus *Fusarium aquaeductuum.* The amount of carotenoids which can be induced by a photon fluence of 4.2×10^{-3} mol photons m^{-2} is given as a function of wavelength. (From Rau, 1967.)

from parsley *(Petroselinum hortense),* can only be characterized experimentally. Parsley cell cultures form large amounts of flavone glycosides under white light of high fluence rate (Hahlbrock and Wellmann, 1970). The spectral responsivity of the light-mediated induction of flavone glycosides in the parsley cell suspension was determined with cutoff filters as well as with a monochromator (Table I). The data indicate that visible light is almost ineffective and that the peak of action is in the UV-B range (320–280 nm). The white fluorescent light which was used by Hahlbrock and Wellmann (1970) was effective because it contains short-wavelength UV. Wellmann showed that the UV-B effect is specific and is not due to nonspecific damage caused by UV-B absorption by proteins or nucleic acids.

The involvement of a specific UV-B effect has been postulated in phototropism of the *Avena* coleoptile (Curry and Jotcham, 1980) and in light-mediated anthocyanin synthesis in the mesocotyl of sorghum seedlings (Drumm-Herrel and Mohr, 1981; Yatsuhashi *et al.,* 1982). The action spectrum for anthocyanin formation in dark-grown sorghum shows a single intense peak at 290 nm and no action at wavelengths longer than 350 nm,

Table I. Estimation of the spectral range active in flavone glycoside accumulation in cells of parsley by means of UV-absorbing cut-on filters.[a]

Filter used $\lambda_H{}^b$ (nm)	Flavone glycosides $A_{380}{}^c$
435	0.13
385	0.19
345	0.31
320	0.73
280	0.90
Without	0.92
Dark control	0.12

[a] The samples were irradiated with white fluorescent light for 2 h and flavone glycosides measured after a further 20 h of darkness. (From Wellmann, 1971.)
[b] Wavelength at which transmittance of the filter is 50%.
[c] Standard extract: 1 g fresh weight/5 ml buffer.

provided that the operation of phytochrome (Pfr) is prevented by a long-wavelength far-red light-pulse treatment following the inductive monochromatic light (Fig. 2). It could be that the UV-B photoreceptor is ubiquitous but has been overlooked so far because of experimental difficulties in working with UV-B.

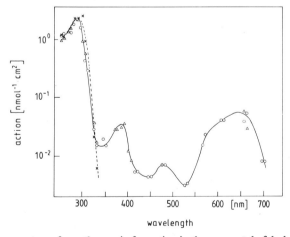

Fig. 2. An action spectrum for anthocyanin formation in the mesocotyl of dark-grown broom sorghum (*Sorghum bicolor* Moench, cv. Acme Broomcorn). Action is expressed by reciprocal of photon fluences required for anthocyanin formation of A 0.02. ——, Action spectrum with monochromatic light treatment alone. – – –, Action spectrum when monochromatic light treatments were followed immediately by a phytochrome-saturating pulse (5 min) with long wavelength (760 nm) far-red light. (From Yatsuhahi *et al.,* 1982.)

C. Phytochrome

Phytochrome is the best known sensor pigment in higher plants for the detection of photosignals from the environment and for making use of these signals to regulate the orderly growth and development of a plant. Phytochrome was predicted by S. B. Hendricks and H. A. Borthwick on the basis of purely physiological experiments (Borthwick *et al.*, 1952). As we know (Rüdiger, 1980), phytochrome is a bluish chromoprotein with photochromic properties. Its function depends directly on its property of photochromicity.

The essential features of the phytochrome system are represented by the model in Fig. 3. Phytochrome is a biliprotein existing in two forms which are interconvertible by light: Pr, which absorbs maximally in the red spectral range at around 665 nm; and Pfr, which absorbs maximally in the far-red spectral range around 725 nm. Pfr has commonly been considered to be physiologically active, and Pr inactive, although this point has been thoroughly investigated in only a few cases (e.g., Hendricks *et al.*, 1956; Drumm and Mohr, 1974; Oelze-Karow and Mohr, 1976; Schmidt and Mohr, 1982). The sum of the amount of Pr and Pfr is called "total phytochrome" or Ptot. In a dark-grown seedling, only Pr is present. Synthesis *de novo* of Pr is a zero-order process which was believed not to depend on light (Schäfer *et al.*, 1972). However, recently Gottmann and Schäfer (1982) found evidence — using the measurement of the activity in an *in vitro* translation system of poly-A-RNA coding for phytochrome apoprotein — that the capacity for phytochrome synthesis in light-grown *Avena sativa* seedlings is much smaller than that of dark-grown seedlings of the same age. After a light–dark transition, the activity slowly reappears, whereas after a dark–light transition, a rapid disappearance could be demonstrated. Pfr originates from Pr through a first-order phototransformation which is photoreversible. Pfr is not stable; it disappears through a first-order destruction process which is independent of light.

Both forms of phytochrome have broad overlapping absorption spectra throughout the visible range. This is the reason why a photoequilibrium is achieved if visible light is falling on a plant or on a phytochrome preparation in the test tube. If the fluence rate of the light is high (in the sense that k_1 and

$$\text{Pr}' \xrightarrow{\ {}^{0}k_s\ } \text{Pr} \underset{k_2}{\overset{k_1}{\rightleftharpoons}} \text{Pfr} \xrightarrow{\ k_d\ } \text{Pfr}'$$

Fig. 3. A model of the phytochrome system as it occurs in mustard seedling cotyledons and hypocotylar hook. The symbols ${}^{0}k_s$, k_1, k_2, and k_d represent the rate constants of *de novo* synthesis of Pr, back and forth phototransformations and Pfr destruction, respectively. (From Oelze-Karow and Mohr, 1976.)

k_2 are much greater than k_s and k_d) a true photoequilibrium of the phytochrome becomes established. The wavelength dependent photoequilibrium is usually defined by $\varphi_\lambda = $ [Pfr]λ/[Ptot]. It is rapidly established in the red and far-red range of the spectrum even at moderate photon fluence rates, e.g., with 5 min red (660 nm) or far-red (756 nm) light at a fluence rate of 7 W m^{-2} (see Schäfer and Mohr, 1980). Virtually only "light pulses" are required to establish photoequilibria of the phytochrome system in long-wavelength visible light.

Another important property of the phytochrome system is that it develops a steady state under continuous light (Schäfer et al., 1973). Based on the model in Fig. 3, any change of total phytochrome can be described by the equation

$$\frac{dPtot}{dt} = {}^{o}k_s - {}^{1}k_d \cdot [Pfr]$$

Under steady state conditions, i.e., "no change of Ptot," the rate of Pr synthesis is equal to the rate of Pfr destruction, ${}^{o}k_s = {}^{1}k_d \cdot$ [Pfr]. Therefore,

$$[Pfr] \text{ steady state} = {}^{o}k_s/{}^{1}k_d$$

In other words, as long as the model in Fig. 3 is valid, the steady state concentration of Pfr is only a function of the rate constants for Pr synthesis and Pfr destruction. This means that the steady state concentration of Pfr does not depend on the wavelength of the light incident on the system (provided that the action of Pfr on Pr synthesis is the same irrespective of wavelength). The essential prerequisite is, however, that the incident light is absorbed by both phytochrome forms to an extent sufficient to establish a steady state (see Schäfer and Mohr, 1974). This property of the phytochrome system offers the opportunity to establish the steady state in red and far-red light, including wavelengths which do not cause significant chlorophyll synthesis or photosynthesis. In this way, one can easily separate photomorphogenetic light effects and photosynthesis even in long-term experiments. For long-term irradiation, it is convenient to use a medium fluence rate (3.5 W m^{-2}) far-red light source which is equivalent, as far as phytochrome is concerned, to the wavelength 718 nm (standard far-red light, see Mohr, 1966; Oelze-Karow and Mohr, 1973).

Some properties of phytochrome will now be described in more detail because the criteria for phytochrome involvement in a photoresponse depend directly on these properties.

(1) Action spectra of photochemical transformations of Pr and Pfr
The respective action spectra of the photochemical transformations with purified phytochrome in solution are shown in Fig. 4. The action spectra in

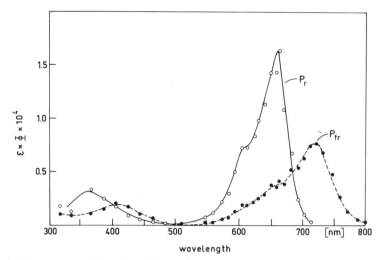

Fig. 4. Action spectra of photochemical transformations of Pr and Pfr. The molar extinction coefficient ϵ is in L mol^{-1} cm^{-1} and the quantum yield Φ is in mol mol photons^{-1}. (From Butler *et al.*, 1964.)

solution are given as the product of the molar extinction coefficient ϵ and the quantum efficiency Φ as a function of wavelength. Since the phototransformations *in vitro* are known to be first order, the product $\epsilon \cdot \Phi$ can be determined at any wavelength from the first-order rate constants. The value of the first-order rate constant has been determined experimentally from the initial slope of a semilogarithmic plot of percent conversion versus time of irradiation.

(2) Photoequilibrium of phytochrome as a function of wavelength
Since the photoequilibrium of the photychrome system at wavelength λ is defined as

$$\varphi_\lambda = \frac{\epsilon_r \Phi_r}{\epsilon_r \Phi_r + \epsilon_{fr} \Phi_{fr}}$$

the data in Fig. 4 can be used to calculate φ_λ. It was found that $\varphi_{660\,nm}$ is close to 0.8. Later *in vitro* estimates (Pratt, 1975, 1978) were slightly lower (0.75). Studies with undegraded phytochrome from different sources have confirmed that φ_{660}, the reference value, is close to 0.8 (Yamamoto and Smith, 1981; W. O. Smith, personal communication). Figure 5 summarizes data about φ_λ values measured *in vivo* by different methods in the red and far-red range. A φ_{660}-nm value of 0.8 was used as a reference. The major result is clear: If the photoequilibrium is established with red light the Pfr/

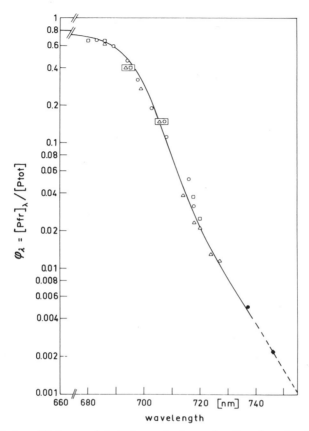

Fig. 5. The photoequilibrium φ_λ of phytochrome *in vivo* as a function of wavelength. The data are adopted from Hanke *et al.* (1969) (*Sinapis alba* hypocotylar hook, spectrophotometric measurements, □), Oelze-Karow and Mohr (1973) (*Sinapis alba,* hypocotylar hook, physiological assay, △), and Quail and Schäfer (1974) (*Zea mays* coleoptile, spectrophotometric measurement, ○). ●, calculated from physiological measurements with *Zea mays* and *Avena sativa* coleoptiles (Schäfer *et al.,* 1975). The value for 746 nm, ◆, was obtained by extrapolation. (From Schäfer *et al.,* 1975.)

Ptot ratio is of the order of 0.8; if the photoequilibrium is established with far-red light (λ_{max} = 720 nm) the Pfr/Ptot ratio is of the order of 0.03; if the photoequilibrium is established with long-wavelength far-red light (λ_{max} = 756 nm) the Pfr/Ptot ratio is less than 0.01.

The situation in the blue waveband is less clear. This is partly due to the low $\epsilon \cdot \Phi$ values in this spectral range (Fig. 4) and partly caused by increased scattering and light attenuation by carotenoid and flavin absorption which make measurements of φ_λ values difficult and less reliable in blue and UV light than in red and far-red light.

II. CRITERIA FOR PHYTOCHROME INVOLVEMENT

A. Action spectra for induction of a photoresponse

Action spectra of photoresponses (see Chapter 6) constitute the most power-ful tool to identify a sensor pigment *in vivo*. A simple mathematical treat-ment shows that an action spectrum (quantum effectiveness as a function of wavelength) represents the extinction (or absorption) spectrum of the photo-receptor pigment involved if the Law of Reciprocity is valid (i.e., the degree of the photoresponse is determined only by the product of fluence rate and time of irradiation) and if several further prerequisites which can be tested are fulfilled.   It is obvious that identical action spectra of different photore-sponses strongly suggest mediation by the same photoreceptor (see also Chapter 6).

The action spectrum for opening of the plumular hook in bean seedlings (Fig. 6) shows that it is mainly light between 550 and 700 nm which induces hook opening.   Red light around 660 nm is the most effective.   Light beyond 700 nm hardly induces hook opening.   At first glance it seems that this long wavelength light is without effect.   However, if a maximum hook opening is induced with a red-light pulse and then a corresponding fluence is applied at a wavelength above 700 nm, e.g., 730 nm, virtually no hook opening occurs.   Clearly, the induction of hook opening by red light can be nullified by a subsequent treatment with 730-nm light, and is thus "revers-ible."   The action spectrum for reversion (Fig. 6) shows light between about

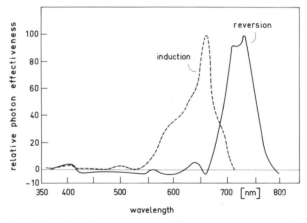

Fig. 6. Typical action spectra of phytochrome-mediated photoresponses, elaborated under induction conditions (Law of Reciprocity valid).   These spectra were determined by Withrow *et al.* (1957) for the light-dependent opening of the plumular hook of the bean seedling.   They are representative of other action spectra of responses induced by phytochrome, lettuce seed germination and anthocyanin synthesis being examples.

700 and 800 nm to be effective (far-red light). The highest quantum effectiveness is around 730 nm. The essential point is that the action spectra in Fig. 6 are explained by the action spectra of photochemical transformations of Pr and Pfr (Fig. 4). This explanation is straightforward if Pfr is assumed to be the active form (the effector molecule) of the phytochrome system. This assumption has been justified repeatedly (see Schmidt and Mohr, 1982). To generalize: if action spectra of induction and reversion of induction of a photoresponse are identical to (or at least similar to) the action spectra shown in Fig. 6, involvement of phytochrome is indicated. Since the elaboration of action spectra is not only time-consuming but also depends on the availability of sophisticated equipment, the criteria for the involvement of phytochrome have been simplified.

B. Criteria for phytochrome involvement in responses caused by light pulses

(1) Etiolated seedlings

The criteria are summarized in Table II. If induction of a photoresponse by a saturating red-light pulse ($\varphi_{red} \approx 0.8$) is fully reversed by following it imme-

Table II. Criteria for the involvement of phytochrome in a photoresponse which can be elicited by light pulses.

Light treatment		Extent of response
Dark control	d	
Red-light pulse	a	$(a \gg d)$
		$(a > b)$
Far-red-light pulse	b	$(b > d)$
Red-light pulse		
+ far-red-light pulse	b	[100% (=full) reversibility][a]
red-light pulse		
+ far-red-light pulse		
+ red-light pulse	a	
red-light pulse		
+ far-red-light pulse		
+ red-light pulse		
+ far-red-light pulse	b	

[a] Percent reversibility is calculated according to the formula:

$$\% \text{ reversibility} = \frac{(a - c)}{(a - b)} \cdot 100$$

where a is the extent of the response obtained with a saturating red light, b is the response obtained with a saturating far-red-light pulse, c is the response obtained with the program: red-light pulse – far-red-light pulse. (From Mohr, 1972.)

diately with a saturating pulse of far-red light ($\varphi_{fr} \approx 0.03$), the response is controlled by phytochrome.   These criteria have been verified in many instances, a typical example being light-mediated anthocyanin synthesis in the mustard seedling (Fig. 7).   The dark-grown mustard seedling does not synthesize anthocyanin.   Five min of red light (applied at 36 h after sowing, 25°C) will induce considerable anthocyanin synthesis.   The effect of the red-light pulse ($\varphi_{red} \approx 0.8$) can be fully reversed by immediately following it with a saturating far-red light pulse ($\varphi_{fr} \approx 0.03$).

The experimenter is advised to determine the time course (kinetics) of the response to ensure that the light pulse operates via phytochrome alone.   In this example the kinetics of anthocyanin synthesis following a red or far-red light pulse should be "similar" (in a mathematical sense) if the only effect of light is to establish a certain amount of Pfr.   A probit transformation of the ordinate shows that the time course of anthocyanin synthesis follows a Gaussian summation curve in both cases (Fig. 8).

Difficulties may arise in those cases where the initial action of Pfr is fast. The term "initial action" designates the action of Pfr on some cell function which is no longer reversible by the removal of Pfr.   The onset of the initial action is defined by the loss of full reversibility (or escape from reversibility).   In many instances escape from reversibility becomes detectable only after 5–10 min or even longer (see Fig. 7 for an example).   Figure 9 shows the kinetics of the escape in the case of a relatively slow initial action.   In this particular case — the action of phytochrome on chlorophyll a and b synthesis in milo seedlings — full reversibility of the effect of a red-light pulse by a far-red (756 nm) light pulse is maintained over a time period of 40 min. Thereafter, escape is quite fast and completed after approximately 2.5 h.

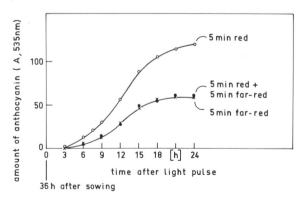

Fig. 7. Time courses of anthocyanin synthesis in the mustard seedling after a brief irradiation ("pulse") with red and/or far-red light (5 min each).   Pulses were given at time zero, i.e., 36 h after sowing. (From Lange *et al.*, 1971.)

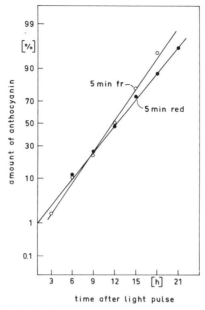

Fig. 8. Another plot of the data in Fig. 7 after a probit transformation of the ordinate. In both cases straight lines are obtained. The different slopes are due to the fact that the time period of synthesis is shorter after a far-red light pulse than after a red-light pulse.

An example of a rapid initial action of phytochrome is the control of grana formation in mustard seedlings (Table III). Phytochrome does not cause appreciable grana formation in the absence of chlorophyll. However, a pretreatment with red light pulses strongly increases the rate of grana development after the onset of a white-light treatment which is saturating with respect to chlorophyll synthesis. Thus a control function of phytochrome becomes detectable as soon as the level of chlorophyll no longer limits the process. In agreement with the criteria for the involvement of phytochrome, the effect of a 15-sec red-light pulse is fully reversible by a 756-nm light pulse. However, a 5-min red-light pulse is no longer reversible. Clearly, the failure to revert the effect of a 5-min red-light pulse by a far-red light does not exclude phytochrome as the sensor pigment involved.

(2) Light-grown plants

An impressive example which dates back to the pioneer days of phytochrome research is flowering of the obligatory short-day plant *Kalanchoe blossfeldiana* (Fig. 10). As in many other short-day plants, a long-day effect can be evoked by brief irradiation in the middle of the dark period, given in addition to a short-day main light period. The *Kalanchoe* plant on the left

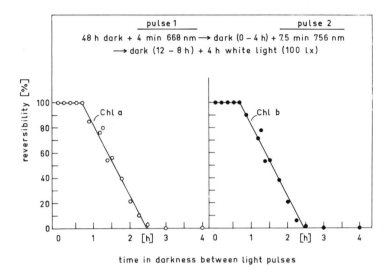

Fig. 9. Time courses of the escape from reversibility. The action of phytochrome on synthesis of chlorophyll *a* and *b* was measured in milo *(Sorghum vulgare)* seedlings. A saturating (4 min) red-light pulse was applied at 48 h after sowing (pulse 1) and followed with a saturating 756-nm light pulse (pulse 2). The two pulses were separated by dark intervals (abscissa). By definition (see Table II), 100% reversibility is obtained if the extent of response induced by a saturating 756-nm pulse equals the extent of response induced by a saturating red-light pulse followed by a saturating 756-nm light pulse. Percent reversibility is calculated according to the formula: % reversibility $= \dfrac{(a - c) \cdot 100}{(a - b)}$, where a is the extent of the response obtained with a saturating red-light pulse, b is the response obtained with a saturating far-red-light (e.g., 756-nm light) pulse, c is the response obtained with the programme: red-light pulse—dark interval—far-red-light pulse. As a response, the amount of chlorophyll *a* or *b* accumulated in saturating white light (0.23 W m^{-2}) between 60 and 64 h after sowing was measured. (From Sawhney *et al.,* 1980.)

was kept under short-day conditions only; it received 8 h of high-fluence-rate white light every day. The plant in the middle received the same short-day treatment but, in addition, received 1 min of red light in the middle of every dark period; the plant remained vegetative. The plant on the right received the short-day treatment plus 1 min of red light in the middle of the night followed immediately by a 1-min pulse with far-red light. The effect of the red-light pulse on flowering was fully reversed, and the criteria for the involvement of phytochrome are fulfilled. This means that photoperiodism in *Kalanchoe blossfeldiana*—and in many other short-day plants—is controlled (in some way or other) by phytochrome.

Table III. Control of grana formation by phytochrome in the palisade parenchyma of the cotyledons of the mustard seedling.[a]

Pretreatment	Number of stroma lamellae[b]	Number of grana[b]
4 × 15 sec 657-nm light[c]	5.6 ± 0.3	19.3 ± 1.0
4 × 15 sec 657-nm light + 2 min 759-nm light	4.2 ± 0.3	12.3 ± 1.2
4 × 2 min 759-nm light	4.4 ± 0.3	12.3 ± 1.1
4 × 5 min red light[d]	5.8 ± 0.2	18.0 ± 0.8
4 × 5 min red light + 5 min far-red light	5.8 ± 0.3	20.4 ± 1.1
4 × 5 min far-red light[d]	4.8 ± 0.3	11.9 ± 0.8
Dark control	4.6 ± 0.3	9.6 ± 0.6

[a] Grana formation was measured after 4 h of white light ($\sim$ 16 W m^{-2}). Onset of white light was 60 h after sowing. (From Girnth *et al.,* 1978.)

[b] Average number per section; control of stroma lamellae formation by phytochrome is given for comparison.

[c] Four red-light pulses equally distributed between 36 and 48 h after sowing.

[d] Broadband standard sources.

(3) Chloroplast movement in Mougeotia cells

Filaments of the green alga *Mougeotia* are made up of long cylindrical cells each of which contains a single, large, flat, plate-like chloroplast (Fig. 11). The chloroplast is able to turn inside the cell. The light-induced movement of this chloroplast is a response to light absorbed by the cytoplasm and not by the chloroplast itself. The profile $\rightarrow$ face position movement of the chloro-

Fig. 10. An example for the controlling function of phytochrome in flower initiation in the short-day plant *Kalanchoe blossfeldiana.* Details of the experiment are described in the text. (From Hendricks and Siegelman, 1967.)

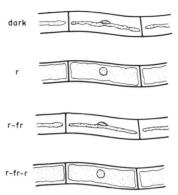

Fig. 11. An experiment which demonstrates the involvement of phytochrome in the low irradiance movement (profile position → face position) of the *Mougeotia* chloroplast. Dark, starting position; r, r-fr, r-fr-r, orientation of the chloroplast about 30 min after brief irradiations (1 min each) with red and far-red light. (From Haupt, 1970.)

plast is a typical phytochrome-mediated photomodulation, as can readily be shown by the usual experiments with a sequence of red and far-red light pulses (Fig. 11). Movement induced by 1 min of red light can be prevented by a subsequent treatment with 1 min of far-red light, etc. The movement itself requires about 30 min for completion (at 20°C) with a lag phase of a few minutes.

C. End-of-day light pulses

This experimental approach was also designed during the pioneer phase of phytochrome research (between 1954 and 1960) to demonstrate the operation of phytochrome during the dark period in a quasi-natural light–dark regime. Control of internodal lengthening during the dark period in light-grown vegetative bean plants is a convenient example (Fig. 12). Conspicuous changes in internode extension were caused by brief treatments with red or far-red light at the end of an 8-h day with white fluorescent light of high fluence rate (main light period). The Pinto bean on the left received no supplementary light, which means that at the end of every main light period approximately 70% of Ptot existed as Pfr in the plant (the Pfr/Ptot ratio established by white fluorescent light is only slightly lower than the photo-equilibrium established by a saturating red-light pulse). The Pinto plant in the centre received 5 min of far-red light at the end of every main light period ($\varphi_{fr} \approx 0.03$), and the plant on the right received 5 min of far-red light followed by 5 min of red light ($\varphi_{red} \approx 0.8$) at the end of every main light period.

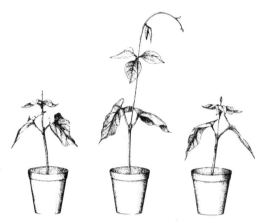

Fig. 12. A demonstration of the controlling function of phytochrome in internodal lengthening in light-grown green bean plants (*Phaseolus vulgaris,* cv. Pinto). Details of the experiment are described in the text. (From Hendricks, 1964.)

These observations demonstrate that phytochrome is involved in the control of internode lengthening during the dark period of a quasi-natural light–dark regime.

D. Criteria for phyotchrome involvement in the control of growth during the light period in a quasi-natural light–dark regime

Two different approaches have been used to demonstrate the involvement of phytochrome.

(1) Simultaneous treatment of green plants with long-term red and far-red light

Green mustard seedlings grown in white light (spectral distribution similar to sunlight, approximately 40 W m^{-2}) for 48 h after sowing were placed under different light sources, and growth of the hypocotyl was followed for a further 24 h. More than 75% of the effect of white light on hypocotyl elongation could be duplicated by low-fluence-rate red light (Table IV). When far-red (756 nm) light was applied simultaneously with red light to decrease the Pfr/Ptot ratio, the effectiveness of the red light was strongly decreased, even though the total fluence rate falling on the plant (and therefore the phytochrome cycling rate, i.e., the rate of Pr ↔ Pfr interconversions) was increased. The findings summarized in Table IV were taken as evidence that it is predominantly the Pfr level which determines the rate of axis elongation in continuous light.

Table IV. Effect of 24 h continuous light on hypocotyl length at 72 h after sowing.[a]

Light treatment	Increase of hypocotyl length (mm)
Dark control (pretreatment terminated with a 756-nm light pulse)	11.6
White light	1.8
Red light	3.9
756-nm light	10.7
Dichromatic light	6.2

[a] Mustard seedlings were pretreated with 48 h of white light (-40 W m^{-2}). Fluence rates used: red light, 0.68 W m^{-2}; 756-nm light, 15.7 W m^{-2}; dichromatic light, red, and 756-nm light applied simultaneously. (From Wildermann *et al.*, 1978.)

(2) Supplementary far-red light, applied in addition to white light during a quasi-natural daylight period

A more sophisticated way to demonstrate the involvement of phytochrome in the modulation of stem-extension rate in light-grown plants is based on the use of a transducer apparatus to follow continuously the growth rate of the stem in individual plants. In elegant experiments, Morgan *et al.* (1980) have shown that supplementary far-red light, given in addition to background white light, caused a rapid increase in stem extension rate in light-grown mustard *(Sinapis alba)* plants (Fig. 13). When the far-red light was turned off, the extension rate decreased again. This response pattern can be explained by a decrease (far-red light on) or increase (far-red light off) of the Pfr/Ptot ratio. This explanation is further justified by the observation that the increase in extension rate caused by supplementary far-red light was reversed by an increase in the fluence rate of supplementary red light (which increases the Pfr/Ptot ratio). The rapid, reversible control of the stem extension rate by phytochrome is an example of the response type called "photomodulation."

A problem in all these experiments is that the total fluence rate received by the plants varies with changing supplementary lights. This necessarily leads to changes of the phytochrome cycling rate for which it is very difficult to compensate. The experimenter must be aware that phytochrome cycling may contribute to the inhibitory effect of light on axis elongation in green plants, although the existence of such cycling-dependent growth inhibition has not yet been demonstrated convincingly.

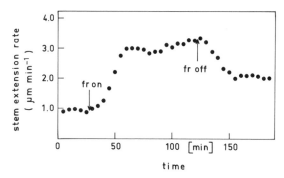

Fig. 13. A demonstration for the effect of supplementary far-red light (fr), in background white light, on the stem-extension rate in light-grown *Sinapis alba* plants. Far-red light, λ_{max} 719 nm, fluence rate 27 μmol m^{-2} sec^{-1}; white light, fluence rate 23 μmol m^{-2} sec^{-1}, Pfr/Ptot ratio in white light $\approx$ 0.65. Pfr/Ptot ratio in simultaneously applied white and far-red light $\approx$ 0.45 (From Morgan *et al.,* 1980.)

E. Long-term, continuous light given to etiolated seedlings

The use of long-term, continuous light has been a popular treatment in those studies which aim at a causal understanding of the biophysical and molecular processes operating during photomorhogenesis. Continuous light has the advantage that a constant signal output of the sensor pigment can be expected over some period of time. If phytochrome is the sensor pigment, a photosteady-state (see above, page 17) will become established after a transition period. The phytochrome photosteady-state ([Pfr] = constant) can be maintained even with wavelengths of light which do not lead to an appreciable accumulation of chlorophyll (e.g., far-red light). Using this method, the molecular events related to photomorphogenesis can be studied without interference from photosynthesis even in long-term experiments (e.g., over 24 h).

In long-term experiments with etiolated seedlings the action spectra show peaks (maxima) in the far-red and in the blue part of the spectrum with a shoulder or peak in the red region (Fig. 14). The ratio of the effectiveness of blue to far-red light varies considerably from action spectrum to action spectrum, although this may be partially due to differences in irradiation time (Mancinelli and Rabino, 1978; Mancinelli and Walsh, 1979). An extreme case is the detailed action spectrum for inhibition of hypocotyl growth in previously etiolated seedlings of *Lactuca sativa* which shows peaks in the far-red and blue spectral range with almost no action in the red region (Fig. 15). Whereas the identity of the photoreceptor responsible for the blue peak of action remains to be demonstrated in each case (because of the

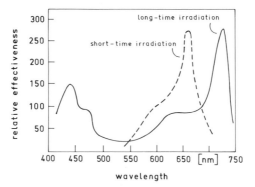

Fig. 14. Characteristics of action spectra of photomorphogenesis for short-time irradiation (for induction conditions, see Fig. 6) and for long-time irradiation of previously etiolated seedlings (From Wagner and Mohr, 1966.)

operation of phytochrome and the blue/UV-A photoreceptor in this spectral range), the action in the red region and the far-red peak of action can be attributed to phytochrome. K. M. Hartmann (1966, 1967a,b) has provided evidence with dichromatic light that the far-red peak of action in the case of

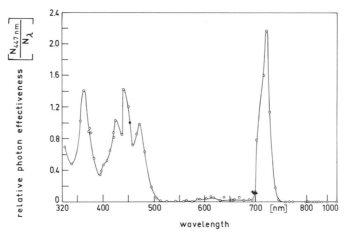

Fig. 15. An action spectrum for control (inhibition) of hypocotyl lengthening in lettuce seedlings (*Lactuca sativa,* cv. Grand Rapids) by continuous light. The action spectrum was elaborated between 54 and 72 h after sowing. During this period lengthening of the hypocotyl is almost exclusively due to cell lengthening. Note that this action spectrum is not for the induction of a response where the reciprocity law is valid (see Fig. 6) but for a long-term irradiation where reciprocity does not hold. However, a steady state situation (i.e., a constant rate of the response) can be assumed for all wavelengths during the period of light treatment. (From Hartmann, 1967a.)

the lettuce action spectrum (see Fig. 15) can be attributed to the action of a photochromic photoreceptor with the properties of phytochrome.

Hartmann's approach was briefly as follows. Etiolated lettuce seedlings were treated simultaneously with 768-nm light, a wavelength which is ineffective if applied separately (see Fig. 15), and with red light (658 nm) at different photon-fluence rates. The red light was also almost ineffective with respect to extension growth of the whole hypocotyl. The fluence rate of the far-red light (758 nm) was kept constant and the fluence rate of the red light was varied. It was found (Fig. 16) that there was a certain fluence-rate ratio of these two wavelengths, which led to an optimum response. On the other hand, the high effectiveness of 717 nm would be nullified eventually by a simultaneous irradiation with 658 nm given at a high photon fluence rate. This kind of result indicates that the strong effect of prolonged far-red light might be due to a favourable Pfr/Ptot ratio. Using the same experimental approach with other pairs of wavelengths, it was found that a maximum response with respect to control of hypocotyl extension always occurred when the Pfr/Ptot ratio was approximately 0.03. If the Pfr/Ptot ratio was above 0.3 or below 0.002 the action of light on hypocotyl extension growth was almost negligible (less than 10% deviation from the dark control).

In long-term irradiations with etiolated seedlings, a strong dependency of the light effect on photon-fluence rate ("irradiance") is generally observed (Fig. 17). (For this reason this light response was called in the pioneer days "High Irradiance Reaction," HIR). This universal and conspicuous "irradiance" (fluence rate) dependency can also be attributed to properties of the

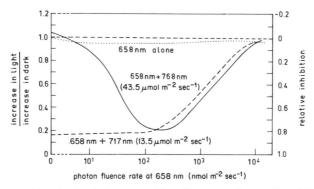

Fig. 16. Hypocotyl lengthening in lettuce seedlings (*Lactuca sativa,* cv. Grand Rapids) under continuous simultaneous irradiation with different photon fluence rates in the red (658 nm) at a constant high photon fluence rate background of different far-red bands (717 and 768 nm). Ordinate: ratio between "hypocotyl increase in light" and "hypocotyl increase in dark," measured between 54 and 72 h after sowing. Before 54 h, the seedlings were kept in complete darkness. (From Hartmann, 1967b.)

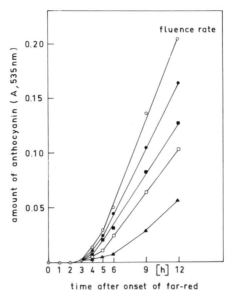

Fig. 17. Time courses of anthocyanin accumulation in mustard seedlings under continuous far-red light as a function of fluence rate. Fluence rate 1000 denotes a standard far-red energy fluence rate (3.5 W m^{-2}). Onset of far-red light: 36 h after sowing. (From Lange *et al.,* 1971.)

phytochrome system (Hartmann, 1967b).   Analogous to the procedure described in connection with Fig. 16, a number of experiments were performed in which red light (658 nm) of increasing fluence rate was applied simultaneously with a far-red wavelength at a determined constant fluence rate.   In this way response maxima were determined with a number of different wavelengths (e.g., 801, 768, and 746 nm) as the partner of 658 nm.   These far-red wavelengths are almost exclusively absorbed by Pfr.   While the Pfr/Ptot ratio for the response maxima is always close to 0.03, the extent of the maximum response is very different, ranging from about 20 up to 80% inhibition of hypocotyl growth.   This is due, as Hartmann (1967b) has shown experimentally, to the fluence-rate dependency of the response. Since then it has become customary to assume that the total absorption of photons in the phytochrome system (i.e., the phytochrome cycling rate = rate of Pr ↔ Pfr interconversions) determines the extent of the response in continuous light once a certain Pfr/Ptot ratio is established (see Mohr, 1972).

Hartmann (1966, 1967a,b), Schäfer (1975), and Johnson (1980) — following a previous proposal by Smith (1970) — have published quantitative models to explain the effect of continuous light on the basis of phytochrome. Even though none of these models is fully satisfactory, there can be little doubt that the seemingly complicated action of continuous red and far-red light, applied to a previously etiolated seedling, is exclusively due to the operation of the phytochrome system (see Mancinelli and Rabino, 1978).

F. Long-term, continuous light given to deetiolated seedlings

The word "deetiolated" describes the state of seedlings which have received a white-light pretreatment before the action spectrum under continuous light is measured. As an example, mustard seedlings *(Sinapis alba)* were kept in white light from sowing onwards for 54 h and then placed in monochromatic light for a further 24 h to determine the action spectrum for inhibition of hypocotyl elongation. The resulting action spectra (Fig. 18) show that the effectiveness of far-red and blue light (see Fig. 14) has almost disappeared. The position of the action maximum in the red spectral range is somewhat shifted to shorter wavelengths in plants containing chlorophyll compared with those containing almost no chlorophyll. The latter fact can readily be explained by screening of the photoreceptor (phytochrome) by chlorophyll. The synchronous disappearance of the far-red and blue peaks of action (compare Fig. 18 with Fig. 14) is probably related to the dramatic decrease of Ptot (by more than 90%) in the course of the white-light pretreatment. On the other hand, the mode of action of the red light is not yet clear. While

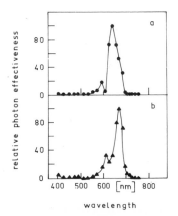

Fig. 18. Action spectra for control (inhibition) of hypocotyl lengthening by continuous light (24 h) in seedlings of mustard *(Sinapis alba)*. (a) green seedlings, pretreated with 54 h of white light; (b) visually white seedlings, pretreated with 54 h of white light in the presence of the herbicide San 9789 which prevents synthesis of coloured carotenoids and accumulation of chlorophylls (see Frosch *et al.,* 1979). (From Beggs *et al.,* 1980.)

hardly any doubt exists that phytochrome is the only photoreceptor, currently available explanatory models are not satisfactory.

III. CRITERIA FOR THE INVOLVEMENT OF A BLUE/UV-A LIGHT PHOTORECEPTOR (CRYPTOCHROME)

A specific blue/UV-A absorbing photoreceptor is clearly indicated in those cases where a strong light action is observed in the blue/UV-A part of the spectrum without any action in the red and far-red range.

A. Photoreception in fungi

As far as we know at present, photomorphogenesis (including light-induced carotenoid synthesis), photomodulation (e.g., light-growth responses) and phototropism in fungi are mediated exclusively by blue/UV light (Rau, 1975; Presti and Delbrück, 1978) (as an example, see Fig. 1). Isolated reports about the involvement of phytochrome in the control of fungal development require confirmation. Intense efforts to demonstrate an interaction between blue light effects and phytochrome in carotenogenesis in *Fusarium aquaeductuum* were without success (W. Rau, personal communication).

B. Phototropism

Except for some fern and moss chloronemata, where phytochrome is involved in the phototropic and polarotropic responses (see Mohr, 1972, Chapter 22), phototropism in the plant kingdom is mediated through cryptochrome. In particular the phototropic responses of higher plants must be attributed to the operation of the ubiquitous blue/UV-A photoreceptor since neither red nor far-red light have ever been found to be phototropically effective. The action spectrum of phototropism, e.g., phototropism of the *Avena* coleoptile, closely follows the established action spectrum for involvement of cryptochrome: three peaks of action (or shoulders) in the blue spectral range and another peak at approximately 370 nm in the near UV-A (see Fig. 1).

C. Methods for separating the concomitant action of phytochrome and the blue/UV-A photoreceptor in higher plants

(1) A method of Meijer (1968)
Elongation growth of dark-grown gherkin hypocotyls was recorded continuously — with the aid of a displacement pickup in combination with a

measuring bridge and a recorder—in darkness and on exposure to blue, red, or far-red light. The data (Fig. 19) strongly indicate that the rapid inhibition of elongation in blue light is not mediated by phytochrome, but that a specific blue-light absorbing photorecepter pigment is involved.

(2) A method of Thomas and Dickinson (1979)

Hypocotyl extension was measured in deetiolated seedlings of lettuce, cucumber, and tomato. Plants were kept under high-fluence-rate light from low-pressure sodium (SOX) lamps, which do not emit any blue light (all of the visible and most of the total output is concentrated in a narrow band centered at 589 nm). Small but highly inhibitory amounts of blue light were added to a high-fluence-rate background irradiation from the SOX lamps Since the added blue light did not alter phytochrome photoequilibrium and its effect was independent of the total fluence rate, it was concluded that both phytochrome and a separate blue-light-dependent photoreceptor pigment are involved in the control of axis elongation. This is a simple and convincing experimental criterion for the operation of a nonphytochrome, blue-absorbing, photoreceptor pigment.

(3) The "light equivalence" test of Schäfer et al. (1981)

Phytochrome (Fig. 3) is a photochromic pigment: $Pr \underset{k_2}{\overset{k_1}{\rightleftharpoons}} pfr$. A photoresponse, mediated through phytochrome, will therefore be a function of both

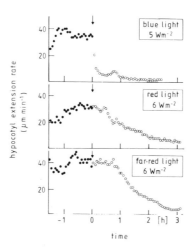

Fig. 19. Hypocotyl extension of dark-grown gherkin seedlings (25°C, 80% air humidity; 3 days after sowing, the length of the hypocotyls then being 25–30 mm) in darkness and on exposure to light (↓). The blue light decreases the growth rate almost instantly while the inhibition by red and far-red light occurs more gradually. (From Meijer, 1968.)

light rate constants k_1 and k_2. Because the photoequilibrium

$$\varphi = \frac{\epsilon_r \Phi_r}{\epsilon_r \Phi_r + \epsilon_{fr} \Phi_{fr}} = \frac{k_1}{k_1 + k_2}$$

(see legend to Fig. 4) and the rate of photoconversion $v = k_1 + k_2$ can be measured directly, it is convenient to describe the light-pigment interaction simply by φ and v. This means that any light will be the same for phytochrome if φ and v — as determined by the light — are the same.

A plot of the extent of a response as a function of φ for constant v in the red – far-red spectral range can therefore be considered as a calibration curve for phytochrome action. Any light treatment should comply with this calibration curve provided that phytochrome is the only photoreceptor involved. Using this approach it was possible to show the involvement of a separate blue-light photoreceptor in the control of hypocotyl growth in several species.

D. Method for demonstrating the obligatory coaction of phytochrome and the blue/UV-A photoreceptor in higher plants

The criteria for defining this complex light response can best be introduced by briefly presenting a case study, namely, light-induced anthocyanin synthesis in the mesocotyl (=first internode) of the mild *(Sorghum valgare)* seedling (Downs and Siegelman, 1963; Drumm and Mohr, 1978). Without light, no anthocyanin synthesis is observed. The principal light effects can be described as follows (Table V). White light has a strong inductive effect. However, if only light above 500 nm is given, no anthocyanin synthesis takes place. Phytochrome can only act once a blue/UV light effect has occurred. On the other hand, the expression of the blue/UV light effect (action peaks in the blue and in the near UV-A around 370 nm) is controlled by phytochrome. It is obvious that we are dealing with an obligatory coaction of cryptochrome and phytochrome. In dichromatic experiments [simultaneous irradiation with two kinds of light to establish very different levels of Pfr (see Fig. 16)] it was found that the blue/UV-mediated photoreaction as such is not affected by the presence or virtual absence of Pfr. This means that the initial action of the blue/UV light is totally independent of phytochrome even though the expression of the blue/UV-light effect is controlled by phytochrome.

It is not clear at present how widely spread in nature this obligatory coaction of cryptochrome and phytochrome, in fact, is, whereas the following category, facultative coaction of cryptochrome and phytochrome, has been documented in many instances.

Table V. Induction of anthocyanin synthesis in the mesocotyl of milo *(Sorghum vulgare)* seedlings by light.[a]

Light treatment (onset, 60 h after sowing)	Amount of anthocyanin (measurement, 87 h after sowing, relative units)
27 h dark	0
27 h white light	185
27 h red light	0
27 h far-red light	0
3 h white light	19
3 h white light + 5 min 756-nm light	6
3 h white light + 5 min 756-nm light + 5 min red light	20
3 h blue-UV	19
3 h blue-UV + 5 min red light	19
3 h blue-UV + 5 min 756-nm light	5
3 h blue-UV + 5 min 756-nm light + 5 min red light	19

[a] In the case of a 3-h light treatment, the seedlings were kept in the dark for 24 h before extraction of anthocyanin. (From Drumm and Mohr, 1978).

E. Method for measuring the facultative coaction of phytochrome and the blue/UV-A photoreceptor in higher plants

Again, the criteria for defining this complex light response can be advanced by introducing a clear-cut case study, namely, anthocyanin synthesis in the hypocotyl of the tomato *(Lycopersicon esculentum)* seedling (Drumm-Herrel and Mohr, 1982b). The tomato seedling shows an intermediate response between mustard (strong effectiveness of single red-light pulses) and the milo seedling (no effectiveness of red light of any kind without a blue/UV pretreatment) since in the tomato hypocotyl long term red or far-red light (operating through phytochrome) leads to a low rate of anthocyanin synthesis, while light pulses are totally ineffective (Table VI). A quantitative gauge for the effectiveness of phytochrome is the "extent of the reversible response," ΔR = response$_1$ minus response$_2$, whereby response$_1$ is obtained if the light treatment is terminated with a saturating red-light pulse, and response$_2$ is obtained if the light treatment is terminated with a saturating

Table VI. Induction (or lack of induction) of anthocyanin synthesis in the hypocotyl of tomato seedlings *(Lycopersicon esculentum).*[a]

Light treatment[b]	Anthocyanin (A_{535})
Dark (D)	0.001
12 h D + 5 min RL ($\varphi_{RL} \approx 0.8$)	0.001
12 h D + 5 min RG 9-light ($\varphi_{RG9} < 0.01$)	0.001
	$\Delta R = 0.000$
12 h RL + 5 min RL	0.020
12 h RL + 5 min RG 9-light	0.009
	$\Delta R = 0.011$
12 h RL_{10} + 5 min RL_{10}	0.032
12 h RL_{10} + 5 min RG 9-light	0.011
	$\Delta R = 0.021$
12 h FR + 5 min RL_{10}	0.025
12 h FR + 5 min RG 9-light	0.010
	$\Delta R = 0.015$
12 h BL + 5 min RL_{10}	0.119
12 h BL + 5 min RG 9-light	0.051
	$\Delta R = 0.068$
12 h UV(PG 218) + 5 min RL_{10}	0.137
12 h UV(PG 218) + 5 min RG 9-light	0.066
	$\Delta R = 0.071$
12 h UV(WG 345) + 5 min RL_{10}	0.116
12 h UV(WG 345) + 5 min RG 9-light	0.066
	$\Delta R = 0.050$

[a] Emphasis is on the extent of the reversible response ΔR as a function of light pretreatment. (From Drumm-Herrel and Mohr, 1982a.)

Programme: 3 days dark — 12 h light (or dark) — 5 min light pulse (saturating red or RG 9-light) — 24 h dark — assay of anthocyanin. ΔR = response$_1$ − response$_2$; response$_1$, response obtained if the light treatment is terminated with a saturating red light pulse (RL_{10}, 6.8 W m^{-2}); response$_2$, response obtained if the light treatment is terminated with a saturating long wavelength far-red light (RG 9-light) pulse (10 W m^{-2}). ΔR is considered a reliable gauge for the responsivity of anthocyanin synthesis to Pfr at the time of the differential light pulses.

[b] Abbreviations: RL_1, red light, 0.68 W m^{-2}; FR, far-red light, 3.5 W m^{-2}; BL, blue light, 7 W m^{-2}; UV(PG218), UV-A with a small amount of UV-B, 12.6 W m^{-2}; UV(WG345), pure UV-A, 9.3 W m^{-2}. (For details of light sources, see Mohr and Drumm-Herrel, 1983).

long-wavelength far-red light pulse. Although single light pulses have no effect on anthocyanin synthesis in the tomato hypocotyl, a surprisingly strong reversible response (ΔR) appears after prolonged light exposure. Blue and UV light are particularly effective in eliciting the reversible re-

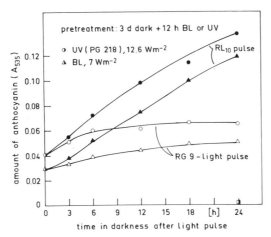

Fig. 20. Anthocyanin accumulation in the tomato hypocotyl in darkness following a light treatment. The seedlings were pretreated either with ultraviolet (UV) or with blue light (BL). The pretreatment was terminated either with a saturating red-light pulse (RL$_{10}$, $\varphi_{RL} \approx 0.8$, see Table VI) or with a saturating RG 9-light pulse ($\varphi_{RG9} < 0.01$, see Table VI). (From Drumm-Herrel and Mohr, 1982a.)

sponse (high values of ΔR). The ΔR values obtained after prolonged red or far-red light are far lower. Figure 20 shows that anthocyanin synthesis after a 12-h blue or UV light exposure continues in darkness at a considerable rate only if Pfr is available. When most of the Pfr is returned to Pr by a saturating long-wavelength far-red light pulse, further anthocyanin synthesis is almost negligible. Clearly, phytochrome is involved in light-mediated anthocyanin synthesis in the tomato seedling, even though single light pulses are totally ineffective. Obviously, a high effectiveness of phytochrome depends on a preceding light absorption in a blue/UV-light photoreceptor.

F. Method for demonstrating the obligatory or facultative coaction of phytochrome and the UV-B photoreceptor in higher plants

Anthocyanin synthesis in the coleoptile of wheat seedlings (*Triticum aestivum*, cv. Schirokko) is a useful case for defining the criteria for this complex light response which may be widespread in nature (Drumm-Herrel and Mohr, 1983). Anthocyanin synthesis in the wheat coleoptile takes place readily in natural sunlight and in fluorescent white light (Table VII). Short- and long-term treatments with red and far-red light are ineffective (Table VII). Surprisingly, blue and pure UV-A have very little effect, while UV light with a small content of UV-B is very effective in eliciting anthocyanin synthesis (Table VIII). Following a UV treatment, the response is con-

Table VII. Induction (or lack of induction) and reversion of anthocyanin synthesis in the coleoptile of *Triticum aestivum,* cv. Schirokko with light of different light qualities.[a]

Treatment	Amount of anthocyanin (A at 546 nm)		
	Response$_1$	Response$_2$	ΔR
5 d D (dark)	0.004		
2 d D + 10 h SL[b]• + 40 h D	0.169	0.043	0.126
2 d D + 12 h WL[b]• + 38 h D	0.092	0.030	0.062
3 d D + 2 h D• + 24 h D	0.004	0.004	0.000
3 d D + 2 h RL$_{10}$• + 24 h D	0.004	0.004	0.000
3 d D + 2 h BL• + 24 h D	0.004	0.004	0.000
3 d D + 2 h UV(WG345)• + 24 h D	0.008	0.004	0.004
3 d D + 2 h UV(PG218)• + 24 h D	0.0	0.00	0.00
3 d D + 2 h UV-B• + 24 h D	0.033	0.012	0.021
2 d D + 12 h FR• + 38 h D	0.005	0.005	0.000
2 d D + 12 h RL$_{10}$• + 38 h D	0.004	0.004	0.000
2 d D + 12 h BL• + 38 h D	0.010	0.004	0.006
2 d D + 12 h UV(WG345)• + 38 h D	0.021	0.005	0.016
2 d D + 12 h UV(PG218)• + 38 h D	0.160	0.071	0.089

[a] At the time indicated by a point (•) seedlings received either a saturating RL$_{10}$ or RG 9-light pulse. Anthocyanin contents were measured at the age of 98 h. For definitions of response$_{1,2}$, ΔR, and abbreviations, see Table VI. (From Mohr and Drumm-Herrel, 1983.)

[b] SL = sunlight (changing fluxes due to clouds). WL = fluorescent white light, ~ 34 W m^{-2}.

trolled by phytochrome (Fig. 21). It appears that the major effect of the UV treatment is to establish responsiveness ("sensitivity") to phytochrome. In contrast to the situation in milo and tomato, UV cannot be substituted for by blue light in its "sensitivity amplification" function in the wheat coleoptile. Thus the function of a separate UV-B photoreceptor (see Fig. 2) is indicated.

IV. INVOLVEMENT OF PHOTOSYNTHETIC PHOTORECEPTORS IN PHOTOMORPHOGENESIS

From time to time, photomorphogenetic effects of long-term light treatments have been related to photosynthesis (or to parts of it such as photosystem I activity or cyclic photophosphorylation). These suggestions have been refuted using quite different criteria (see Mohr, 1972, Chapter 4; Mancinelli and Rabino, 1978). An argument refuting the role of chlorophyll or photosynthesis is illustrated in Fig. 18. The similarity in the responses of

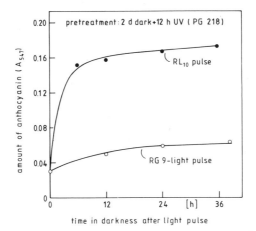

Fig. 21. Anthocyanin accumulation in the wheat coleoptile in darkness following a UV pretreatment. The UV is predominantly UV-A with a small amount of UV-B, total fluence rate 12.6 W m^{-2}. The pretreatment was terminated either with a saturating red light (RL$_{10}$, $\varphi_{RL} \approx 0.8$, see Table VI) or with a saturating RG 9-light pulse ($\varphi_{RG} < 0.01$, see Table VI). (From Mohr and Drumm-Herrel, 1983.)

herbicide-treated plants, which contain only traces of chlorophyll and no functional chloroplasts (Frosch *et al.,* 1979), and their normally green counterparts provides strong evidence that chlorophyll and photosynthesis are not involved in the photomorphogenetic effects of continuous light.

REFERENCES

Beggs, C. J., Holmes, M. G., Jabben, M. and Schäfer, E. (1980). *Plant Physiol* **66**, 615–618.
Borthwick, H. A., Hendricks, S. B., Parker, M. W., Toole, E. H. and Toole, V. K. (1952). *Proc. Natl. Acad. Sci. U.S.A.* **38**, 662–666.
Butler, W. L, Hendricks, S. B. and Siegelman, H. W. (1964). *Photochem. Photobiol.* **3**, 521–528.
Curry, G. M. and Jotcham, J. R. (1980). *Book Abstr. Am. Soc. Photobiol. Annu. Meet.* pp. 120–121.
Downs, R. J. and Siegelman, H. W. (1963). *Plant Physiol.* **3**, 25–30.
Drumm, H. and Mohr, H. (1974). *Photochem. Photobiol.* **20**, 151–157.
Drumm, H. and Mohr, H. (1978). *Photochem. Photobiol.* **27**, 241–248.
Drumm-Herrel, H. and Mohr, H. (1981). *Photochem. Photobiol.* **33**, 391–398.
Drumm-Herrel, H. and Mohr, H. (1982a). *Photochem. Photobiol.* **35**, 233–236.
Drumm-Herrel, H. and Mohr, H. (1982b). *Photochem. Photobiol.* **36**, 229–233.
Frosch, S., Jabben, M., Bergfeld, R., Kleinig, H. and Mohr, H. (1979). *Planta* **145**, 497–505.
Girnth, C., Bergfeld,R. and Kasemir, H. (1978). *Planta* **141**, 191–198.
Gottman, K. and Schäffer, E. (1982). *Photochem. Photobiol.* **35**, 521–525.
Hahlbrock, K. and Wellmann, E. (1970). *Planta* **94**, 236–239.

Hanke, J., Hartmann, K. M., and Mohr, H. (1969). *Planta* **86**, 235–249.
Hartmann, K. M. (1966). *Photochem. Photobiol.* **5**, 349–366.
Hartmann, K. M. (1967a). *Z. Naturforsch.* **22B**, 1172–1175.
Hartmann, K. M. (1967b). *Book Abstr. Eur. Photobiol. Symp. Hvar (Jugoslavia)* pp. 29–32.
Haupt, W. (1970). *Physiol. Veg.* **8**, 551–563.
Hendricks, S. B. (1964). *Photophysiology* **1**, 305–331.
Hendricks, S. B. and Siegelman, H. W. (1967). *Compr. Biochem.* **27**, 211–234.
Hendricks, S. B., Borthwick, H. A. and Downs, R. J. (1956). *Proc. Natl. Acad. Sci. U.S.A.* **42**, 19–26.
Johnson, C. B. (1980). *Plant, Cell Environ.* **3**, 45–51.
Lange, H., Shropshire, W. and Mohr, H. (1971). *Plant Physiol.* **47**, 649–655.
Mancinelli, A. and Rabino, I. (1978). *Bot. Rev.* **44**, 129–180.
Mancinelli, A. and Walsh, L. (1979). *Plant Physiol.* **63**, 841–846.
Meijer, G. (1968). *Acta Bot. Neerl.* **17**, 9–14.
Mohr, H. (1966). *Z. Pflanzenphysiol.* **54**, 63–83.
Mohr, H. (1972). "Lectures on Photomorphogenesis." Springer-Verlag, New York.
Mohr, H. and Drumm-Herrel, H. (1983). *Physol. Plant.* **58**, 408–414.
Morgan, D. C., O'Brien, T. and Smith, H. (1980). *Planta* **150**, 95–101.
Oelze-Karow, H. and Mohr, H. (1973). *Photochem. Photobiol.* **18**, 319–330.
Oelze-Karow, H. and Mohr, H. (1976). *Photochem. Photobiol.* **23**, 61–67.
Pratt, L. H. (1975). *Photochem. Photobiol.* **22**, 33–36.
Pratt, L. H. (1978). *Photochem. Photobiol.* **27**, 81–105.
Presti, D. and Delbrück, M. (1978). *Plant, Cell Environ.* **1**, 81–100.
Quail, P. H. and Schäfer, E. (1974). *J. Membrane Biol.* **51**, 393–404.
Rau, W. (1967). *Planta* **72**, 14–28.
Rau, W. (1975). *Ber. Dtsch. Bot. Ges.* **88**, 45–60.
Rüdiger, W. (1980). *Struct. Bonding (Berlin)* **40**, 101–140.
Sawhney, S., Oelze-Karow, H., Sawhney, N. and Mohr, H. (1980). *Photochem. Photobiol.* **32**, 787–732.
Schäfer, E. (1975). *J. Math. Biol.* **2**, 41–56.
Schäfer, E. and Mohr, H. (1974). *J. Math. Biol.* **1**, 9–15.
Schäfer, E. and Mohr, H. (1980). *Photochem. Photobiol.* **31**, 495–500.
Schäfer, E., Marchal, B. and Marmé, D. (1972). *Photochem. Photobiol.* **15**, 457–464.
Schäfer, E., Schmidt, W. and Mohr, H. (1973). *Photochem. Photobiol.* **18**, 331–334.
Schäfer, E., Lassig, T.-U. and Schopfer, P. (1975). *Photochem. Photobiol.* **22**, 193–202.
Schäfer, E., Beggs, C. J., Fukshansky, L., Holmes, M. G. and Jabben, M. (1981). *Book Abstr., Eur. Symp. Light-mediated Plant Dev., 1981* Abstract 9.14.
Schmidt, R. and Mohr, H. (1982). *Plant, Cell Environ.* **5**, 495–499.
Senger, H. (1980). "The Blue Light Syndrome." Springer-Verlag, Berlin and New York.
Smith, H. (1970). *Nature (London)* **227**, 665–668.
Thomas, B. and Dickinson, H. G. (1979). *Planta* **146**, 545–550.
Wagner, E. and Mohr, H. (1966). *Photochem. Photobiol.* **5**, 397–406.
Wellmann, E. (1971). *Planta* **101**, 283–286.
Wellmann, E. (1974). *Ber. Dtsch. Bot. Ges.* **87**, 267–273.
Wildermann, A., Drumm, H., Schäfer, E. and Mohr, H. (1978). *Planta* **141**, 211–216.
Withrow, R. B., Klein, W. H. and Elstad, V. (1957). *Plant Physiol.* **32**, 453–462.
Yamamoto, K. T. and Smith, W. O. (1981). *Plant Cell Physiol.* **22**, 1159–1164.
Yatsuhashi, H., Hashimoto, T. and Shimizu, S. (1982). *Plant Physiol.* **70**, 735–741.

3

Light Sources

M. G. Holmes

I. INTRODUCTION

Because light is the causative factor in photomorphogenetic responses, it is critical that the light sources are chosen, constructed, and used correctly. The primary requirements are a light source which is efficient in the relevant waveband and a filter combination which provides maximum transmission in that waveband and maximum rejection of unwanted wavelengths. Of equal importance, however, are simplicity, cleanliness, regular maintenance and regular monitoring of the light output.

Where possible, discussion is restricted to the practical aspects of choosing and operating light sources. Theoretical treatment of the design and construction of equipment is limited to where brief mention will aid practical application. Selected references to theoretical aspects are included at the end of the chapter. It is impossible to list all firms which supply components for light sources. Where appropriate, specific major firms have been mentioned; in many instances there are other reputable manufacturers which supply comparable equipment.

The construction and use of light sources for the 300–800-nm waveband is discussed and the following abbreviations are used for radiation: UV = ultraviolet; B = blue; G = green; R = red; FR = far-red; IR = infrared; PAR = photosynthetically active radiation = 400–700 nm. For convenience, the UV, visible, and IR regions are defined here as the 300–380, 380–770, and greater than 770-nm wavebands, respectively.

TECHNIQUES IN PHOTOMORPHOGENESIS
0-12-652990-6

II. BASIC LIGHT SOURCES

A. Incandescent filament lamps

(1) Lamp type
Vacuum lamps (i.e., lamps which are partially evacuated of air to decrease the rate of filament oxidation) are generally not suited to photomorphogenesis research applications because of their limited wattage. The low pressure and lack of convection currents within the bulb limit the temperature at which the filament may be operated. The exception, however, is the tubular lamp, which is available in various lengths and in ratings from 35 to 150 W. This low-luminance lamp is an excellent source for large FR-light fields and provides a very uniform light output (see Section IV).

General service lamps with ratings of 40 W and higher usually contain argon and nitrogen gases at slightly below atmospheric pressure to reduce evaporation of tungsten. A gradual decrease in light output takes place throughout the life of these lamps because of deposition of tungsten on the bulb (blackening) and a decrease in the operating temperature of the filament. This decrease in output varies according to lamp type and application, but a depreciation of 15% from new to the end of the rated life is typical. In situations where high irradiances are not required, these lamps offer a cheap source of radiation for FR sources (Section IV) when the space they require is not restricting; regular monitoring of output is necessary, however.

For higher outputs of light, lamps operating on the tungsten–halogen principle should be used. These lamps use rarer gases such as iodine, the pressure of the gas depending on the size and construction of the glass envelope and the use to which the lamp is put. The halogen serves to combine with the evaporated tungsten near the bulb wall. This combination is returned to the filament by convection where the tungsten is released, thereby eliminating blackening of the bulb wall and maintaining high light output. There is a large difference, however, between hard-glass halogen lamps and quartz halogen lamps. The former have the same output and rated life as normal lamps, the halogen merely preventing blackening. The quartz halogen lamps offer much higher light output because quartz glass can withstand higher thermal shock and heat loads, so the envelope can be much closer to the filament and therefore smaller. The more compact construction of these lamps allows the use of higher gas pressures, thereby increasing both the temperature at which the filament can be operated and its longevity.

(2) Lamp output and life
Lamp life depends primarily on the temperature at which the filament is operated. As filament temperature increases, the evaporation rate of tungsten also increases, thereby sooner reaching the stage at which the filament becomes so thin that it fractures. A balance must therefore be found between light output and lamp life. Whereas general lighting service and linear tungsten–halogen lamps have a long rated life (1000–2000 h), higher-output lamps such as those used in a projector have a relatively short life (25–50 h). The lamp life data provided by the manufacturer should be treated as a rough guide only. Even when operated strictly within the manufacturer's limits, a wide variation will be experienced in true lamp life on either side of the rated value.

(3) Voltage variation
The best compromise between light output and lamp life will usually be obtained at the rated voltage. High-wattage lamps are more efficient than low-wattage lamps of the same voltage and life rating. Also, low-voltage lamps are more efficient than high-voltage lamps of the same wattage and life rating. This is because the filaments of high-wattage–low-voltage lamps are thicker in diameter than low-wattage–high-voltage types; the increase in filament diameter allows higher operating temperatures and therefore greater light output in the 300–800-nm range without excessive evaporation of tungsten.

Substantially greater light output can be obtained by running lamps at higher than rated voltages. Although the extra energy consumption costs may be negligible, the costs of greatly decreased lamp life and the possibility of experiments ruined by lamp failure must be considered. As an example, increasing voltage by 10% will typically increase light output by around 40% but will reduce lamp life to only 25% of its rated value. As a rule-of-thumb guide, lamp life is approximately halved by every 5% increase in constant operating voltage above that recommended by the manufacturer. Within limits, over-voltage running can be practical, especially when planned replacement is used to avoid the effects of lamp failure during experiments; the final choice depends primarily on the application. For growth cabinets, environmental simulators and standard FR sources, the number and type of lamps should be based on running at manufacturer's rated voltage. For monochromatic sources using filters (e.g., for action spectra) adequate fluence rates can rarely be reached, especially in the blue waveband, with incandescent lamps run at recommended voltage. Over-voltage operation in these cases may avoid the need to resort to specialised sources.

(4) Ambient operating temperature
Under normal conditions, temperature has little effect on the performance and life of the incandescent lamp. Two situations may arise, however, in which the temperature may become high enough to cause lamp failure. The first results from inadequate ventilation. Such conditions cause general lamp overheating which can result in rapid blackening of the bulb, followed by failure. Also, the cement between the base and the bulb may disintegrate, the solder in the base can soften or melt, and cable insulation may be damaged. The second cause of overheating is incorrect focusing. If the image of the filament is focused on the bulb wall instead of the filament plane, softening and blistering of the wall may lead to early failure. This is particularly important in tubular lamps.

(5) Operating position
Although most general lighting service lamps can be operated in any position, certain specialised lamps (e.g., tubular tungsten–halogen double-ended lamps and projector lamps) must be burned only in the positions designated by the manufacturer. Failure to observe this will result in envelope blackening and extremely short lamp life.

(6) Vibration
Vibration can be a major factor in reducing lamp life and should be minimized wherever possible. The high temperatures reached result in the filament becoming softer and more pliable. If there is a great deal of vibration (e.g., from ventilating fans), vibration-absorbing fixtures should be used, especially when the vibrations are of high frequency. Fortunately, where high light outputs are required, a low-voltage and high-wattage lamp will be used; lamps of this type offer greatest resistance to vibration because of their inherent greater filament thickness. Coil type is also important; single-coil filaments are much more resistant to vibration than coiled-coil filaments.

(7) Switching
Incandescent lamps offer the simplst form of switching available, and repeated switching has negligible effect on the life of lamps run at rated voltage. Repeated switching of lamps run at over the rated voltage (e.g., projector lamps used as a pulsed source) will reduce lamp life substantially because of the excessive current surge through the cold filament. The main problem with switching tungsten–halogen lamps is that tungsten will be deposited on the lamp envelope if the lamps are unable to reach full operating temperature. As a rule-of-thumb guide, the lamp should run continuously and a shutter system be used if repeated irradiations of less than 30 sec are required.

(8) Cleanliness and general handling
Quartz-envelope lamps should be protected from contamination with dirt or grease because this will cause fine surface cracks to develop during operation and eventually lead to envelope failure.   The bulb should never be touched with the fingers, but handled with the paper sleeve or polythene glove supplied with the lamp.   Should the lamp become contaminated, it should be thoroughly cleaned with alcohol.   Despite their strong construction, quartz–halogen lamps can explode because the internal gas pressure is significantly higher than atmospheric.

B. Fluorescent lamps

Fluorescent lamps are low-pressure mercury discharge lamps.   The mercury vapour is excited by electrons emitted from the electrodes positioned at each end of the tube; on de-excitation the mercury emits UV light at 253.7 nm which is absorbed by the phosphor coating on the tube wall and is converted to visible light.   The spectral quality of this light is determined by the phosphor used.   Fluorescent tubes offer high efficiency, stable light output, a long life and low FR and IR output.   At the same time, however, all discharge lamps require more complex circuitry than the incandescent lamp.

(1) Lamp type
Choice of tube type depends on the spectral quality of the light required and availability of lamps which physically fit within the confines of a lamp or filter housing.   A wide variety of phosphors have now been developed. Lamps emitting in a fairly narrow waveband (half-power bandwidth ⁓ 50 nm) are useful for basic monochromatic sources between 350 and 750 nm (see Section IV).   Standard white fluorescent lamps emit mainly between 400 and 700 nm.   Appropriate types for generating monochromatic light are considered in Section IV and for polychromatic light in Section V.

(2) Lamp output and life
Light output decreases by about 5–10% during the first 100 h of operation. After this period, the deterioration rate is much lower and dependent on lamp type and operating conditions.   It is therefore advisable to let new tubes run for about 100 h before use.

   Under most conditions, fluorescent tubes can be run for at least 10,000 h before a substantial decrease in output necessitates replacement.   Lamps normally fail when the electron-emitting material on the electrodes becomes depleted below a certain level.   Although this emissive material is gradually depleted when the lamp is running, a large amount is lost during starting.

Frequent switching (<3 h) should be avoided. A lamp operating for only 30 min per start will have half the average life of a lamp operating for 3 h per start. Daily switching has negligible effect and the lamp will have a similar life to one running continuously.

Light output depends on the wattage per unit length of tube and on tube length. The use of reflecting surfaces within a filter housing or in growth cabinets and correct spacing of lamps will usually ensure adequate light output. Where lack of space does not allow the use of external reflectors, reflector tubes can often be used. These have an internal reflective coating around approximately two-thirds of the internal surface. Light is therefore emitted only over a 120° angle, resulting in an increased output of around 60% in the direction required. Tubes in which only a 60° angle is free from an internal reflective coating (aperture tubes) are also available.

(3) Voltage variation

Although less sensitive to changes than in incandescent lamps, decreasing or increasing line voltage alters lamp output, the extent depending mainly on the type of circuitry used. The line voltage can be decreased by between 20 and 50%, depending on lamp type, before they are extinguished. More important is that both over- and under-voltage operation results in decreased lamp life, decreased light output, and accelerated end discolouration.

(4) Ambient operating temperature

Although ambient temperature has little effect on lamp life, it is an extremely important factor in obtaining optimal light output. Most tubes operate best at an ambient temperature of $\sim 25\,°C$ and a decrease or increase from this value reduces output. When lamps are operated in enclosed areas the air exchange rate must be high to ensure optimal performance. Allowing the air temperature to reach 50°C for example, will result in an almost 30% decrease in light output. Indium-amalgam tubes provide higher energy levels than other tubes if the temperature in the lamp housing cannot be held below 50°C.

(5) Humidity

When separated from the growing space, little difficulty in starting due to humidity is liable to be encountered. Under extreme conditions, however, such as when a temperature increase in a cabinet is synchronised with switching on the lamps, excessive moisture may condense on the tubes. The use of silicone-coated tubes should assure reliable starting under most high humidity conditions because any film of moisture is dispelled into droplets.

(6) Switching

Devices for operating fluorescent lamps in a flashing mode are available and are suitable for "Rapid Start" type lamps.   However, for flashing rates of a few minutes (e.g., 5 min on, 5 min off), lamp life is severely reduced and end discolouration will be rapid.   Most fluorescent tubes can be dimmed, but not to the same extent as in incandescent lamps.   Excessive dimming will cause the lamp to be extinguished and it will not restart until high enough voltages are applied.   Reduction of the arc current also reduces cathode heating, thereby making restarting difficult.   Cathode-heating dimmer controls which provide a high voltage starting pulse or full voltage to the electrodes at all times are available, but the manufacturer's advice must be sought concerning suitable control gear for the type of tube being used.

(7) End discolouration

End discolouration in fluorescent tubes is usually a sign of old age.   There are various forms of discoluration, however, and knowledge of the commonest of these can provide easy identification of the cause.

A general blackening which extends toward the centre of the tube usually indicates the imminent end of lamp life, and the tube should be replaced.   If the blackening occurs early in lamp life, incorrect circuitry or line voltage should be suspected.   If blackening is restricted to only a few millimeters at the ends of a cold lamp, this will probably be a mercury deposit which will evaporate when the lamp is running.

Dark spots may be present in new lamps owing to mercury condensation. These will normally disappear with time and have negligible effect on light output.   Mercury condensation may also be evident as dark, lengthwise streaks on the lower side of the tube owing to the lower side being substantially cooler than the upper.   Rotation of the tube by 180° will normally enable the mercury to evaporate.   Brownish-black rings or elongated spots extending around the tube 20–60 mm from the base may appear during lamp operation.   This is a common occurrence, and such tubes can still be used unless light output is unacceptably low.

C. Short-arc discharge lamps

When high fluence rate monochromatic sources are required, incandescent lamps may provide inadequate energy, especially at shorter wavelengths. Under these circumstances it is necessary to use short-arc or compact-arc high-pressure discharge lamps.   The most suitable are short-arc versions of the xenon, mercury, or mercury–xenon discharge lamps.   Several other arc lamps are available; these include the carbon-, deuterium-, hydrogen-, tungsten-, sodium-, and zirconium-based systems. However, they all suffer from individual disadvantages which make them poorer choices than xenon- or mercury-based lamps for generation of light in the 300–800-nm waveband. The xenon and mercury short-arc lamps provide excellent point sources, from an arc which is typically 1–10 mm in length. The bulbs, 10–70 mm in diameter, are quartz and filled with gas under pressure. The majority operate on direct current, which aids stability and lamp life. Light output can be maintained constant by careful control of temperature and current.

The xenon arc offers a more continuous spectrum than the mercury arc, but suffers from a fall in output below 400 nm. The mercury arc has a very high output in the UV and blue wavebands. The decrease in energy above 600 nm makes the latter source little better than a high-wattage quartz halogen lamp in the 600–800-nm waveband. Compared with the xenon-arc lamp, mercury-arc lamps require more care in use, the arc is less stable and the lamps have a relatively short life. For the 300–500-nm waveband, however, mercury lamps offer a high light output without the very high levels of IR produced by the equivalent xenon lamp. If the user is restricted to only one type of lamp, the xenon offers the best compromise.

(1) Short-arc xenon discharge lamps

Short-arc xenon discharge lamps provide approximately constant light output between 400 and 800 nm with the exception of several small peaks in the 400–500-nm waveband. There is substantial falloff in emitted radiation below 400 nm. The spectral output in the visible waveband remains fairly uniform throughout the lamp's life and is not significantly dependent on supply voltage or current. The proportion of UV radiation emitted does decrease with age, however. These lamps have a very high output in the IR waveband.

Rated lamp life varies between ∼400 and 2000 h; the majority have a life of ∼1500 h, after which time output in the visible waveband will have decreased by ∼25%. There is some deposition of electrode material on the bulb during use, the rate of which increases with shorter running periods. This deposit decreases UV emission far more than visible light. Unlike mercury-arc lamps, xenon arcs may be switched on when hot (albeit sometimes with difficulty) and require virtually no warm-up time, although the arc may flicker for a few minutes. Frequent switching reduces lamp life, and the start button should never be pressed for more than 0.5 sec since this can damage the lamp.

Low-wattage lamps (250 W or less) will only operate satisfactorily at the recommended voltage. The output of higher-wattage lamps can be adjusted by varying the current. Operation above the recommended range will severely reduce lamp life, whereas operation below it will reduce both lamp life and the stability of the light output.

Most lamps are designed to operate within 20° of vertical. Electrode material usually deposits itself on the upper side of the envelope in horizontally burning lamps, but turning the lamps by 180° after 50% of their rated life will aid uniformity of light output. Correct focusing is extremely important both for optimising light output and avoiding damage to the lamps. Because they operate with direct current, the cathode (negative) is smaller than the anode (positive) to aid heat dissipation. The inverted image of the arc must be focused on itself and not on either of the electrodes.

When changing lamps, special care should be taken to ensure correct polarity and that the electrical contacts are in good condition. Failure to observe either of these pecautions will result in lamp failure within a few hours. Adequate forced cooling of the entire lamp is important. The socket temperature should not be allowed to reach 230°C, and care must be taken to ensure that the cooling is uniform. Unlike mercury discharge lamps, ventilation should continue for at least 5 min after the lamp is switched off. Water-cooled lamps are more complicated to operate and are of limited use because their arc length is usually too large and their output is relatively low.

(2) Short-arc mercury discharge lamps

The mercury-arc discharge lamp produces a discrete line emission superimposed on a relatively low-level continuous spectrum. The size and distribution of the mercury lines and the proportion of energy emitted as a continuous spectrum depends primarily on the gas pressure used. The life of these lamps is typically 200–400 h, although this is strongly dependent on the frequency of switching; a loss in light output of approximately 20% must be expected during this period. Lamp life is usually given for repeated 2-h burning periods; using a burning period of only 20 min will reduce lamp life by approximately half. Stable light output is not achieved until the end of the warm-up phase, and during this period a voltage increase and current decrease will be observed. The lamps should never be burned for less time than it takes to reach full operating temperature, which is usually between 5 and 15 min, depending on lamp type and running conditions. A thin film of evaporated electrode material deposits itself on the inner surface of the bulb during use, the rate of deposition increasing with the frequency of switching.

Lamps operating on direct current offer much longer lamp life, better starting and warm-up characteristics, and greater arc stability than those operating on alternating current. Even direct current lamps, however, do not offer complete reliability in starting. When difficulty is experienced with hand-started equipment, long pauses ($\sim$60 sec) should be allowed between each attempt. If the lamp has just been running for 15 min or longer, wait at least 5 min. The igniter button may be safely pressed for much longer ($\sim$5 sec) than with xenon-arc lamps.

Correct focusing is extremely important. Light output is symmetrical, except where the electrodes fall within the output angle. In older three-electrode lamps, the starting electrode also reduces the solid angle in which light is emitted. The electrodes in alternating current lamps are the same size and the highest light output is next to both electrodes. The light distribution within direct current arcs is asymmetrical and these lamps should be focused in the same way as the direct-current xenon arc. Lamps are usually burned in a vertical position and the positive end must be at the bottom in direct current lamps.

Mercury-arc lamps require different ventilation from that of xenon-arc lamps. The ventilation should be arranged so that the bulb itself does not receive a direct blast of cooling air. At the same time, however, the temperature of the sockets shuld not be allowed to exceed $\sim$230°C. The ventilation should be wired to switch off at the same time as the lamp.

(3) Safety precautions

The danger of explosion because of the high running pressures used in high-pressure discharge lamps cannot be overemphasised. The likelihood of an explosion increases with lamp age, so lamps should be replaced at the end of their rated life. Apart from the danger to personnel, an explosion will damage expensive internal projector equipment, such as mirrors. The lamp must be operated only when all lamp housing covers are in place. Although the possibility of an explosion is greatest when the lamp is hot, a face shield and gloves must be worn when handling the lamps. If a protective casing is supplied with the lamp, it should be removed after fitting the lamp and attached again before removing it. The quartz bulbs must never be touched with the fingers, although accidental grease marks can be removed with alcohol.

Careful attention should be paid to the specific ventilation requirements of the lamps. Apart from the great heat output of high-wattage lamps, the possibility of a fire following lamp explosion must be considered. The large quantities of ozone produced by some lamps must be expelled from the building. All ventilation openings must be constructed so that flying glass produced by an explosion is contained within the housing. The direct-current voltage used to start these lamps can be as high as 50 kV, and a potentially lethal current flows during normal operation, so suitable precautions must be taken to avoid electrocution.

The arc must never be viewed directly. Apart from the dangers of high

UV output, the high emission in the visible waveband can cause permanent damage to the eyes in the same way as looking at the solar disc. Even a brief glimpse of the source may cause the unpleasant and often painful sensation similar to that of having sand trapped under the eyelids; this appears some 8–12 h after viewing the arc and lasts for several hours. Excessive exposure to radiation below approximately 330 nm can promote skin cancer and cause keratosis of the cornea. The lips and cornea are very susceptible to serious damage by UV radiation. If extensive adjustments have to be made near an unfiltered xenon or mercury high-pressure lamp, then gloves and a face mask fitted with a UV-absorbing window should be worn. Care should be taken to ensure that the eyes are also protected at the sides because light may strike the cornea without the operator actually seeing it.

When observing the performance of equipment near bright light sources which contain little or no UV radiation, neutral-density or coloured-glass filters can be used for brief periods. Alternatively, a piece of cardboard with a pinhole (diameter ⁓ 1 mm) can be used; the most convenient method is to curl the index finger to form a pinhole gap between finger joints and to view through this.

III. FILTERS

A wide variety of filter types are available for selecting a portion of the radiation emitted by a light source. The choice of a correct filter is determined by the waveband required, resolution needed, size of the area to be irradiated, cross-sectional area of the light beam, heat load, resistance to humidity or liquid water, and cost. Most filters operate on the basis of selective absorption and transmission of the incident light, whereas the optical interference filter selectively interferes with the incident light, thereby reflecting unwanted wavelengths and transmitting the waveband required. The selective-absorption filters can be subdivided into cutoff types, which transmit most of the radiation on one side of a certain wavelength and little or none on the other, and bandpass filters, which transmit mainly in one or more distinct wavebands.

It is useful to look at filter transmission curves on a logarithmic scale because many filters have important transmission characteristics which are not immediately evident from a linear scale (see Fig. 1). Inadequate blocking can be very important in action spectroscopy, for example, where the integrated energy outside the main passband may be relatively high and give spurious results if the major passband of the filter is far away from the actinic wavelength.

Bandpass filters are primarily characterized by their *peak transmission* or

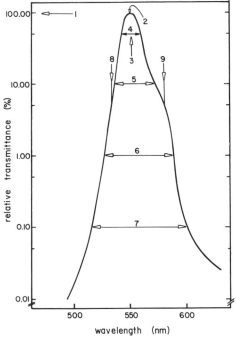

Fig. 1. Transmission characteristics of a bandpass filter. 1 = peak transmission or transmission maximum; 2 = peak wavelength or wavelength maximum; 3 = central wavelength; 4 = half bandwidth, half-power bandwidth, or bandwidth; 5 = $\frac{1}{10}$ power banwidth; 6 = $\frac{1}{100}$ power bandwidth; 7 = $\frac{1}{1000}$ power bandwidth; 8 = cut-on (5%); 9 = cutoff (5%).

transmission maximum, which is their maximum percentage transmission, and by their *peak wavelength* or *wavelength maximum* (λ_{max}), which is the wavelength at which maximum transmission occurs. Because many filters are not symmetrical, it is often necessary to measure the *central wavelength* (λ_o); this is the wavelength at the midpoint of the half-power bandwidth (see below); if the bandpass is perfectly symmetrical, then $\lambda_{max} = \lambda_o$. The *half-power bandwidth* (sometimes given as *half bandwidth* or *bandwidth*) describes the bandwidth at 50% of peak transmission. The half-power bandwidth is sometimes expressed as a percentage of the central wavelength; this is the *percentage bandwidth*. A filter with a central wavelength of 550 nm and a half-power bandwidth of 5.5 nm, for example, has a 1% bandwidth. Because of the possible confounding effects of long tails in the transmission of filters mentioned above, it is often necessary to give more information on bandwidth than the half-power bandwidth alone. Supplying information on the *tenth-power* and *hundredth-power* bandwidth is usually adequate; with filters with extremely shallow slopes, the *thousandth-power bandwidth* should be given.

It is sometimes useful to indicate where the major transmission band of a filter begins and ends. The *cut-on* is the point on the transmission curve, going from zero at lower wavelengths to maximum transmission, at which the transmission is a certain percentage (usually 5%) of peak transmission. The *cutoff* is the point on the same curve, going from maximum to zero transmission at higher wavelengths, where the transmission is a certain percentage (usually 5%) of peak transmission. The percentage transmission at cut-on or cutoff should be cited. Note that "cutoff filters" usually refer to filters which absorb lower wavelengths and transmit higher wavelengths. The *blocking range* is the range over which wavelengths other than those required are reduced by an acceptable amount. The extent of the blocking is given by the *degree of blocking,* which can be expressed as a percentage or as absorbance. Some manufacturers supply a *rejection ratio* or *degree of rejection,* which is the ratio of peak transmission to the transmission outside the passband; it is usually expressed as a percentage.

A. Liquid chemical filters

Liquid chemical filters contain salts or dyes in solution. They provide a relatively cheap means of filtering radiation but are expensive if a circulation system is needed to aid cooling. Many combinations can be prepared to act as either cutoff or bandpass filters in the UV, visible, and IR regions. Water is a useful absorber of unwanted IR (see Section V). Generally, however, the disadvantages outweight the advantages; almost all of these filters deteriorate with time, and this process is often irradiance-dependent, so regular monitoring of transmittance is necessary.

B. Gelatin filters

A typical example of gelatin filters is the Kodak "Wratten" filter (Kodak). The great majority are formed by mixing synthetic organic dyes in gelatin. These filters are cheap and have relatively sharp absorption bands. Apart from being fragile, and therefore needing to be supported in a glass or Perspex sandwich, their main disadvantages are their sensitivity to moisture and temperature. Contact with high humidity or liquid water causes gelatin films to swell and become cloudy; this, and their maximum usable temperature of ~ 40°C, makes them unsuitable for many applications.

C. Cinemoid filters

"Cinemoid" filters (Rank Strand Electric) have an acetate sheeting base which is more flexible and much stronger than gelatin. Although only a few provide wavelength cutoffs as sharp as those available with the equivalent

Wratten filter, they are more resistant to moisture and offer a useful but cheap material for the construction of basic monochromatic sources (see Section IV). Like gelatin filters, their physical flexibility demands some form of support if they are to be used over large areas.

D. Plastic filters

Plastic filters (e.g., ICI Perspex, Röhm Plexiglas) are excellent for filtering large areas of light because they are fairly cheap and are rigid enough to be self-supporting. They are therefore useful for light fields (Section IV) and for filtering additional FR radiation in environmental simulators (Section V).

The heat tolerance of plastic filters is good, distortion starting above ~110°C. Operation above 80°C, however, may produce permanent changes in transmission properties. Their transmission characteristics are otherwise relatively stable. Because they absorb IR radiation effectively, only energy levels should be used unless heat-absorbing glass is placed between the light source and filter. Plastic filters can be used for narrow-waveband sources in projector systems (Section VI) but they do not have the high abrasion and heat resistance of glass filters.

E. Glass filters

Owing to their high heat tolerance and durability, glass filters (e.g., Schott) provide the best form of cutoff filter where high irradiances are used; within limitations, they are also good bandpass filters. Glass filters are expensive and normally available only in small sizes, so they are more suitable for use in projector systems than for producing basic monochromatic light sources.

Cutoff filters are available for absorbing radiation above or below a chosen wavelength. They are especially useful for eliminating unwanted orders transmitted by diffraction-grating monochromators and interference filters. Single glass cutoff filters which transmit only radiation above a certain wavelength (long-pass filters) are available for the UV, visible, and IR wavebands. These filters have a sharp transition between rejection and transmission and a very high internal transmittance. Single glass filters which transmit below a certain wavelength and reject above it (short-pass filters) are less satisfactory. They have a much shallower transition from blocking to transmitting and often have a second waveband of transmission. This can be partially overcome by using a combination of filters. Placed separately, approximately 8–10% of the incident light is lost per filter (4–5% at the air–glass interface and 4–5% at the glass–air interface). This can result in a substantial loss of light if several filters are used.

Combinations of filters that are cemented together can be used, thereby improving transmission.   However, their construction usually limits their use to situations where filter temperature does not rise above 70°C during continuous use, although 100°C is permissible for short periods if the temperature rise is not too rapid.   Despite their construction, composite cutoff filters with an acceptably sharp transition from minimum to maximum (i.e., from 1% to maximum transmission over less than 30 nm) rarely have a transmission in the required waveband as high as 80%, whereas a single glass cutoff filter usually has a transmission of over 90%.

By using suitable combinations of coloured glass filters, bandpass filters can be constructed.   They can also be purchased as precemented combinations.   The same precautions apply as with cutoff filters; the use of several separate filters will cause a large reduction in fluence rate, but cemented filters have lower heat tolerance.   The face which should be directed toward the light source is usually marked.   Cemented filters also suffer from the same humidity restrictions as interference filters and should be handled with equal care (see Section III. F).   Although fairly narrow wavebands can be achieved with glass filters, their low transmission usually makes the interference filter a better alternative.   As a rule-of-thumb guide, accounting for both cost and suitability, glass bandpass filters are superior for wide bandwidths (HW > 50 nm) and interference filters for narrower bandwidths.

Some glass filters fluoresce under UV light.   The extent differs, and the emission wavelength can be in various positions of the visible spectrum. When high UV emitting sources are being used, it is important that the fluorescence properties of the filters are checked.   Prolonged irradiation with sources emitting high levels of UV radiation may also alter the transmission characteristics.   This normally results in an increase in absorption, particularly in the B waveband.

Sensitivity to temperature of individual glasses varies, but 250–300°C is the usual maximum before permanent damage results; some may withstand 500°C.   Most glasses are resistant to thermal shock if elementary precautions are taken (e.g., inserting a heat filter between light source and coloured glass, using forced ventilation which switches off at the same time as the light source, and not placing water and coloured glass together when the glass is closest to the light source).   Glasses with a high coefficient of thermal expansion have a low resistance to thermal shock.   Some manufacturers offer "hardened" glasses; these are preferable if high-output light sources are used.   A tight-fitting holder should not be used, especially if the glass has a high thermal expansion coefficient.   Temperature has a large effect on the cutoff wavelength.   Although the extent also depends on the thickness of the glass, increasing the temperature by 100°C will typically result in a 15–20-nm shift of the cutoff to longer wavelengths.

F. Interference filters

Interference filters are constructed from several layers of semitransparent surfaces which, depending on the spacing of the layers, cause multiple reflection and constructive interference of the required wavelength which passes through the filter. The majority of the other wavelengths interfere destructively and are reflected back out of the filter. Although not providing the spectral purity of a correctly used double monochromator system, interference filters have the advantages of relative ease of handling and low expense. The most important point to bear in mind with interference filters is that many have secondary transmission peaks which must be eliminated by using appropriate cutoff filters and light sources.

Like cemented coloured glasses, interference filters cannot withstand high temperatures. Heat must be removed from the light before it reaches the filter (see also Sections V. B. 5 and VI); the visually most reflective surface of the filter should face the radiation source. For continuous use, 70°C is the maximum allowable temperature, although intermittent use at 90–100°C should not result in any damage. Using the filters above ~110°C will result in permanent alteration (primarily in the central wavelength) of the optical characteristics, or even destruction. It is important that filter temperature by changed slowly (during both heating and cooling) because of the different coefficients of thermal expansion of the various layers. No damage should ensue if temperature is changed at a rate of less than 10°C per min. The central wavelength increases with increasing temperature; the amount varies from filter to filter but is usually less than 2.0 nm and typically less than 1.0 nm when changing from 25 to 75°C. Temperature also affects bandwidth and peak transmission values, but this is normally by a negligible amount.

The hygroscopic dielectric layers must be protected from excess moisture. Most filters are produced with a moisture-resistant sealing cement around the outside rim. The thickness of the seal varies, but if it is chipped by misuse in handling, this will allow the ingress of moisture which will ruin the filter. As an extra precaution, filters should be stored with a desiccant. If they must be used constantly under humid conditions, special encased filters should be used. When extremely high fluence rates are needed, it is possible to use the filters in flowing water if the edges are kept dry (see Section VI).

The angle of incidence of the light at the filter is extremely important, mainly because the apparent thickness of dielectric or metallic layers in an interference filter changes with the angle of incidence. Ideally, the filters should be used in collimated light arriving normal to their surface. Placing the filter at off-normal angles results in a shift of the transmission peak to lower wavelengths and increases the bandwidth of the transmitted light.

The extent of this shift is dependent on the refractive index of the spacer layer(s) and the wavelength maximum (λ_{max}) of the filter. The wavelength shift per degree the filter is angled away from normal incidence increases with increasing λ_{max}. At 800 nm, a 10° off-normal positioning is unlikely to cause a greater than 6% shift to lower wavelengths, and 3% is more typical. The exent of the shift depends on the construction of the filter. With parallel light in the 300–800-nm waveband, an entry angle of 10° and an exit angle of 20° can be considered to have negligible effect. Done correctly, off-normal positioning of the filter can be a useful method for finely adjusting to the required λ_{max}. It should not be forgotten, however, that turning the filter also broadens and flattens the transmission curve and that the effect is especially marked in narrow band-pass filters. The effect of a divergent or convergent incident beam is less than that of an off-angle collimated beam because the angle of incidence in the former two situations is averaged.

G. Polarising filters

Although there are several forms of polarisation, linearly polarised light is that which is usually required in photomorphogenetic research. Two general types of polarisers are appropriate. One operates by dichroism (e.g., iodine-based sheet polarisers) and the other by double refraction (i.e., polarising prisms). The final choice depends on the degree of extinction required, price, and the fluence rate which must be reached.

Sheet or film polarisers are useful where high extinction ratios are not required. They have the advantage of relative cheapness and are available in large sizes. Some filters cover the entire 300–800-nm waveband, but their efficiency varies greatly with wavelength. When both the UV and visible waveband have to be polarised, it is usually preferable to use more than one filter type if optimal transmission and extinction are sought. The angle of incidence of the light must be kept as close as possible to 90° in order to reduce transmission of unpolarised light. Purity can be improved by operating filters in tandem. Physical protection of sheet and film polarisers varies; whereas some are enclosed in a glass or quartz sandwich, some are made of unprotected plastic. Those which are intended to polarise the far UV can consist of a soft coating on quartz; this surface must never be touched. None of these filters can withstand temperatures higher than ~80°C, and with some the maximum is only 60°C. Most filters cannot withstand continued or high fluence rates, and polarisers for the visible waveband are usually very sensitive to UV light.

If high purity of polarisation is required, prism polarisers must be used; a single pair of prisms should provide an extinction ratio of better than 1×10^{-5}. They are manufactured with air spaces between the elements or with cemented prisms. Those with air spaces have a narrow light-acceptance angle (typically ∼ 10°, wider angles of incidence allowing transmittance of unpolarised light). Cemented prisms offer wider acceptance angles, but this is at the expense of some loss of transmission, especially in the UV waveband. Care should be taken with cemented types to avoid overheating.

Polarisers should be mounted in rotators. Rotators for prism polarisers are more precise (and expensive) than those for films but should be used for maximum performance. Although obvious, it should be remembered that polarised light will become depolarised to various extents by reflection and scattering by objects placed in the light path. Filters, lenses or other optical elements, when needed, should be placed between the light source and the polariser, and not between the polariser and the irradiated object.

H. Neutral-density filters

Neutral-density filters are used to reduce fluence rate by fixed amounts. Despite their names, all filters absorb selectively, the exent varying from manufacturer to manufacturer and with wavelength. Major deviations from their nominal transmission value must be expected in the UV waveband unless special UV filters are used. A marked increase in transmission above 700 nm is often found in higher-density filters. As with coloured glasses, care should be taken to discern between internal transmission and percentage transmission of the incident light; the manufacturer's transmission curves normally present the former information, while general specification tables usually present the latter. Densities ranging from 0.15 to 5.0 (∼ 71 – 0.001% transmission) are commonly available.

The heat resistance of neutral-density filters is similar to that of coloured-glass filters, although special care should be taken with high-density filters. Thermal shock should be avoided and the filters must be kept clean and stored in dry, clean containers. The spectral transmission characteristics of each individual filter must be known so that the appropriate filter can be inserted into the optical system to produce a known change in fluence rate. Density measurements from the glass melt used for each batch are usually supplied. It is useful to have a wide range of filters to enable instant production of any fluence rate required. Whereas this method avoids the need to remeasure the actinic light every time a change in fluence rate is made, it is necessary to recalibrate the filters occasionally and to mark each filter individually.

The most suitable form of neutral-density filter for covering large areas is aluminum gauze. It has negligible effect on the spectral quality of the transmitted radiation and is heat and corrosion resistant. Several mesh sizes are available, but gauze intended for use as mosquito netting (1.4×1.4-mm grids; 0.28-mm wire diameter) is easily obtained and several layers can be used to produce the desired transmittance.

IV. BASIC MONOCHROMATIC LIGHT SOURCES

The UV, B, G, R, and FR monochromatic light sources are the most important pieces of basic equipment for photomorphogenesis research. For this reason, special care should be taken to ensure that only the desired wavelengths are present. Common difficulties are the elimination of FR radiation from UV, B, and G light sources and the suppression of IR radiation from all five. Because there are no ideal light source and filter combinations, the general basic requirements and some commonly used suitable combinations (Table I) are presented here. The filters in the table have been limited to those produced by manufacturers most often cited in the literature. Others are available, but their transmission characteristics should be measured to ensure that they are suitable.

If meaningful comparisons are to be made between the effects of different

Table I. Basic monochromatic light sources.[a]

Number	Waveband	Light source	Filters
1	B	Fluorescent	Cin 20 (×2)
2	B	Fluorescent	PG 0248
3	B	Fluorescent	Ros 863
4	G	Fluorescent	PG 303 + PG 0248
5	G	Fluorescent	PG 303 + PG 627
6	G	Fluorescent	PG 303 + PG 701 (×2)
7	G	Fluorescent	PG 303 + PG 701
8	G	Fluorescent	Cin 39 (×2)
9	G	Fluorescent	Ros 874 + Ros 877
10	G	Fluorescent	Cin 46 + Cin 1 + Cin 20
11	G	Incandescent	Cin 39 + BG 18 (2 mm)
12	G	Incandescent	BG 18 (2 mm) + GG 495 (3 mm)
13	G	Incandescent	Cin 24 (×3)
14	G	Incandescent	Cin 39 (×3)
15	R	Fluorescent	Cin 14 + Cin 1
16	R	Fluorescent	Cin 6 + Cin 1
17	R	Fluorescent	PG 501
18	FR	Fluorescent	Ros 823
19	FR	Incandescent–fluorescent	Cin 5a + Cin 20
20	FR	Incandescent–fluorescent	Cin 5a (×2) + PG 627 (×2)
21	FR	Incandescent–fluorescent	PG 501 + PG 627 (×2)
22	FR	Incandescent–fluorescent	FRF 700 (3.2 mm)
23	FR	Incandescent–fluorescent	ICI 400 + ICI 600 (3.2 mm)

[a] BG and GG = Schott coloured glass filters; Cin = Rank Strand Electric "Cinemoid"; PG = Röhm & Haas "Plexiglas"; ICI = Imperial Chemical Industries "Perspex"; Ros = Rosco Laboratories Inc.; FRF = Westlake Plastics Company far-red filter. The recommendations are for 3-mm thick Plexiglas; the thickness for other rigid filters is given in parentheses. Unless otherwise stated (i.e., "×2" or "×3"), only one layer of each filter is required. Further details are presented in the text, using the filter numbers given in the table.

monochromatic light sources, the IR output must also be the same. Total elimination of heat is impractical, but it can be reduced to an acceptable level by using appropriate heat filters and adequate ventilation of the lamp housing. Whenever fluorescent tubes are used, a 50-mm band of black tape should be wrapped around the ends of the tubes to absorb the FR and IR radiation emitted by the cathodes. An indication of the variation in heat output between the sources can be obtained with a precise mercury thermometer with a blackened bulb. Absolute measurements should be made with a thermal detector (see Chapter 4).

A. Ultra-violet light

Fluorescent tubes with phosphors which emit primarily in the UV-A waveband provide convenient UV sources. Tubes made of black glass ("Woods glass") should be used and not the tubes designed for UV sun lamps, since the latter also emit in the B waveband. 20-W and 40-W lamps which fit standard fluorescent lamp control gear are commonly available. Small tubular (150 mm long, 26 mm diameter) 6-W tubes which fit standard lamp sockets (screw type, E27) can be used for irradiating small areas. As a general rule, however, the longest tubes (1200 mm) are most efficient, partially because 50 mm of the tube adjacent to the cap has to be covered by black tape to reduce FR emission. A high pressure mercury lamp with a black glass envelope can be used to obtain maximum output near 365 nm, although if narrow waveband sources are required, interference filters are usually more suitable. The light source should be separated from the plants by a UV-transparent barrier such as Plexiglas 233, 606, or 612 (3 mm) or Cinemoid 14 or 33. Regular monitoring of light output is necessary. Additional care should be taken to monitor redistribution of the phosphor within the tube, which results in transmission of high levels of FR light. This is most easily done visually by observing the tubes through an FR transmitting filter. These light sources excite fluorescence in the finger nails, teeth and vitreous humour of the eye. Direct inspection of the tubes is harmless, and the misty appearance of the tubes and surrounding objects caused by fluorescence in the eye ceases as soon as the eyes are averted.

B. Blue light

A pure B light source is difficult to achieve because most B filters also transmit FR light. Suitable FR-absorbing filter combinations are shown in Table I. These filters also have the function of removing the relatively high levels of UV radiation emitted by some B fluorescent tubes. B fluorescent tubes are better than white fluorescent tubes, and incandescent lamps must never be used with the filter combinations shown.   Note that Plexiglas 627 should not be used since it transmits FR light.   An alternative B source is the gallium–arsenide discharge lamp used for photographic purposes.

C. Green safelight

A G light is used because if offers a compromise between good vision and low effectiveness in causing photomorphogenetic responses.   The human eye reacts maximally to light near 555 nm.   Higher wavelengths are increasingly more effective in photoconverting Pr to Pfr, while lower wavelengths start to encroach on the absorption waveband of the B-light photoreceptor(s).   A survey of action spectra for photomorphogenetic and phototropic responses indicates a compromise wavelength of 520–530 nm for these three factors. The final choice depends on the objective and needs of the study; no G light can be regarded to be safe unless this is proved to be the case.   Extremely low levels of G light can be detected by a plant, so the so-called safelight should be used at the lowest practical fluence rate and for the briefest period possible.

Removal of radiation above 680 nm is difficult without expensive filters, so all safelights for irradiating large areas should be based on the G fluorescent tube.   Care should be taken to choose a tube which emits over a narrow bandwidth; current Philips TL 40W/17 tubes, for example, are preferable to Osram D 40W/235V for this application.   Filter combinations (4) and (5) provide very effective safelights with a λ_{max} near 525 nm.   If visually brighter sources are needed, source (6) or, brighter still, source (7) may be more suitable; the bandwidths of these sources are greater than for (4) and (5), and the λ_{max} is slightly higher (near 530 nm).   If Cinemoid or Roscolene filters are preferred (they can be conveniently wrapped around fluorescent tubes), combinations (8) (λ_{max} near 535 nm) and (9) may be used to cover most applications; greater spectral purity, a lower λ_{max} ($\sim$ 505 nm), and a visually dimmer source is provided by combination (10).   Combinations (11), (12), (13), and (14) are for use with portable sources (i.e., torches), only.   The G495 glass fluoresces and must therefore be placed between the light source and the BG18 glass.   Combinations (11) and (12) are superior to (13) and (14).

D. Red light

R light is produced by filtering the light from white fluorescent tubes through R plastic, which absorbs all radiation below 580 nm (Table I).   The correct choice of fluorescent tube is important.   Most white fluorescent tubes have an emission maximum near either 580 or 660 nm; the latter type is preferable.   The final choice is then made on the basis of which tube emits the lowest amount of FR radiation.   If high fluence rates are required, mono-chromatic R fluorescent tubes can be used, although those from some manu-facturers emit disproportionly high levels of FR light.   In the examples presented, filters (16) and (17) transmit over 10% more radiation at 660 nm than filter (15), but the 50% cut-on wavelength is approximately 25 nm shorter at $\sim$610 nm.

E. Far-red light

The FR filters differ primarily in the wavelength and steepness of the cut-on curve and the amount of IR radiation which is transmitted.   Increasing the cut-on wavelength decreases the photostationary state established by the source.   If the source is also to be used for dichromatic (Section VII) or polychromatic irradiation (Section V), it must be borne in mind that longer cut-on wavelengths reduce the fluence rate of radiation capable of convert-ing Pfr to Pr.   When describing the fluence rate produced by an FR source, the upper wavelength limit should be provided, since many light detectors measure the IR radiation produced by some of these FR sources.

V. ENVIRONMENTAL SIMULATORS

A. Basic requirements

The primary objective of using environmental simulators is to be able to apply defined light treatments which are representative of the natural envi-ronment.   Exact replication of natural conditions is less important than the ability to provide reliably uniform, reproducible, and definable radiation.   In many instances, the additional facility of providing unnatural light spec-tra (e.g., night-breaks with monochromatic light) is of advantage.

The construction of a sophisticated, artificially lit "controlled environ-ment" is not justified if factors other than the actinic light cannot be held constant.   It is most important that the responses of the plans are due to incident light alone and are not caused by secondary factors such as carbon dioxide, temperature, liquid water, ethylene, humidity, mineral nutrition, and air pollutants.   Depletion of carbon dioxide in the air is a major problem when working with green plants.   Rapid air circulation is the best solution; an air exchange rate of five times per min (preferably downward) will over-come both carbon dioxide and temperature problems.   Care should be taken to avoid overcrowding, since this can limit air movement.   Conduc-tion of heat can cause large variations in temperature within an environmental simulator. Wood and hardboard are the cheapest building materials available and offer a high degree of insulation. The additional use of foamed polyurethane or expanded polystyrene of at least 60-mm thickness will reduce temperature problems owing to conduction to negligible proportions. Floors are liable to receive excess water and should therefore be constructed from perforated aluminium. Since the floor is used for correct air circulation, it does not need to act as an insulator. Most environmental simulators will require a cooling system to maintain air temperature. Cooling systems cause drastic reductions in humidity and therefore necessitate the use of a humidifier.

B. Simulation of daylight

High fluence rates are best achieved with high-pressure xenon long-arc lamps which have a colour temperature approaching 6000 K, which is similar to sunlight. Although it can also provide high fluence rates, the high-pressure mercury-arc lamp is less representative of sunlight, consisting of a basic continuous spectrum with several discrete peaks. This lamp also has several operational disadvantages when compared with the xenon source, the major of which are its shorter life, switching difficulties, and the need for water cooling in some of the higher-power lamps.

Fluorescent tubes are cheaper, simpler, and safer than the high-pressure arc lamps, but have the disadvantages that very high fluence rates are difficult to attain and that their spectra do not represent daylight. If ⁓ 30% of the rated wattage is supplied by incandescent lamps, however, plants representative of those grown in natural daylight will be obtained. If a bank of white fluorescent tubes is used, some of these can be supplemented with colour tubes to obtain the required spectral balance. Obtaining major spectral quality changes which are representative of those occurring in the natural environment requires construction of special growth cabinets. The principal components of an environmental simulator are a background of PAR and a source of FR light which can be superimposed on the PAR. The various components of an environmental simulator are represented schematically in Fig. 2.

(1) Lamps for PAR

Although "normal" growth is obtained with light sources containing some FR radiation, a PAR light source may be required which enables maximum possible photoconversion of phytochrome into the Pfr form. For this purpose, fluorescent lamps which contain the minimal amount of FR and a high output around 660 nm should be used.

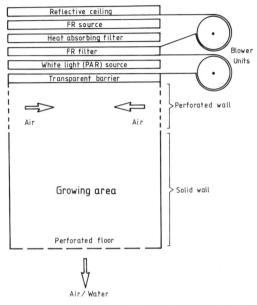

Fig. 2. Schematic layout of a plant growth cabinet which provides light approximating the spectral quality of daylight.

(2) High-output FR sources

The most appropriate light sources for FR radiation are the high-pressure xenon arc, quartz–halogen bulbs, and household incandescent bulbs. If the cabinet is also going to be used as a broad-waveband WL source, then the xenon arc is the best alternative because of its high light output of a spectrum representing daylight. If the lamps are only going to be used as a source of FR radiation, then quartz–halogen bulbs are preferable. These lamps have the advantages of being cheaper, simpler to install, and having a more extensive dimming range than that of xenon-arc lamps. Tubular (linear) halogen lamps are available in various ratings up to 2 kW with an average life of ∼ 2000 h. Although halogen bulbs of up to 10 kW can be purchased, their rated life is relatively short. The use of several low-wattage lamps produces a more homogeneous light distribution across the growing area than a few high wattage lamps.

(3) Reflective surfaces

Reflective surfaces are necessary both to minimise light loss and to provide homogeneous light distribution. Incorrect installation or design of reflectors in the lamp compartment can result in worse light distribution across the growing surface than when no specific reflector is used. For the lamp compartment, polished aluminium provides a highly reflective surface at reasonable cost. Aluminised or silvered plastic adhesive film will not withstand the heat generated by the lamps. For the growth compartment, aluminised "Mylar" or reflecting silvered plastic film ("Melinex") are cheap and suitable but tear easily when abraded. If lamps must be used within the growing compartment, it should be borne in mind that even low levels of heat can cause discolouration and peeling. Greater resistance to abrasion and heat is offered by specular aluminium (e.g., "Alzak") which is also fairly corrosion resistant.

(4) FR filter

The cheapest form of filter is constructed with acetate sheets (e.g., Cinemoid). Although acetate sheets are fairly durable and are moisture and heat resistant, their thinness necessitates that they be supported by a sandwich of glass or Perspex to avoid warping. A suitable self-supporting filter can be made with plastic sheeting such as Perspex or Plexiglas. Both the acetate and plastic FR filters transmit a small amount of PAR above 650 nm, the exact amount depending on the combination used. The removal of this radiation, however, results in the concomitant removal of a disproportionally large amount of FR radiation. When very high photon fluence rates of transmitted FR are required, a combination of one layer of 3.2-mm thick R 400 Perspex and one layer of 3.2 mm thick G 600 Perspex provides an excellent FR filter combination.

(5) Heat-absorbing filter

Although both window glass and the colour filters each absorb a small amount of IR radiation, a specific heat-absorbing layer is necessary to reduce the heat load on the plants to an acceptable level. Two basic methods for removing the heat produced by the lamps are available, but both have their disadvantages.

Heat-absorbing glass is the best form of IR filter because of its compactness and the lack of the obvious dangers of using water and electricity together. It is, however, extremely expensive. Heat absorbing glass is available only in limited sizes; a balance must be struck between the number of sections used and the expense of replacements. Many small plates necessitate extensive supporting grids between sections, thereby causing shadows, but the replacement of broken plates is relatively cheap. A few large sheets produce fewer shadows, but are more expensive to replace.

Transmission curves of heat-absorbing glass should be studied carefully when choosing the type and thickness. Choosing a glass which removes the maximum amount of IR radiation will result in a disproportionally large loss of FR light. A glass which transmits less than 10% of the incident IR radiation is usually adequate because the filter and window glass used between the light source and the plants also absorb a high proportion of the heat. It is important that a high-speed air current blows across the heat-absorbing glass to conduct the heat away before it is reradiated.

A water filter is an effective absorber of excess heat. A 100-mm layer of water removes over 90% of the radiation above 900 nm. A greater depth is not recommended because of the weight of the water and the high absorbance of FR wavelengths. As a lower limit, 30 mm of water will remove ~ 75% of the radiation above 900 nm. The addition of ferrous ammonium sulphate to the water results in a disproportionally high loss of FR radiation. A further disadvantage is that it is necessary to add 1 or 2% sulphuric acid to decrease the rate of oxidation of the ferrous ammonium sulphate solution. This in turn makes it necessary to use a totally corrosion-resistant system.

Ideally, a water heat-absorbing filter system should be of the closed type, being circulated and cooled with an air–water or water–water heat exchanger. Where this is not possible, a stirrer should be employed to decrease evaporation rate from the warmer upper layers. Under heavy heat loads, evaporation may still be excessive, but this can be overcome by coils above the water surface containing flowing cold water to condense the water vapour. In a "through-flow" type system, an antiscale agent will be necessary in most areas. If a closed system is used, distilled water will avoid the formation of scale, but an algicide must be used to avoid the rapid buildup of algae under the almost perfect growing conditions.

(6) Transparent barrier
A transparent barrier should separate the plants from the lamps. This aids continuous light output by reducing the accumulation of dirt and by allowing optimal air-flow rates for lamp cooling. A barrier also prevents convected heat from the lamps from reaching the plants.

C. Operation and adjustment

(1) Uniformity of light distribution
Owing to their low luminance, fluorescent tubes used for background PAR radiation provide fairly homogeneous light distribution over the plant growth area. The supplementary FR light produces a less uniform light distribution. This problem is best reduced at its source by using several low-wattage lamps rather than only one or two of high wattage to obtain the same fluence rate. Fluence rate is usually lowest at the edges of the growth area; this can be partially overcome by using slightly higher-wattage lamps near the walls than at the centre, but individual experimentation is required

to determine the best positioning of lamps. If floor height is adjusted, the distribution of the light across the floor must be remeasured, although the correct use of reflective surfaces helps to reduce spatial variations in fluence rate. All efforts to obtain uniformity can be defeated by mutual shading by the plants or by using containers whose walls partially shade the seeds or seedlings. The increase in variability results in a need to have more replicates to achieve the same degree of confidence.

(2) Uniformity of light output

Fluorescent lamps are very sensitive to changes in temperature. Light measurements should therefore not be made until the temperature within the lamp compartment has equilibrated. Although dimmer units are available for fluorescent lamps, they are fairly expensive and fluence rate can be readily adjusted by altering the number of tubes which are switched on and by adjusting floor height. A greater range of dimming is possible with xenon-arc lamps and, more especially, with incandescent lamps. The output of fluorescent lamps decreases with hours of use and number of starts. The decrease in output is often greater than that claimed by the manufacturers and should be checked regularly. Details of these and other factors affecting light output are covered in Section II.

(3) Supplementary lighting

Because an environmental simulator is normally equipped with automatic time-switching clocks, the inclusion of a facility for automatically controlled "end-of-day" or "night-break" irradiations presents negligible extra cost. For broad-band monochromatic light in the 300–750-nm waveband, several types of coloured fluorescent tubes are available. A suitable number can replace some of the tubes used for PAR if these can be switched separately. Alternatively, a separate circuit and fittings can be used within the lamp compartment, or the lamp can be placed within the growth compartment itself if only brief periods of irradiation are required. If an end-of-day R or FR treatment is given as a supplement to the PAR irradiation, it is important that both light sources switch off simultaneously.

(4) Breakdowns and their prevention

The possibility of breakdowns or significant changes in any environmental factor is greatly reduced by preventative maintenance. To avoid sudden lamp failure, it is advisable to change all lamps after 80% of the rated life, this rating being based on the conditions under which the lamp was used. Several factors affect lamp life (see Section II). A minimum requirement with environmental simulators is the continuous monitoring of fluence rate and temperature. If possible, carbon dioxide concentration and humidity should also be monitored.   Since many experiments may take several days, remote operating alarm systems can save an experiment.   If a power cut is known to be imminent during long experiments, the hiring of an auxillary generator may be worth the cost involved.

VI. MONOCHROMATOR SYSTEMS

The simplest and most convenient equipment for generating monochromatic light is a projector and narrow-band filter.   The principal advantages of filter monochromators as compared to prism or grating monochromator systems are their relative ease of operation and adjustment.   Whereas a filter monochromator system offers homogeneous areas of monochromatic light which are easy to characterise, the spectrograph produces a continuum requiring more space, providing difficulties in light quality definition and difficulty in avoiding scattering of energy from one wavelength station to the next.   A spectrograph has unequal distribution of energy across the spectrum and, although this may not be a major problem, the ability to use equal fluence rates at all wavelengths allows direct comparison of plant responses within each experimental run.   These problems can be avoided by the use of a series of single exit slit monochromators, but their expense, limitations at high fluence rates, and relative complexity in operation all point to the filter monochromator system as being the most suitable.

Many types of commercially available projectors are suitable for adaptation as monochromatic light sources.   For low levels of light, slide projectors fitted with prefocus quartz – halogen lamps are cheap and convenient optical systems.   For the UV and B wavebands, it is usually necessary to have projectors fitted with small xenon or mercury short-arc discharge lamps. For higher light outputs, such as those required to obtain saturating responses with narrow-band interference filters, large xenon or mercury short-arc discharge lamp projectors such as those used in cinemas are necessary. These need little modification and produce a condensed rectangular light output.   Theatre spotlights offer a relatively cheap means of achieving high fluence rates, although they are not suitable for use with interference filters. Units are commonly available in the range from 0.5 to 10 kw.

The exact specifications of a filter monochromator system depend on the type of projector used.   Since projector design varies widely between manufacturers, only the general principles for the choice, design, and operation of a generalised optical system are discussed.   The numbers in parentheses refer to the items in Fig. 3.

The optics of lower-output projectors are usually based on a concave mirror (A) behind the lamp and an aspherical lens in front (B).   High output

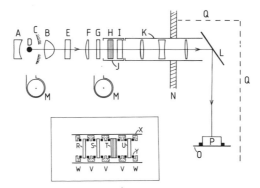

Fig. 3. Schematic representation of a monochromator system. Individual components are described in the text. Inset: longitudinal section of a water-cooled filter jacket. The glass (R), heat-absorbing filter (S), interference filter (T), and coloured filter (U) are sandwiched between brass spacing rings (V). The brass retaining rings (W) have threaded rims and screw into the jacket (X). Water enters and leaves through pipes in the spacing rings. Rubber rings (Y) ensure a watertight seal.

sources usually have auxillary mirrors (C) to capture forward radiation which would otherwise be lost. Cold projector mirrors (heat transmitting) are available, which reflect most of the radiation between 400 and 700 nm and transmit 80–90% of radiation above 800 nm, thereby removing much heat from the light source. The amount of radiation in the 300–400-nm and 700–800-nm wavebands which is reflected varies between manufacturers. Unfortunately, those mirrors which reflect well between 300 and 400 nm usually have poor reflection in the 700–800-nm waveband. Conversely, good reflectors of 700–800-nm radiation often perform poorly between 300 and 400 nm.

Various types of light sources (D) can be used (see also Section II). With incandescent lamps, which have a grid-filament in one plane, it is important that the inverted image of the individual filament segments from the miror is focused into the spaces between the segments. Failure to do this results in a large loss of light, a decrease in filament life, and a decrease in homogeneity at the irradiated surface. A major cause of light loss is dust; poor earthing increases dust accumulation. Short-arc discharge lamps are correctly focused when both the arc and its mirror image are of the same size, are on the same vertical plane, and overlap by ⁓ 50%. Care should be taken to avoid focusing the mirror image on the electrodes or the glass envelope. Incorrect adjustment leads to poor light output and possible lamp damage.

The use of a quartz aspherical condenser lens (B) is rarely justified for transmitting radiation above 300 nm. They are required only when the minimum of light loss can be tolerated below 350 nm. Borosilicate crown glass gives satisfactory transmission down to ∼ 350 nm. Pyrex glass is superior if wavelengths between 300 and 350 nm are required, although the cutoff is very sharp around 300 nm.

Another major cause of light loss is the use of an excessive number of lenses in the optical system. Each lens causes a loss of approximately 8% of the incident light beam throughout the 300–800-nm waveband. This loss can be reduced to less than 2% by a simple antireflective coating; multilayer coating can be used to reduce reflective loss to around 0.3% but is disproportionally more expensive. The lens manufacturer should be consulted because the change in reflectance is wavelength dependent and may cause an increase in reflection at certain wavelengths.

The heat output of the light sources can be removed with heat-absorbing glass or with a hot or cold mirror or with a water filter. For wavelengths between 300 and 600 nm, heat-absorbing glass is an excellent heat sink. Internal transmission of this type of glass within the 300–600-nm waveband is typically between 80 and 90%, which represents a true loss of between 20 and 30% of the incident energy. Above 600 nm, transmission falls off rapidly. Care should be taken to choose a filter of the correct thickness and with suitable transmission characteristics. The maximum temperature at which these filters can operate is usually between 400 and 600°C; "hardened" glass is more resistant to large changes in temperatures and is recommended if high-output light sources are used. If heat-absorbing glass is placed in a water jacket cooling system, it must be cooled on both faces.

A hot mirror is placed at 45° to the optical axis. These mirrors reflect IR radiation and transmit the visible waveband. Using a mirror has the advantage of obviating the re-irradiation of heat which is accumulated by heat-absorbing glass. With many hot mirrors, transmission in the desired wavebands can be further increased by altering the angle of light incidence without significantly increasing the amount of IR radiation which is transmitted. Although these mirrors are useful up to 800 nm, it should be noted that their transmission falls off rapidly below 450 nm, and therefore heat-absorbing glass is better in the UV waveband.

Cold mirrors may be used to advantage in the 700–800-nm waveband if the projector is used at 90° to the optical axis of the filters and objective lenses. With this system, unwanted IR light is largely transmitted, but the visible and FR waveband is reflected into the filters and objective.

As with the condenser lens, it is rarely necessary to use quartz for the projection lens (F). Practical aspects of choosing and using neutral-density glass (G), bandpass (H) and cutoff filters (I) are covered in Section III.

The filter holder (J) should act as a collimator, preventing stray light from reaching the filters. A slight lip or overlap is necessary around the rim of the face of an interference filter because they are not always silvered up to the edge. The width of this rim varies between filter types but is typically only 1–2 mm in modern filters. Although excess IR radiation should be removed before the actinic light reaches it, the filter should fit loosely enough to allow for some expansion. Filters also vary greatly in thickness, so a spring-loaded holder can be used to accommodate a wide range of thicknesses.

An adjustable objective lens system (K) allows image size to be changed, and with it, the fluence rate. The focal length of the system controls the size of the projected image; the shorter the focal length, the larger the irradiated area at a given distance.

A standard flat glass aluminum-based mirror (L) provides almost 90% reflection between 360 and 700 nm. The reduction to around 75% reflection between 700 and 800 nm is normally acceptable because of the relative ease of achieving high-incident energy fluxes in this waveband. Special, hard overcoats are available at extra cost. To obtain satisfactory reflection down to 300 nm, the more expensive magnesium fluoride overcoat will be necessary. Only mirrors with special, hard overcoats should be wiped with lens tissue when cleaning. The best method of cleaning is to use compressed, filtered air. If minimal light loss is required over a short period of time (i.e., months) bare aluminum offers excellent reflection from 300 to 800 nm. Mirror faces must never be touched with bare fingers.

It is a useful precaution to ensure that the fan (M) switches off at the same time that a lamp fails, because too rapid cooling can damage interference filters.

Ideally, the projectors and the material to be irradiated should be in separate rooms which are divided by a wall or partition (N). The isolation of the plant material from the projectors has several advantages. Projectors produce a great deal of heat which must be removed separately if overloading of the cooling system is to be avoided. A partition also avoids the possibility of plants receiving stray light; even if enclosures are used for the plants, it is difficult to eliminate stray light from several wavelength stations when monitoring fluence rate during an experiment. If xenon- or mercury-arc lamp systems are used, the ozone which they emit must be kept isolated from the plant material.

A correctly adjusted monochromator system will have less than 10% variation in energy flux across the irradiated surface (O), but a variation as low as 5% is unlikely to be achieved. The area to be irradiated should be marked by a physical contour which can be felt, because this allows exact positioning of the plant material (P) in darkness or when using invisible wavelengths.

Scattering of light between wavelength stations can be prevented with black curtains (Q), but care must be taken to permit free air flow. All reflective surfaces in the irradiation cabinet or room should be painted matt black.   The amount of radiation absorbed by black coatings varies widely, its effectiveness depending primarily on its composition and on how it was applied.   Where large areas such as floors, walls, and ceilings must be covered, cost will usually play a major factor in deciding which type to use. Carbon-based absorbing surfaces are generally superior to those based on black metallic oxides or black organic dyes.   Care should be taken in applying black paint.   Each coat must be allowed to dry thoroughly to keep reflectance to a minmum.   Increased reflectance will be observed if the surface is moist or contaminated with grease.   To reduce reflection from objects in rooms, the use of anodised surfaces and black curtains can offer useful alternatives to painted sufaces.   When a black coating is being used to prevent the transmission of light to an object, organic dyes should be avoided because they all transmit FR radiation to a certain extent.   Indeed, some types of black plastic sheeting act as excellent deep FR filters.

VII. DICHROMATIC IRRADIATION

Several methods can be used to produce dichromatic light.   The method used depends primarily on the required fluence rates and bandwidths of the sources.   The most important requirement for dual wavelength irradiations is that the light distribution at the target be homogeneous.   This is because the most important factor is the ratio of the fluence rates of the two wavelengths.

Diffuse light sources (e.g., fluorescent or incandescent tubes) offer homogeneous light distribution.   This can be enhanced by increasing the spacing near the centre of the light bank [Fig. 4(a)] or by increasing the distance between source and target in the centre of the light bank(s) [Fig. 4(b)].   If a narrow bandpass or high fluence rate is required for one of the wavebands, a projector plus filter system (Section VI) can be used in conjunction with a diffuse light source [Fig. 4(c)].   If high fluence rates or narrow wavebands are required for both wavelengths, two projector-based monochromator systems will be preferable.   The simplest method [Fig. 4(d)] involves the use of two projectors, placed one above the other, and simple mirrors (see also Section VI).   The angle of incidence of the actinic radiation must be arranged to obviate possible phototropic responses.

Beam splitters can be used to irradiate one or more targets with two light sources.   They function by transmitting part of the incident radiation and reflecting part of it at 90°.   Before purchasing a beam splitter, the spectral selectivity and the absorption loss should be ascertained since these facts are

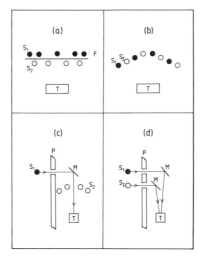

Fig. 4. Dichromatic irradiation sources. F = filter; M = mirror; P = lightproof partition; S_1 = first monochromatic source; S_2 = second monochromatic source; T = target.

not always clearly stated by the manufacturers. A wide range of reflection/ transission ratios is available.

Some beam splitters absorb almost half of the incident light, so care must be taken to choose the appropriate type if minimal light loss is required. Whereas high-output light sources can partially compensate for this, rapid changes in temperature will damage splitters which absorb strongly. Dielectric types are generally more efficient transmitters and reflectors than those with metallic coatings, although the former perform poorly in the UV waveband. Coated pellicle beam splitters give excellent efficiency in both the UV and visible waveband. However, because they consist of a very thin plastic membrane, they are easily broken (the surface must never be touched) and cannot withstand temperatures above ~80°C.

VIII. PULSED LIGHT

Pulsed light can be obtained either by using a specific pulsed-light source or by interrupting the light path of a continuous output source. The latter method has the advantage of being cheaper if it is incorporated into preexisting lighting systems and is adequate for most applications. For pulses of nanosecond or picosecond duration, the reader should consult manufacturers of lasers and gas plasma discharge radiators.

A. Pulsed-light sources

(1) Xenon-arc discharge lamp

The best high performance flashing source is the xenon discharge lamp which can be repeatedly discharged by high voltage capacitors. The duration of the flash depends on how long the current is passed through the lamp and is limited primarily by the cooling rate of the gas; sources providing flash durations of a few μsec are commonly available. A wide range of flash frequencies can be obtained, although most types increase greatly in price if frequencies above ~ 500 Hz are required because additional control equipment is needed.

These lamps can produce extremely high fluence rates of light with a spectral distribution which is similar to daylight. Their output in the blue waveband is proportionally greater than that of other xenon lamps, and adequate physical protection should be taken against the high UV output. Lamp life varies with size. Small lamps have a typical working life of 100–200 h and large lamps of around 500 h, although the flashing rate of the large lamps is slower. Lamp life is increased by decreasing the output or increasing the rate of flashing.

(2) Photographic flash guns

Photograpic flash guns provide a relatively cheap means of producing msec duration flashes if a recycling time of between 2 and 10 sec is acceptable. Some flash guns can achieve up to three pulses per second, but the output is drastically reduced. Flash guns which can be connected directly to the mains supply will ensure constant light output and recycling time.

B. Pulsed-light accessories

There are two basic methods for interrupting the light path between a projector light source and the object to be irradiated.

(1) Shutters

Iris diaphragm lens shutters enable irradiation times down to ~ 1.0 msec. A sliding shutter operated by a solenoid may be more suitable if a large area of light must be cut off, but the duration of the pulse will be longer.

(2) Rotating sectors and filter wheels

The rotating sector consists of a motor-driven disc (e.g., 2-mm thick aluminium) from which radial sectors are cut. A wide variety of pulse durations and frequencies are obtained by varying the speed of revolution and the size and spacing of the cutout angles. The quality of the light and the fluence rate is determined by static filters. Flash durations of 1.0 msec are easily achieved.

As its name implies, the motorised rotating filter wheel has one or more filters placed at predetermined intervals near its perimeter. It is used where long duration pulses (several seconds) are required and has the advantage that several types of filter can be placed in the wheel. This system is ideal for automatic R – FR reversibility tests.

IX. MICROBEAMS

Irradiation of small areas of plant tissue can be achieved by using either metallic light guides or fibre-optic light guides.

A. Metallic light guides

Metallic light guides are hollow tubes whose inside surfaces are highly polished and then plated with gold. They provide an extremely efficient means of transmitting light (less than 20% loss over a 1-m length, the transmission varying only slightly with wavelength). Coupling units which can pivot about a perpendicular axis are used for reflecting the light and cause less than 10% light loss. Beam splitters are also available and tapered guides can be used to concentrate the light beam. If only one or two sections are required, metallic light guides provide a good method for producing microbeams. They have the major disadvantages of being rather cumbersome and extremely expensive.

B. Fibre-optic light guides

Fibre-optic light guides consist of extruded glass or man-made fibres bound in an outer casing. The casing has a different refractive index from that of the fibre(s). Within the range of acceptance angles determined by the refractive indices, the light entering the guide undergoes multiple total internal reflection along its length. Some light is lost; this is partially due to reflectance at the air – guide and guide – air interfaces (up to 30% loss) and partially due to scattering and absorption within the tube.

(1) Glass-fibre systems

Glass-fibre systems are made of seveal fibres (typically 20 to 80 μm in diameter) to form a bundle. The commonest bundle diameters are between 1.0 and 15 mm, but sizes outside this range are available. The comonest lengths are between 100 and 1000 mm. The fibres are held together within a stainless steel or PVC casing, and the ends are optically polished to maximise transmission. The ends are usually cemented (with an epoxy resin) or may be fused together while the rest of the bundle is free to flex. Opposite ends can be any shape as long as their areas are the same.

Glass-fibre bundles transmit maximally in the 500–600-nm waveband. There is a steep falloff in transmittance below ∼400–450 nm; the relative falloff toward 800 nm is not so marked. In absolute terms, a 300-mm bundle passes ∼65% of the incident light, whereas a 4000-mm bundle lets through only 30% at the same wavelength (550 nm). Increasing length therefore, results in a strong decrease in transmittance. Special UV transmitting fibres are also available. The majority of the output from a glass bundle is restricted to a 60–70° angle, with a sharp dropoff outside this range. Focusing heads are available and tapered bundles can be used to focus the emitted light.

(2) Plastic-fibre systems
Plastic-fibre systems are generally cheaper and easier to make and are less brittle than glass fibres. They are commonly produced in 0.25 to 1.5-mm diameters. In contrast to bundles of glass fibres, plastic guides are usually produced as one fibre. The light-transmitting tube is covered with similar material of lower refractive index. In the G region, loss of light is greater than with equivalent glass-fibre systems. Light loss at the ends (entrance – exit) can be reduced to ∼5% if they are well cut and polished.

(3) Choosing the appropriate system
When deciding whether to use plastic- or glass-fibre systems, several points should be borne in mind. Plastic fibres are cheap and resilient, whereas gradual breakage of individual fibres is a major problem with light sources using glass fibres. Up to 5% of the glass fibres are broken during manufacture. Many of the dark spots at the ends of new guides are glued spots, and not broken fibres. Plastic fibres are available in long sections, and these can be cut to the exact size required with a razor blade. The cut ends are easily polished with commercial rubbing compounds; this is an important factor in reducing light loss.

Although plastic fibres are suitable for the spectral range 400–1100 nm, glass fibres are superior for the shorter wavelengths and silica fibres are better for the UV waveband. If high fluence rates are required, the fact that plastic fibres cannot withstand temperatures higher than ∼75°C and that diameters greater than 1.5 mm are difficult to obtain may decide which type must be used. Whichever type is chosen, minimum light loss is obtained by using the minimum possible length, the minimum number of connections (air – conductor interfaces), and by use of well-polished ends.

The flexibility of fibre-optic light guides depends primarily on the diameter and the type of protective sheath. As a rule-of-thumb, a radius of 10 times the diameter of the light transmitting bundle can be safely used. PVC sheaths are more flexible than metal piping but are less protective. Most manufactureres can provide guides with tighter, but permanent, inbuilt angles. Alternatively, the exit of the guide can be cut and polished at an oblique angle to obtain limited diversion of the exit beam.

Light sources for use with fibre-optic light guides must be aligned correctly. It is simplest to use commercial instruments which are available with quartz–halogen or xenon- or mercury-arc discharge lamps. In combination with a Schott KG1 heat-absorbing filter, a typical quartz–halogen-based source produces an emission maximum from the tube at ⁓640 nm with a relatively sharp dropoff on either side. This value is reduced to half at ⁓540 and ⁓750 nm. At 400 and 800 nm, the relative fluence rate is less than 1% and 10% of maximum, respectively. The more expensive xenon-arc source provides a broader waveband coverage. The mercury arc is useful for irradiation in the B and G wavebands. The advantages of commercial instruments are that various interchangeable fibre systems can be used, the light sources are compact, and the problems of optical alignment have already been overcome. Standard systems provide a variety of guide thicknesses and models with up to 10 arms are available.

SOURCES OF FURTHER INFORMATION

Bickford, E. D. and Dunn, S. (1972). "Lighting for Plant Growth.". Kent State Univ. Press, Kent, Ohio.

Calvert, J. G. and Pitts, J. N. (1966). "Photochemistry." Wiley, New York.

Canham, A. E. (1966). "Artificial Light in Horticulture." Centrex Publ. Co., Eindhoven, Holland.

Downs, R. J. (1980). *Bot. Rev.* **46,** 447–489.

Heathcote, L., Bambridge, K. R. and McLaren, J. S. (1979). *J. Exp. Bot.* **30,** 347–353.

Jagger, J. (1967). "Introduction to Research in Ultraviolet Photobiology." Prentice-Hall, Englewood Cliffs, New Jersey.

Koller, L. R. (1965). "Ultraviolet Radiation." Wiley, New York.

Tate, R. L. C. (1978). "Technical Handbook," Publ. No. 424. Thorn Lighting Ltd., London.

Withrow, R. B. and Withrow, A. P. (1956). *In* "Radiation Biology" (A. Hollaender, ed.), Vol. 3, pp. 125–258. McGraw-Hill, New York.

4

Radiation Measurement

M. G. Holmes

I. INTRODUCTION

Although light is the elective factor in photomorphogenetic responses, the importance of its accurate measurement is often neglected. Errors introduced during the measurement of the actinic light may be incorporated into the error inherent to the plant response measured. The accuracy of light measurements is determined by the choice of appropriate equipment and the ability of the operator to use the equipment correctly.

Three basic requirements must be fulfilled if accurate and precise measurements are to be made. First, the acceptance angle of the receiver should be compatible with the angle of incidence of the light and the geometry of the irradiated tissue. Second, the spectral sensitivity of the detector should be appropriate for the light source. Third, the error involved in making the measurement should be known and definable. Having made the light measurements, the question arises of how much information should be provided for the reader. Whereas excessive detail is a waste of publishing space, enough detail should be provided to ensure that the experiment can be repeated, and that the response of the plant material to the actinic radiation can be fully assessed.

The following abbreviations are used in the text: UV = ultraviolet; R = red; FR = far-red; IR = infrared; PAR = photosynthetically active radiation = 400–700 nm. The UV, visible, and IR regions are defined here as the 300–380-, 380–770-, and greater than 770-nm wavebands, respectively.

TECHNIQUES IN PHOTOMORPHOGENESIS
0-12-652990-6

II. UNITS AND NOMENCLATURE

The choice of appropriate units and nomenclature requires a clear, exact and simple terminology which is readily comprehensible. Until recently, a major distraction in the literature was the use of photometric units and terminology. Repetitive reminders of their uselessness and irrelevance has resulted in their virtual absence from current photobiological publications. There is still disagreement, however, on which radiometric units and terminology should be used in the plant sciences.

Knowledge of the quantum properties of the actinic light is of great importance because the Stark–Einstein law of quantum equivalence defines that, in the primary photochemical process, each molecule is activated by the absorption of one photon. It follows that one mole of photoreceptor should be excited by the absorption of one mole of photons. In practice, this one to one relationship is unlikely to be observed because of the complexity of the physical conditions. Nevertheless, it is preferable to describe actinic light in terms of the wavelengths under consideration and the number of photons arriving at the irradiated object. Actual light measurements are normally made of the energy content of the radiation rather than the particle content. Both descriptions are correct and applicable and are meaningful if the spectral characteristics of the radiation are provided.

The Système International (SI) units are nomenclature used here are based on the recommendations of several international organisations: the Association Internationale de Photobiologie, the Bureau International des Poids et Mesures, the Commission Internationale de l'Eclairage, the International Commission on Radiation Units and Measurements, and the Union Radioscientifique Internationale. The Shorter Oxford English Dictionary and Webster's International Dictionary were used to verify etymology.

A. Definitions

Units and terminology are summarised in Table I. To describe the amount of energy carried by radiation, we can consider the *radiant energy* (in joules, J) or the *number of photons*. The latter quantity is dimensionless unless it is expressed by Avogadro's number of photons (mole, mol).

The amount of energy carried by radiation per unit time is the *energy flow rate* (J sec^{-1} or watt, W). The analogous quantity is the *photon flow rate* (sec^{-1} or mol sec^{-1}).

The amount of energy or number of photons traversing a unit surface area normal to the direction of propagation is described, respectively, by the *energy fluence* (J m^{-2}) or *photon fluence* (m^{-2} or mol m^{-2}). The *energy*

Table Ia. Terminology and units for describing the radiation arriving at a flat surface. The acceptance angle of the radiation detector is 180° or less and a cosine correction is applied to radiation arriving at angles which are not normal to the receiver surface.

Term or quantity	Unit
A. Energy content	
Radiant energy	joule (J)
Energy flow rate[a]	J sec^{-1} *or* watt (W)
Energy applied[b]	J m^{-2}
Energy flux	W m^{-2}
B. Photon content	
Number of photons[c]	dimensionless
Avogadro's number of photons	mol
Photon flow rate[a]	sec^{-1} *or* mol sec^{-1}
Photons applied[b]	m^{-2} *or* mol m^{-2}
Photon flux	m^{-2} sec^{-1} *or* mol m^{-2} sec^{-1}

[a] The term *flow rate* is preferable to *flow* because flow does not in itself imply rate.

[b] The energy applied, or number of photons applied, refer to the radiation measured. In analytical photobiology, it is usually necessary to determine the quantity of radiation absorbed by the photoreceptor; in this situation, the appropriate terms are the *energy absorbed* and the (number of) *photons absorbed,* respectively.

[c] The term photon is preferable to quantum because the photon is specifically a quantum of electromagnetic radiation. Other types of quanta exist; for example, the phonon is the quantum of a lattice vibration. Quantum is acceptable if it is clear that electromagnetic radiation is being referred to.

Table Ib. Terminology and units for describing radiation arriving at a point. The radiation detector has a spherical acceptance angle of 360° and the measurement is the integral of radiation from all directions arriving at that point.

Term or quantity	Unit
A. Energy content	
Radiant energy	joule (J)
Energy flow rate	J sec^{-1} *or* watt (W)
Energy fluence[a]	J m^{-2}
Energy fluence rate	W m^{-2}
B. Photon content	
Number of photons	dimensionless
Avogadro's number of photons	mol
Photon flow rate	sec^{-1} *or* mol sec^{-1}
Photon fluence[a]	m^{-2} *or* mol m^{-2}
Photon fluence rate	m^{-2} sec^{-1} *or* mol m^{-2} sec^{-1}

[a] *Energy fluence* and *photon fluence* refer to the amount of energy or number of photons applied to an object. In analytical photobiology, account will be taken of screening factors to determine the *absorbed energy fluence* or *absorbed photon fluence,* respectively.

fluence rate is the energy fluence per unit time (J m^{-2} s^{-1} = Wm^{-2}), and to express the photon fluence per unit time, the *photon fluence rate* is used (m^{-2} sec^{-1} or mol m^{-2} sec^{-1}).

The amount of energy or number of photons arriving at a flat surface (typically cosine-correlated) are described by the *energy applied* (J m^{-2}) or *photons applied* (m^{-2} or mol m^{-2}), respectively. The energy applied per unit time is the *energy flux* (J m^{-2} sec^{-1} or W m^{-2}) and the photons applied per unit time are described as the *photon flux* (m^{-2} sec^{-1} or mol m^{-2} sec^{-1}).

When dealing with the spectral composition of radiation, the energy fluence per unit time and wavelength is described by the *spectral energy fluence rate* (W m^{-2} nm^{-1}) and the photon fluence per unit time and wavelength is described by the *spectral photon fluence rate* (m^{-2} sec^{-1} nm^{-1} or mol m^{-2} sec^{-1} nm^{-1}). The equivalent terms when using a cosine correction are the *spectral energy flux* (W m^{-2} nm^{-1}) and the *spectral photon flux* (m^{-2} sec^{-1} nm^{-1} or mol m^{-2} sec^{-1} nm^{-1}).

There is a fundamental difference between the terms flux and fluence rate. Flux describes the radiation arriving at a flat surface; a proportionality factor (cosine correction) is applied to radiation arriving at other than normal incidence because the flux arriving at a flat surface decreases as a function of the angle of incidence. The term fluence rate is defined as the radiant energy per unit time which has entered a small sphere surrounding a specific point in space, divided by the sphere's cross-sectional area. Fluence or fluence rate, therefore, considers radiation entering an imaginary three-dimensional object (a transparent sphere) and is a function of the sphere's cross-sectional area. An ideal collector for measuring fluence or fluence rate (which is a spherical diffusing object) responds equally to radiation at any point on its surface. This contrasts with an ideal collector for measuring flux, which is typically a flat surface with an inbuilt cosine-correction factor.

The practical differences between measurements of flux and of fluence rate can be pictured by imagining two light sources, A and B. If 100 light particles from source A arrive at normal incidence to a plane surface and 100 arrive from source B at 60° from normal incidence, the cosine-weighted flux is 100 plus 50, or 150 particles per unit time per unit area. This is because the proportionality factor $\frac{1}{\cos} \theta$ (where θ is the angle between the plane of incidence of radiation and the normal to the surface), i.e., 0.500, is applied to the radiation from source B. If, however, a collector for measurement of fluence rate is placed in the same position as that used to measure flux, the detector will measure 100 particles from source A plus 100 particles from source B, i.e., 200 particles per unit area per unit time. Flux and fluence rate are only identical in the case of a single parallel beam of radiation traversing a surface perpendicular to the beam. In practice, there is only a small (less than 4%) difference between the two definitions within the limits of a solid incident angle of about 15% which is applicable to most projector-based sources. For measurements in most growth cabinets and under natural conditions, the difference is large.

Simplification can be achieved by expressing light measurements in instantly recognisable and meaningful units. The use of mm^{-2} or cm^{-2} to describe the base unit m^{-2} is superfluous, merely adding to the number of quantities which need to be easily recognised and understood. Simplification of the radiation units for the energy or photon content arriving at a surface is less straightforward. For the energy content, it is usually unnecessary to add any prefix such as micro- or milli- to the basic units joule or watt. The majority of light sources used in photomorphogenesis research produce an energy fluence rate or flux of between approximately 1 and 100 $W\ m^{-2}$; this is a simple range of values which should provide instantly recognizable quantities. In line with this argument, the photon fluence rate or flux should be expressed as $mol\ m^{-2}\ sec^{-1}$. In this case, however, the values normally encountered require the expression of data as $\mu mol\ m^{-2}\ sec^{-1}$.

B. Exceptions

Whereas the terminology presented here represents the recommendations of the major international bodies responsible for the simplification and clarification of nomenclature and units, there are terms, both SI and non-SI, whose use occurs in the literature. Most of these units remain in use for historical reasons. Many reference sources, such as those used for meteorological data, continue to use units based on non-SI nomenclature. The calorie, einstein, erg, langley, and micron are particularly common. Conversion of data presented in this way to modern units by the institutions involved will take a long time and in some cases may never happen. Nevertheless, their inclusion in new photobiological publications will not speed up the process. Conversion factors are given in Table II to facilitate interconversion of the most commonly occurring units in the literature.

C. Common misnomers

The almost complete elimination of photometric units from the literature has led to the loss of photometric terms such as luminous energy, luminous flux, illumination, and illuminance. Several radiometric terms still remain which are either misused or are irrelevant to modern terminology. The commonest error is the misuse of the term "intensity" (which refers to the radiation emitted per unit solid angle) to describe the fluence rate or flux. The term "dose" is commonly misused because although it applies to ab-

Table II. Conversion factors for commonly occurring units.

To convert from	to	multiply by
angstrom	nanometre	1.000×10^{-1}
calorie (thermochemical)	joule	4.184
calorie cm^{-2} sec^{-1}	watt m^{-2}	4.184×10^{4}
calorie cm^{-2} sec^{-1}	mole m^{-2} sec^{-1}	$3.499 \times 10^{-4}\lambda^{a}$
calorie cm^{-2} sec^{-1}	quanta m^{-2} sec^{-1}	$2.107 \times 10^{20}\lambda$
calorie cm^{-2} min^{-1}	watt m^{-2}	6.973×10^{2}
calorie min^{-1}	watt	6.973×10^{-2}
calorie sec^{-1}	watt	4.184
electron volt	joule	1.602×10^{-19}
erg	joule	1.000×10^{-7}
erg	quanta	$5.036 \times 10^{8}\lambda$
erg cm^{-2} sec^{-1}	mole m^{-2} sec^{-1}	$8.362 \times 10^{-12}\lambda$
erg cm^{-2} sec^{-1}	quanta m^{-2} sec^{-1}	$5.036 \times 10^{12}\lambda$
erg cm^{-2} sec^{-1}	watt m^{-2}	1.000×10^{-3}
erg sec^{-1}	calorie min^{-1}	1.434×10^{-6}
erg sec^{-1}	watt	1.000×10^{-7}
joule	mole	$8.362 \times 10^{-9}\lambda$
langley	calorie cm^{-2}	1.000
micron	nanometre	1.000×10^{3}
mole m^{-2} sec^{-1}	quanta m^{-2} sec^{-1}	6.022×10^{23}
mole m^{-2} sec^{-1}	watt m^{-2}	$1.196 \times 10^{8}\lambda^{-1}$
quanta m^{-2} sec^{-1}	mole m^{-2} sec^{-1}	1.661×10^{-24}
quanta m^{-2} sec^{-1}	watt m^{-2}	$1.987 \times 10^{-16}\lambda^{-1}$
watt cm^{-2}	watt m^{-2}	1.000×10^{4}
watt m^{-2}	mole m^{-2} sec^{-1}	$8.362 \times 10^{-9}\lambda$
watt m^{-2}	quanta m^{-2} sec^{-1}	$5.036 \times 10^{15}\lambda$

[a] λ = wavelength (nm).

sorbed radiation, practically all experiments in photomorphogenesis are interpreted in relation to the light arriving at the irradiated object. The terms "energy fluence" and "photon fluence" and the terms "energy applied" and "photons applied" describe the treatment adequately and thereby avoid confusion.

Problems often arise from misuse of the term "flux" to describe "flow rate"; in physics and engineering flux refers to the transference of volume, mass, or energy per unit area (usually normal to the direction of flow) per unit time. It also follows from this definition that when "flux" is used correctly, the term "density" is superfluous.

Although recommendation of use of the term "irradiance" is becoming less frequent, the word has earned international acceptance and respectability and its use is technically correct if restricted to measurement of light incident at a plane surface. Nevertheless, it is to be hoped that only the terminology in Table I will be used in order to promote consistency and uniformity.

III. CALIBRATION

All light-measuring equipment requires periodic calibration. The regularity with which this is done depends on the type of equipment (see Sections V and VI) and on the conditions under which it is used. Generally, spectroradiometers should be calibrated before and after each study, and other detectors should be calibrated at least biannually. Calibration can usually be carried out by the manufacturer. Although this can save time, the cost may often be prohibitive and the fragile and sensitive nature of some equipment may result in some loss of calibration during shipment. For larger laboratories, especially those using several light detectors, purchase of a standard lamp and the auxiliary calibration equipment can be an economical proposition and has the added advantage of encouraging regular recalibration.

A. Basic equipment

The room used for calibrating light equipment should be kept clean at all times to avoid contamination of the optical equipment. A room kept at a constant temperature is essential and all equipment should be thermally adapted. This is particularly important when using thermoelectric detectors because sources of heat, such as the operator's body and electrical power supplies, can produce spurious observations. Extraneous light can be minimised by painting all walls, ceilings, floors, and tables matt black and ensuring that no light enters through cracks in doors, etc. Black cloth can also be used to cover reflective surfaces and to reduce unwanted light from electrical power supplies and readout instruments. Stray light from the standard lamp can be reduced substantially by placing a series of blackened screens with apertures of appropriate size between the source and the detector. Using these precautions, stray light can be reduced to less than 0.1%.

An optical bench enables exact positioning of the source and the receiver and therefore must be sturdily constructed and accurately graduated. A bench scale accuracy of 0.5 mm in 1.0 m (0.05%) is readily achieved. Horizontal positioning of the lamp and receiver can be determined with a plumb line. In many cases, a marked lamp housing is supplied, thereby making accurate positioning easier. Unless a micrometer drive is used for adjustments, horizontal alignment is unlikely to be better than $\pm 0.5\%$, but this is less important than the lamp to sensor distance.

B. The standard lamp

All standard lamps require a precisely regulated power supply. A supply with an output current accuracy of better than 0.25% and voltage stability of 0.15% should be sought. Most standard lamp power supplies operate well within this range but if large variations in supply voltage are expected, a constant output voltage transformer will also be necessary. The ability to adjust the output current smoothly and the provision of a ramping circuit to protect the lamp during switching is advantageous.

Comercially available standards are usually tertiary standards which have been calibrated against secondary standards possessed by the supplier. The supplier's secondary standards have been calibrated against primary standards possessed by bodies such as the National Bureau of Standards in the United States or the National Physical Laboratory in England. Despite this extensive transfer of data, the accuracy of the purchasable (i.e., tertiary) standards is very high. The lamp purchased from the supplier is expensive and has a limited accurate life, usually, of 50 h. This lamp should be used as a shelf standard against which one or more working standard lamps are calibrated. Uncalibrated working standards are cheap in comparison with the calibrated lamp. The working standard should be checked periodically against the shelf standard. It is extremely important that all lamps are stored and handled carefully. The accuracy of all light measurements depends on the accuracy of the standard and the (minimum) amount of error introduced by the user during calibration.

C. Calibration procedures

(1) The receiver

Unless accurate details are given by the manufacturer, the directional sensitivity of the receiver must be calibrated. To do this it is necessary to use a fixed-position light source whose spectral quality can be varied. The emitted radiation must be collimated to restrict the incident angle of the incident radiation at the receiver to $\sim 5°$ (maximum 10°). Using this system, it is possible to measure the spatial and spectral sensitivity of the receiver and the inherent error of normal incidence, cosine response, and spherical response receivers.

The receiver holder is constructed to allow rotation of the receiver in the horizontal and vertical planes, these being calibrated in degrees. To measure the cosine response of a receiver, the output produced by radiation arriving at normal incidence is measured. The detector output is then measured as the angle of incidence is moved away from normal incidence in 5° stages. To ensure that no significant variation exists at other azimuth angles, this procedure is repeated after the receiver has been turned about its own axis in 30 or 45° steps. The receiver's error is the percentage deviation in the observed response from that expected with a true cosine response. The spherical response is measured in the same way except that it is necessary to rotate the receiver through 360°, rather than the 180° used for cosine response receivers.

A system must be used which permits monitoring of the output of the light source at the beginning and end of each series of measurements. A calibrated detector is usually most suitable because it is normally necessary to repeat these measurements at several wavelengths to account for the spectral selectivity of the receiver.

(2) Broad band detectors

To determine the spectral responsivity of broad waveband detectors, the output from a broad waveband light source (it need not be a standardised source, but should operate from a stabilised supply) is passed through a wavelength-calibrated monochromator (see Section VI. C). If the receiver surface is large, it will usually be necessary to disperse the radiation emitted by a monochromator. A less satisfactory method is to use a series of narrow bandpass interference filters. The output from the monochromator is measured at selected wavelength intervals with the detector. These measurements are then repeated with a calibrated detector (e.g., thermopile) to determine the true spectral response of the detector under test. The three primary requirements of this procedure are that the radiation is uniformly distributed over the receiver surfaces, that the whole surfaces of the receivers are irradiated, and that the system assures exact repositioning of the optical system, reference detector, and tested detector between alternate measurements.

To calibrate detectors which have a uniform spectral response, such as thermopiles with blackened receivers, a standard of total "irradiance" should be used. Data for irradiance at three operating currents are usually supplied, but the range of calibration can be extended by applying the inverse square law or using low light level techniques.

(3) Spectroradiometers

A standard of spectral irradiance (i.e., energy flux) is required to calibrate a spectroradiometer. Both quartz–halogen and hard-glass–tungsten filament standards are available. Accurate calibration in the UV waveband requires a high-wattage quartz–halogen lamp. Calibration data are normally supplied for operation at only one current rating, but calibrations for other ratings can usually be provided on request. The calibration report provides the spectral energy flux at a given distance (often 0.5 m) from the lamp. This information is used to calculate the spectral photon flux. The specific conditions and procedures for using each type of standard lamp are provided by the supplier.

Discharge lamp emission lines and dye lasers offer the most precise means of calibrating the wavelength accuracy and wavelength repeatability of monochromator systems. A calibrated monochromator is a good alternative, but allowance must be made for its inherent inaccuracy. Didymium and holmium oxide filter glasses can be used but are only suitable if purchased in calibrated form. As a last resort, narrow band interference filters may be used, although it should be borne in mind that the wavelength of peak transmission supplied by the manufacturer is usually inaccurate, that re-measurement of this peak involves an inaccuracy, and that the peak is often very broad, or even double.

(4) Linearity

The linearity of a detector's response is described by its ability to follow accurately gradual changes over a wide range (several decades) of radiation levels. Deviations from linearity may also be caused by inaccurate range factor switching.

To measure linearity, two light sources, x and y, are aimed at the detector. The radiation output from x is increased in stages from zero to the maximum amount rate which is to be measured. The output from y is held constant and a measurement of x taken alternately with y switched on and with y switched off. Perfect linearity exists if the signal difference is constant. Alternatively, the addition technique may be used; here, the signal from x + y together should be equal to the sum of x alone + y alone.

(5) Error

The primary sources of error during calibration are the accuracy of the calibrated lamp at the time of calibration, the accuracy of the lamp and receiver alignment, the accuracy of the lamp to receiver distance, the elimination of stray light, and the accuracy of lamp current measurement. Further details are included in Section VII.

IV. RECEIVER GEOMETRY

Three factors influence the choice of correct receiver geometry. These are the angle of incidence of the actinic light with the receiver, the orientation of the irradiated plant material, and the proportion of reflected or scattered light arriving from outside the acceptance angle of the receiver (Fig. 1). An example of the error which can be introduced by using the wrong type of receiver is given in Table III.

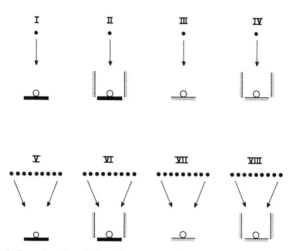

Fig. 1. Factors influencing the choice of correct receiver geometry. Detectors with a very restricted acceptance angle ("normal incidence") are only suitable for measuring point sources (I) in the absence of reflection or scattering. If the plant material scatters this light and a significant amount can be reflected from the surroundings, then a cosine (II) or spherical response (III, IV) receiver will be required.

With diffuse light sources, a cosine-corrected receiver is appropriate if the angle of incidence is $\leq 180°$ (V and VI). If a significant proportion of light arrives from outside this angle (VII and VIII), a spherical response receiver will be required.

A point source is considered here as a light source whose cross section is less than 10% of the distance from the source to the irradiated object.

Key: ● = point source (e.g., projector), ●●●●● = diffuse source (e.g., fluorescent tubes, daylight), ▬▬▬▬ = nonreflective surface (e.g., soil, black filter paper), ∕∕∕∕∕∕ = reflective surface (e.g., white paint or filter paper), ○ = irradiated object.

Table III. Dependence of radiation measurements on the type of receiver used and reflectivity of the surroundings.[a]

	Black background		White background	
Receiver	μmol m^{-2} sec^{-1}	Spherical receiver response, white background (%)	μmol m^{-2} sec^{-1}	Spherical receiver response, white background (%)
Normal incidence	1.0	13	1.2	15
Cosine corrected	2.8	36	2.9	37
Spherical	4.1	53	7.8	100

[a] The radiation was produced by a horizontally positioned 1.1 × 1.2-m monochromatic red light source ($\lambda_{max} = 658$ nm) composed of Philips TL40W/15 fluorescent tubes filtered through a 3-mm thick sheet of No. 501 Plexiglas suspended 0.7 m above a black background (black cloth) or a white background ("Pearl White" formica). Radiation measurements made with cosine response, spherical response, and normal incidence response (acceptance angle = 9.5°) receivers are compared. (From Hartmann, 1977.)

Some detectors, for example, many thermopiles, have a very restricted acceptance angle or field of view. These detectors are suitable for applications where the incident radiation arrives within the nominal aperture angle. This infers that reflection from surroundings lying outside this angle must be negligible. The other requirement is that the irradiated tissue develops predominantly in a plane at right angles to the incident beam. If the irradiated object receives diffuse light (e.g., natural daylight, radiation from fluorescent tubes, growth cabinet lighting, or the radiation within a vegetation canopy), a cosine-corrected or spherically corrected receiver should be used.

The flux arriving at a flat surface decreases as a function of the cosine of the angle of incidence. Consequently, a proportionality factor must be applied to radiation arriving at other than normal incidence. This factor is $\frac{1}{\cos}\theta$, where θ is the angle between the plane of incidence of the radiation and the normal to the surface. Detectors should therefore be compensated so that a cosine response is obtained (Fig. 2). For a perfect receiver, the response at 45° off-axis would be the cosine of 45°, i.e., 0.707 of the response, and at 60°, 0.500 of the response, etc.

Good cosine correction (i.e., less than 5% deviation) at off-normal angles greater than 80° is difficult to achieve. However, significant errors will be introduced only if the proportion of radiation arriving at these extreme angles is substantial (see Section VII for calculation of error) in comparison with that arriving perpendicular to the receiver. Thus the cosine-corrected receiver is suitable when the plant material is irradiated from one side only with either a point source or diffuse light, and when there are no reflective surfaces on the opposite side of the plant material to the light source. Ideally, the irradiated material should be developing primarily in two dimensions and be orientated in one parallel plane at approximately right angles to the incident radiation. If the photosensitive region of the tissue is known and it lies predominantly at other than normal incidence to the actinic beam, then the base of the cosine-corrected receiver should be angled to lie parallel with the surface of the photosensitive region. This procedure compensates for the reduced incident flux per unit tissue area.

Several methods of converting photodetectors to a cosine response can be used (Fig. 2), some of which are simple to construct or apply. Placing a sheet of white chromatography or filter paper or cloudy white Perspex or Plexiglas over the photodetector provides an acceptable cosine response for light arriving at angles of up to approximately 65° from normal incidence. A slightly superior result is obtained with flat opal glass placed within or over the entrance aperture; rough-surfaced opal glass is preferable to a polished surface. Although a flush-fitting opal glass sheet provides an acceptable cosine response up to approximately 70° from normal incidence, reduced

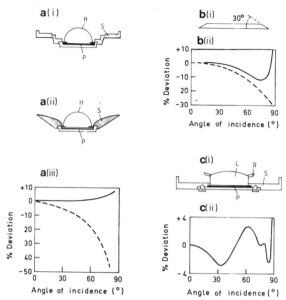

Fig. 2. Examples of cosine-corrected receivers. [a(i)] and [a(ii)] Receivers corrected by metal screens. H = hemispherical opal glass bowl, P = photocell, S = light metal screen. [a(iii)] Deviation from the cosine law of a photocell without (broken line) and with (continuous line) the corrected receiver. (Redrawn after Eckhardt, 1965.) [b(i)] Opal glass plate cut to the shape of a cone frustrum. [b(ii)] Deviation from the cosine law of a photocell without (broken line) and with (continuous line) the correction plate. (Redrawn after Kubín, 1971.) [c(i)] Receiver corrected by a screen and a convex lens. L = lens, S = screen, R = ring attached to the lens, P = photocell. Deviation from the cosine law of the combined photocell and correction system. (Redrawn after Dlugos, 1958.)

error at larger angles can be obtained by altering the height of the glass relative to the aperture neck and using a carefully shaped screen to prevent light arriving at extreme angles. Further improvement can be gained by shaping the glass plate into a cone frustrum or using a hemispherical globe of opal glass.

When the incident radiation arrives at a solid angle of greater than 180° or the object being irradiated develops predominantly in three planes, a spherical response receiver should be used (Fig. 3). The simplest form consists of two photocells placed back to back; one photocell measures the downwelling radiation and the other the upwelling radiation. Two approaches can be used with this type of receiver. One method is to give each receiver a cosine-corrected response; clearly, one cosine-corrected receiver can be used to make two individual measurements. An instrument of this type is suitable for measuring radiation which propagates primarily in two opposite

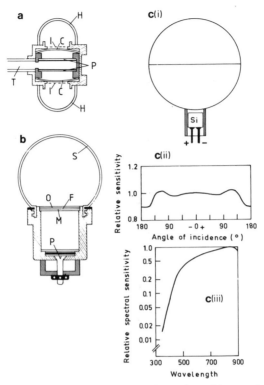

Fig. 3. Examples of spherical response receivers. (a) Receiver with two selenium photocells. H = hemispherical opal glass domes, C = convex opal glasses, I = iris diaphragms, P = photocells, T = tube which serves as handle through which the photocell wires go to the measuring instrument. (Redrawn from Wassink and Van der Scheer, 1951.) (b) Receiver with single dome and photocell. S = opal glass sphere, O = flat opal glass, F = Schott BG 21 (2 mm) filter, M = mosaic spectral sensitivity correction filter, P = photocell below protective clear glass plate. (Redrawn from Šetlík and Kubín, 1966.) [c(i)] Cross section of spherical response receiver constructed from two white Plexiglas hemispheres. [c(ii)] Relative spatial sensitivity. [c(iii)] Relative spectral sensitivity. (Redrawn from Hartmann, 1977.)

directions. Alternatively, each half can be constructed to provide an approximately uniform response to all incident angles. A spherical receiver can be constructed by cementing the bases of two hemispherical white Perspex or Plexiglas bowls together. By placing a photocell within a tube which is directly attached to the sphere, an excellent spherical receiver is obtained, although the sensitivity of the photocell is severely reduced. The spherical error can be decreased by judicious application of a black coating opposite the photocell and in a narrow band where the hemispheres were joined. Reflective correction stripes may also be used.

Self-shading presents a major problem when making measurements with spherical radiation receivers. This problem is best overcome by reducing the size of the receiver to a minimum. Most commercially available spherical receivers are too large for many applications, although relatively small receivers can be obtained (e.g., Li-Cor, Inc.; Biospherical Instruments Inc.). Small receivers can be made from preformed objects (e.g., white table-tennis ball, nominal diameter = 38 mm). The use of encased reinforced fibre-optics allows the detector to be placed remote from the receiver, thereby further reducing shading effects.

When measuring very low levels of radiation, it should be borne in mind that corrective receiver surfaces substantially reduce the sensitivity of the detector. It will also be necessary to recalibrate the spectral sensitivity of the detector because all of the above-mentioned diffusers show slight spectral selectivity in their transmission characteristics. In addition, the inherent spatial error of the receiver should be known or determined. Most manufacturers supply accurate estimates of spatial error. Where it is necessary to construct purpose-made receivers, these errors should be measured directly. The appropriate procedures are presented in Section III. C. 1.

V. DETECTORS

The detector converts absorbed radiant energy into a form of energy which is readily measurable. There are two main types of detectors: thermal radiation, and photoelectric or quantum detectors. The thermal radiation detectors, which are normally used in photobiology, are of the thermoelectric type; these convert heat energy to electrical energy. Nearly all of the wide variety of photoelectric detectors have an application for light measurement in the plant sciences.

A. Thermoelectric detectors

One of the greatest advantages of thermoelectric detectors is their nonselective response to different wavelengths. Their extremely broad waveband of responsivity makes them suitable for many applications, although it is often necessary to restrict the range to be measured by using bandpass or cutoff filters. The thermoelectric detectors offer a fairly linear response to a wide range of radiation levels. Most thermoelectric detectors have a relatively slow response and are therefore not always suitable for measuring small and rapid variations in radiation.

(1) Thermopile
Thermopiles consist of two or more thermocouples with alternately blackened junctions which are usually connected in series. When irradiated, a

heat differential is established between the hot and cold junctions, which produces a measurable electromotive force (emf). The thermopile is one of the most accurate instruments for light measurement and is capable of giving highly reproducible measurements. Thermopiles possess the additional advantage of requiring only periodic recalibration. They are particularly suitable for the calibration of other light-measuring equipment, for light measurement in action spectroscopy if all conditions for suitable receiver geometry are met (Section IV), for measuring the output of monochromators, and for measuring light emitted from lasers. Whereas thermopiles with relatively large acceptance angles are available, many of those designed for laboratory use have a restricted field of view. Whichever type of thermopile is used, the acceptance angle(s) should be known to ensure compatibility with the light environment to be measured (see Section IV).

An important characteristic of thermopiles is their almost uniform response to the UV, visible, and near-IR wavebands. The uniformity range depends partially on the transmission characteristics of the window. Various types of window materials are available to minimise the effects of air circulation, which is a major cause of spurious results in temperature-controlled rooms. A crystal quartz window is adequate, having a fairly uniform transmittance from ~ 250 to over 3000 nm. Because thermopiles also measure IR radiation, all sources of heat are detected, including heat emitted by the operator. Measurements should be made with and without a glass filter which absorbs visible radiation but transmits IR; a 50% transmission cut-on at ~ 800 nm is ideal. The IR flux can then be subtracted to obtain the value for the waveband of interest. Care should be taken in choosing an appropriate filter because some IR-transmitting filters also transmit UV radiation and may therefore lead to inaccurate measurements. As with the thermopile itself, the filter must be totally thermally adapted to the environment in which it is used. This is a common source of operator error and should be checked as a matter of course; if thermally adapted, the filter will cause no meter deflection when placed over the open detector in darkness.

The maximum flux which thermopiles can measure is limited by the specific design characteristics and the period over which measurements are made. Thermopiles are capable of measuring extremely high levels of radiation. Only at energy fluxes of ~ 3×10^3 W m^{-2} is a significant departure from a linear response liable to be observed. If measurements need to be made over lengthy periods (min) the maximum flux measurable is reduced by a factor of approximately 3. The low sensitivity at low radiation levels is more likely to be a limiting factor and varies according to the type of detector. Vacuum cases offer slightly greater sensitivity and a steadier zero, with an increase in response time. If fluxes of less than 1.0×10^{-2} W m^{-2} need to be measured, it will be necessary to use a photoelectric detector.

Various basic thermopile receiver designs are available for measuring radiation in the laboratory. The circular type is suitable for most general applications and many have a large aperture angle. Receivers designed specifically for measuring the continuous output of lasers are available. These have a very restricted acceptance angle and usually employ water cooling. Linear receivers are designed to measure elongated areas of radiation and are therefore especially suitable for measuring the output of a monochromator. If robustness and resistance to shock is a primary consideration, wire-wound thermopiles are preferable to the Coblentz type. With wire-wound instruments, however, the whole area of the receiver must be exposed to the incident radiation. When both portability and ruggedness are important factors, a photoelectric detector may be both more suitable and cheaper.

The emf generated by the thermopile can be measured with direct readout, potentiometric, or electronically amplifying instruments. Details of the required input impedance and sensitivity are supplied by the manufacturer. If the thermopile and readout instrument have to be portable, a robust microammeter or microvoltmeter is preferable to a sensitive and delicate galvanometer. In the past, one of the main disadvantages of thermopiles was the need to use conversion factors to convert from voltage to units of radiant energy. This has been greatly simplified by the introduction of direct readout instruments. These instruments are usually supplied calibrated without a window on the thermopile. If a window is used, the system must be calibrated with the window or a correction factor must be used.

(2) Bolometer

Bolometers are resistance thermometers which measure the change in temperature of the radiation detector. Thermistor bolometers offer only slightly less sensitivity than thermopiles. Like thermopiles, they are very sensitive to extraneous heat but the complications involved in removing error from this source are greater than with thermopiles. This, the requirement for an accurately controlled external power supply, and the technical difficulties in attaching spatial receivers to measure diffuse radiation reduce the popularity of the bolometer as a thermoelectric detector.

B. Photoelectric detectors

Photoelectric detectors differ from thermoelectric detectors mainly in their strong wavelength dependency. They should therefore be rigorously calibrated for all wavelengths at which they are to be used. By using appropriate filters, the spectral responsivity can be tailored to specific requirements within a limited waveband. A major advantage is their lack of sensitivity to small changes in ambient temperature. Their ruggedness varies according to construction, but most types offer a convenient and portable instrument for making rapid and accurate radiation measurements. Unlike thermoelectric cells, however, all photoelectric cells produce a "dark current," or apparent signal, when they are not receiving light. Photoelectric detectors may be conveniently divided into three categories: photoconductive, photoemissive and photovoltaic.

(1) Photoconductive detectors

Photoconductive detectors are not suitable for most applications in photomorphogenesis. They operate on the basis that the resistance of certain semiconductors changes greatly when irradiated; this change in resistance can be used to measure the incident energy flux . The maximum sensitivity lies mainly in the IR waveband and only a few semiconductors provide proportionally high enough sensitivity at shorter wavelengths.

(2) Photoemissive detectors

Photoemissive detectors have very fast time constants and are very sensitive to low light levels, but this is offset by their narrow waveband of spectral sensitivity.

(a) Phototube or photodiode

The phototube or photodiode consists of an anode and a cathode enclosed in a glass or quartz envelope. The envelope may either be evacuated or filled with an appropriate gas. The spectral responsivity depends primarily on the composition of the cathode. An external amplifier is usually required to measure the current. Despite this, phototubes are not as sensitive as photomultipliers, and the noise levels can be unacceptably high. A nonlinear response may be experienced at high fluxes, especially in gas-filled detectors.

(b) Photomultiplier tube

The photomultiplier tube consists of a series of dynodes which amplify the signal within the tube itself. A gain of one million or more is typical. Their extremely high sensitivity makes them a suitable choice for measuring very low light levels. They also offer excellent linearity over a wide range. It is important that detector stability be good if measurement and calibration are separated by a substantial period, but the relatively low stability of photomultipliers does not make a significant contribution to error under normal measuring conditions. Care is needed with new photomultipliers, however, because they exhibit exaggerated aging. There is a large and rapid decrease in sensitivity during the first few hours of use, but this levels off to an acceptable rate ($\sim$ 1% per 100 h) after about 25 h of use. With a new tube, therefore, it is recommended that the instrument be allowed to run for at least 50 h before any measurements or calibrations are undertaken.

Although permanent damage does not usually result, very high fluxes should not be allowed to reach the photomultiplier when it is operating at near maximum voltage since this may result in temporary fatigue (change in sensitivity). Full recovery is normally achieved after a few minutes with no light input. After extreme overloading, it may also be necessary to reduce the applied voltage and switch the equipment off for a few minutes.

(3) Photovoltaic detectors

Photovoltaic or barrier layer cells consist of a low internal resistance semi-conductor layer sandwiched between a thin transparent metallic film (cath-ode) and a solid metal base (anode). Incident light generates an emf which is measured by a microammeter. This emf is high enough for the instrument to be self-powered at higher fluxes.

The most commonly used semiconductor materials are selenium and silicon. The selenium cell is often used for photometric instruments be-cause of the similarity of its spectral response to the average luminous effi-ciency curve of the human eye. Some selenium cells have a secondary peak of sensitivity in the R – FR waveband, making them suitable for measure-ments of up to 800 nm. The selenium cell suffers from fatigue and a non-linear response at high fluxes. Stability can also be a problem, and these detectors should be calibrated regularly, especially if they are often used at high fluxes. Variations in selenium-type detector temperature can cause large measurement errors. This error increases greatly with increasing re-sistance in the measuring circuit.

The silicon photocell offers the benefits of small size and high efficiency. Its photocurrent is proportionally higher than that produced by the selenium cell and also has little dependence on temperature. The silicon cell has the further advantages of superior linearity and is less susceptible to damage or fatigue from high levels of radiation although neutral-density filters may need to be used in order to stay within the meter range.

The spectral sensitivities of these two semiconductor detectors varies with the method of construction. The selenium cell offers high sensitivity down to 300 nm and has an upper limit of either $\sim$ 700 or $\sim$ 800 nm, according to type. On the other hand, the silicon cell offers high sensitivity up to and beyond 800 nm, but its accuracy below 400 nm will be severely restricting unless a suitably constructed UV-enhanced detector is used. Such detectors offer high sensitivity down to and slightly below 300 nm.

C. Detectors for specific wavebands

Two methods can be used to define the amount of energy within specific wavebands of a polychromatic light source. The first and only completely satisfactory method is to measure the spectral photon fluence rate or spectral photon flux with a spectroradiometer (Section VI) and then integrate the energy within the required waveband. The second method involves covering the detector with a combination of filters which transmit only the wavelengths of interest; this procedure is less exact because the cut-on and cutoff slopes lead to problems of definition. The full transmission characteristics of the filter (see Chapter 3, Section III) should be provided to enable meaningful interpretation of data measured in this way. Sensors with an approximately quantum-corrected response to the 400–700-nm waveband and negligible response to radiation below 380 or above 720 nm are commercially available. They offer a satisfactory description of the PAR produced by most polychromatic sources. However, unless they are specifically recalibrated, these detectors are not suitable for measuring monochromatic light or the PAR content of low-pressure discharge lamps with strong emission lines. They are also not capable of measuring all radiation absorbed by phytochrome.

VI. SPECTRORADIOMETRY

A. Introduction

If only monochromatic light needs to be measured, a simple detector with a calibrated broad-waveband response will be adequate if the spectral characteristics of the light sources can be defined. For these purposes the transmission properties of filters may be measured with a spectrophotometer, and the spectral quality of the transmitted light can be calculated using the emission characteristics of the light source supplied by the manufacturer, bearing in mind emission changes due to aging (see Chapter 3). Within limitations, the same approach can be used for polychromatic sources, such as those used for simulating natural daylight conditions. This situation is more complex, however. If several filters are used, the additive errors of many filter transmission measurements may be significantly greater than those involved when using the appropriate instrument for broad waveband sources, i.e., a spectroradiometer.

Although the choice is limited, several instruments are available which will provide useful measurements if operated correctly. These range from instruments which provide only the basic facilities to sophisticated equip-

ment which is self-calibrating, measures automatically, and then converts
the output to requested units before either storing the data or presenting it as
a plot or digital printout. At the same time, however, spectroradiometry
presents more potential sources of error than any other form of light measurement used in photomorphogenesis research. The accuracy and precision of measurements depend on the type of equipment used and on the
operator maintaining error at a minimum. A spectroradiometer can be
divided into four major components. These are the radiation receiver, an
optical system for dispersing the radiation into discrete wavebands, a detector which converts the radiation input into an electrical signal, and an output
system which communicates the signal to the operator.

B. Receiver geometry

The importance of using appropriate receiver geometry is considered in
Section IV. The same arguments apply for spectroradiometric measurements, although very few manufacturers offer more than one type of receiver. The cosine-corrected receiver is often supplied and is suitable for
many applications. Miniature receivers attached to fibre-optic light guides
are useful for making light measurements in confined areas. The output
from the light guide and the entrance to the dispersion system should be
correctly matched and aligned. The use of fibre-optics does have the disadvantages of decreased sensitivity and reduced stability because of the inevitable gradual breakage of individual fibres with time (see also Chapter 3,
Section IX).

C. The dispersion system

The dispersion or monochromator system separates the polychromatic radiation input into monochromatic components for analysis. Several methods
are available for dispersing light but not all permit satisfactory measurements. The three major considerations when choosing and operating a
dispersion system are the amount (percentage or ratio) of stray light, the
bandpass, and the accuracy of the wavelength scale. Interference filters are
generally unsuitable. Unless they are specially made they have an unacceptably large variation in bandpass. Calibration is difficult and accurate
measurement of certain line emission sources is virtually impossible. The
major mechanical disadvantages of interference filters are their need for a
bulky changing mechanism and the long time required to switch through all
the filters. Wedge interference filters permit the construction of more compact and relatively cheap instruments but the minimum spectral bandwidth
is unacceptably large for most applications.

Prisms offer a generally better performance than interference filters, although they suffer from a nonlinear wavelength scale which results in systematic variation in the bandpass throughout the wavelength range. Because prism instruments progressively compress the scale at longer wavelengths, this property can be a major disadvantage for phytochrome studies. The diffraction grating, on the other hand, is more or less linear in wavelength and bandpass but has the disadvantage of passing unwanted higher orders of light. If the manufacturer has not accounted for these, they can be removed by appropriate filters. A single diffraction rating should be adequate for most purposes, stray light being typically less than 0.1%. A combination of prism and grating provides very low stray light and no unwanted orders. A similarly high performance is offered by a single holographic concave grating whose construction reduces stray light to approximately a tenth of that produced by normally rated gratings.

The bandpass should be constant throughout the waveband to be measured and should be as narrow as possible to enhance spectral purity but not so narrow that the amount of light reaching the detector is a limiting factor. The ability to operate at a nominal bandpass of ~ 10 nm or less will enable precise measurements of most light sources. A monochromator with inherent high dispersion (i.e., ~ 5 nm mm^{-1}) has the advantage of permitting the use of relatively large slit widths without exceeding the required bandpass.

Although the wavelength scale of all dispersion systems is temperature dependent, this dependency is unlikely to be large enough to warrant consideration under normal measurement conditions. Calibration is necessary, however, preferably using low-pressure emission line sources. It should be noted that wavelength accuracy (typically ± 2 nm) and wavelength repeatability (typically ± 0.5 nm) are two separate entities; both should be checked.

Dust and other atmospheric contaminants stain mirrors and cause light scattering. For this reason, slit openings should always be covered. In spectroradiometers with exchangeable diffraction gratings, neither the monochromator nor spare grating should be left uncovered, even for short periods of time. Mirrors and gratings must never be touched. Under no circumstances should an attempt be made to clean these parts mechanically. Should it be necessary to remove dust, a jet of dry air may be used, but one should not try simply to blow the dust away because the high moisture content of the breath will cause staining.

D. The detector

Several types of detectors have been used in spectroradiometers but the majority of modern instruments use photomultipliers or silicon photodiodes. The photovoltaic silicon photodiode has the advantages of good stability, robustness, simplicity, and a broad waveband of spectral sensitivity. The photomultiplier, on the other hand, is more sensitive but only within a restricted waveband. If low light levels are to be measured over a very broad waveband, interchangeable detectors can be used.

Detector stability is important if measurement and calibration are separated by a substantial period. This period should be determined for the specific instrument and conditions under which measurements are made. An internal reference source greatly overcomes this problem.

E. Information output

Basic spectroradiometer systems provide a digital display of detector output which can be plotted and recorded, using an XY recorder. If large quantities of data have to be measured, it is more convenient to have data stored on magnetic tape or disc. Many modern spectroradiometers can be equipped with electronic calculators or computer interfaces which enable automatic printout at specified wavelength intervals in either relative or calibrated units. The integrated flux over a previously demanded waveband can also be presented. Sophisticated calculators are available which not only process the output data but also control the measurement system; although useful and time-saving, their practicality must be weighed against their expense.

F. Special situations

(1) Artificial light sources
The spectral purity of the light emitted by the exit slit of the monochromator and measured by the detector is determined by the slit widths and the dispersion of the system. Lowering the dispersion (i.e., the reciprocal dispersion) or decreasing the slit width provides a narrower bandpass. Although very narrow bandpasses provide high spectral purity, it is usually necessary to use a compromise because reducing the slit width also reduces the amount of radiation reaching the detector.

The bandpass should be appropriate for the light source. If a light source which emits a completely discontinuous spectrum must be measured, the bandpass can be opened up to 1 or 2 nm less than the separation of the peaks and 100% spectral purity will be achieved. If the source comprises a low-level background continuum superimposed on the line spectra (such as a medium-pressure discharge arc), the purity remains high because of the relatively high emitted intensity of the lines. As the proportional contribution of the background continuum increases, the bandpass should be reduced accordingly. When continua alone, or continua with relatively small emission lines are to be measured, perfect purity cannot be attained. Under these conditions, bandpass must be reduced to the minimum which provides adequate transmission. When purchasing a spectroradiometer for measuring several types of light sources, it is clearly important that the bandpass can be adjusted, either by interchangeable fixed slits or by a single variable slit.

When single-beam monochromators are used, higher orders of light may be present, and these must be removed with appropriate cut-on or cutoff filters. The visible waveband is overlapped by second orders from the UV region. A clear glass filter cuts off below 350 nm and therefore eliminates false orders up to 700 nm. To clear the region from 700 to 800 nm, a filter which cuts off below 400 nm must be used. The main false orders (with their origins in parentheses) are: 507.4 nm (253.7 nm × 2), 593.4 nm (296.7 nm × 2), 626.4 nm (313.2 nm × 2), 730.0 nm (365.0 nm × 2) and 761.1 nm (253.7 nm × 3). The order-sorting filter can be placed at either the entrance or the exit slit, but the same position must be used for both measurement and calibration.

(2) Natural light measurement

Measurement of light in the field demands several requirements above those already needed in the laboratory. The instrumentation needs to be more robust and to be protected from excessive moisture and dirt. Ingress of dirt and moisture to the optics, circuit boards, and electrical connections can cause inaccurate measurements or even instrument failure. If the instrument needs to be used for long periods under damp conditions, a purpose-made outer casing will help protect against dirt and dust and enable provision to be made for dehumidifying the air around the optics and electrics.

If there is no power supply, a few instruments are available which can be powered by rechargeable batteries. If many measurements have to be made, it will be necessary to have several reserve batteries. A suitable alternative is a small portable generator which provides a stable output. Any variations in this output must be known and their effects checked during calibration.

Small variations in the total flux during each spectral measurement are virtually inevitable, even on apparently cloudless days. Total energy should be measured separately during each spectral measurement and any recordings made during which the variation in total energy exceeds specified limits should be rejected. In practice, a variation of better than ± 3% is difficult to achieve under overcast conditions and ± 5% must be expected during twilight. These restrictions are most easily obeyed by using fast scanning rates and by avoiding making measurements on windy days. Although it is possible, it is not advisable to normalise the spectral measurement at each wavelength stage to the total energy measurement because most changes in total energy, such as those caused by fast moving clouds, are also associated with a change in spectral quality.

Characterisation of natural daylight requires more accurate definition than artificial light sources. To enable full interpretation of the data, the following information should be given: latitude, longitude, date, time, solar elevation, solar azimuth angle, and cloud cover. Additionally, it is important to know whether the sun is obscured by cloud, i.e., whether direct and/or diffuse radiation is being measured.

(3) Measuring low light levels

When measuring low light levels, certain basic precautions can be taken which will greatly aid the accuracy of measurements made at light levels approaching the limits of the equipment's sensitivity. The primary objectives under these conditions are to maximise the signal input and to minimise the amount of noise. Both of these factors are improved by ensuring that the detector is suitable for the waveband to be measured. Under extreme conditions it may be necessary to use more than one detector.

To maximise the signal, it is important that the amount of radiation reaching the detector is not unnecessarily reduced. The largest possible collecting surface should be used, bearing in mind that collector geometry must still be appropriate. Fibre-optic light guides can severely reduce light input especially if many of the fibres are broken; also, their use often necessitates a small receiver surface. Sensitivity can often be improved by increasing the monochromator slit width within the limits imposed by the type of light source which is to be measured (see also Section VI. F. 1). Noise can be reduced by increasing the sampling time. The signal to noise ratio improves with the sampling time, so very slow scanning rates can be used; this has the advantage of allowing long photometer response times which also reduce the noise level.

When operating photomultipliers at extremely low light levels, thermionic emission of electrons from the photocathode or dynodes may become evident; because dark current increases with temperature, this problem can be reduced by cooling the photomultiplier. Although increasing the voltage applied to the photomultiplier increases its sensitivity, very high applied voltages increase the dark current substantially. Operation within 250 V of the maximum rating should be avoided. A high-pitched whine at high applied voltages is a normal occurrence and may be ignored.

Dust and other atmospheric contaminants on the surfaces of the optical system cause scattering of light which results in a decreased signal and increased noise. All external contacts should be cleaned regularly with alcohol because poor contacts owing to dirt or moisture lead to increased dark current or leakage.

VII. SOURCES OF ERROR

The most serious sources of error can usually be traced back to absolute calibration error or to operator error. The major causes of error during calibration (i.e., standard lamp inaccuracy, lamp current inaccuracy, poor alignment, and excessive stray light) are considered in Section III. In practice, calibration error is usually caused by lack of calibration rather than careless calibration. Operator error can derive from many sources and may be so serious as to make light measurement data meaningless. The major sources of operator error are using inappropriate equipment, using uncalibrated equipment, using a different system for calibration from that used for measurement, shading or reflection of radiation by equipment or the operator, incorrect positioning of the receiver, incorrect interconversion of units, dust or moisture modifying the true radiation measurement, and inadequate user control of input voltage, zero drift, dark current, and temperature. Operator error can be reduced to negligible proportions if the user understands the limitations of the available equipment.

In addition to calibration and user error, there are several individual forms of error which can be minimised by relatively simple precautions. These errors can be divided into three areas: receiver, detector, and indication or readout error. Receiver error has been considered in detail in Section IV. Assuming that the correct receiver has been chosen for the light source and irradiated object, it is important that it should perform as closely as possible to its designed purpose. The extent of the error varies with the type of detector being used (see Section V). Detectors should be checked for error due to fatigue (slow loss of sensitivity during exposure), nonlinearity (inability to follow progressive changes in fluence rate or flux), response time (inability to follow rapid changes in fluence rate or flux), and temperature. Relative error in the spectral response is the deviation in the spectral sensitivity of the detector from its designed response. This is an especial problem with detectors which are fitted with filters designed to have a tailored spectral response within a specific waveband. Here, the error may become unacceptably large when measuring light sources with strongly pronounced emission lines. Indication or readout error is the difference between the indicated value given by the readout instrument and the true value. This error can be minimised by using a stable line or battery input voltage, ensuring good electrical connections, avoiding high humidity and temperature variations, and by careful control of zero drift and changes in the dark current.

The total error is calculated directly from the various sources of error which are present in a specific measuring system. As an example, a simplified hypothetical case can be considered which has the following individual errors: absolute error = 3%, receiver error = 5%, stability error = 2% and indication error = 1%. The total error of this system is given by the square root of the sum of squares for each individual error, i.e.,

$$\text{Total error} = \sqrt{3^2 + 5^2 + 2^2 + 1^2} = 6.2\%$$

At worst, this system will measure a $3 + 5 + 2 + 1\% = 11.0\%$ over- or underestimate. Operator error has not been included in this hypothetical case because it is usually impossible to define numerically. The best definition of operator error is given by a comprehensive description of the equipment and measuring techniques.

SOURCES OF FURTHER INFORMATION

Bell, C. J. and Rose, D. A. (1981). *Plant Cell Environ.* **4**, 89–96.
Bickford, E. D. and Dunn, S. (1972). "Lighting for Plant Growth." Kent State Univ. Press, Kent, Ohio.
Commission Internationale de l'Éclairage (1970). "International Lighting Vocabulary," Publ. No. 17, C.I.E., Paris.
Dlugos, H. G. (1958). *Lichttechnik* **10**, 565–567.
Eckhardt, G. (1965). *Lichttechnik* **17**, 110A–113A.
Geist, J. and Zalewski, E. (1973). *Appl. Opt.* **12**, 435–436.
Hartmann, K. M. (1977). *In* "Biophysik" (W. Hoppe, W. Lohmann, H. Markl and H. Ziegler, eds.), pp. 197–222. Springer-Verlag, Berlin and New York.
Hartmann, K. M. and Cohnen-Unser, I. (1972). *Ber. Dtsch. Bot. Ges.* **85**, 481–551.
Holmes, M. G., Klein, W. H., and Sager, J. C. (1984). *HortScience* (in press).
Incoll, L. D., Long, S. P. and Ashmore, M. R. (1977). *Curr. Adv. Plant Sci.* **9**, 331–343.
International Organization for Standardization (1973). "Quantities and Units of Light and Related Electromagnetic Radiations," Int. Stand. ISO 31/VI.
International Organization for Standardization (1979). "Quantities and Units of Light and Related Electromagnetic Radiations," Draft Int. Stand. ISO/DIS 31/6.
Jagger, J. (1967). "Introduction to Research in Ultraviolet Photobiology." Prentice-Hall, Englewood Cliffs, New Jersey.
Kubín, Š. (1971). *In* "Plant Photosynthetic Production. Manual of Methods" (Z. Sestak, J. Catsky and P. G. Jarvis, eds.), pp. 702–765. Junk, The Hague.
Mohr, H. and Schäfer, E. (1979). *Photochem. Photobiol.* **29**, 1061–1062.
National Bureau of Standards (1978). *NBS Tech. Note (U.S.)* **910-2**.
Rupert, C. S. (1974). *Photochem. Photobiol.* **20**, 203–212.
Rupert, C. S. and Latarjet, R. (1978). *Photochem. Photobiol.* **28**, 3–5.
Šetlík, I. and Kubín, Š. (1966). *Acta. Univ. Carol., Biol., Suppl.* **1/2**, 77–88.
Wassink, E. C. and Van der Scheer, C. (1951). *Meded. Landbouwhogesch. (Wageningen)* **51**, 175–183.
Withrow, R. B. and Withrow, A. P. (1956). *In* "Radiation Biology" (A. Hollaender, ed.), Vol. 3, pp. 125–258. McGraw-Hill, New York.

5

Action Spectroscopy

Eberhard Schäfer and Leonid Fukshansky

I. INTRODUCTION

Action spectroscopy is dosimetry with light. It provides a nondestructive procedure for analysing the properties of a functional pigment *in vivo* and *in vitro*. Action spectroscopy has been used frequently in photobiology for qualitative and quantitative analysis of photoreceptors since the time of Engelmann's classic qualitative experiments in 1882 (Engelmann, 1882).

General methods in action spectroscopy have been described in a series of publications (see Shropshire, 1972). Additional recent publications discussing action spectroscopy as related to photomorphogenesis are those by Shropshire (1972), Hartmann and Cohnen-Unser (1972), Hartmann (1977), and Schäfer *et al.* (1983a).

The original aims of action spectroscopy were to obtain either absolute or apparent cross sections for the excitation of a pigment to enable the characterisation and purification of the pigment.· The cross section characterises the interaction between a pigment and light and is proportional to the extinction coefficient. We have to distinguish between the excitation and conversion cross section (the latter taking the quantum yield for the conversion into account) and the absolute and apparent cross section [the latter including optical distortions (see Section II)].

The original aims of action spectroscopy are often no longer relevant in photomorphogenetic research. It is known that these responses are usually either under the control of phytochrome or under the control of a blue–near-UV light-absorbing pigment, which some authors refer to as cryptochrome (see Senger, 1980 and Chapters 2 and 12). To prove that a photoresponse is under the control of phytochrome, much easier approaches than action spectroscopy are available (see Chapter 2). It should be mentioned that neither a definitive identification of cryptochrome—flavin versus carotenoid—nor its purification has yet been achieved (see Song,

TECHNIQUES IN PHOTOMORPHOGENESIS
0-12-652990-6
109

Copyright © 1984 by Academic Press, London.
All rights of reproduction in any form reserved.

1980; Shropshire, 1980; de Fabo, 1980; and Chapter 12). On the other hand, it should be mentioned that, especially under continuous irradiation, involvement of other photoreceptors acting directly on photomorphogenesis (Ng *et al.,* 1964; Vanderhoef *et al.,* 1979; Holmes and Schäfer, 1981), or an interaction between photosynthesis and photomorphogenesis, has to be expected (see McCree, 1976). Therefore, action spectroscopy will still be needed in photomorphogenesis, especially for light-grown plants.

An analysis of photogravitropism in *Phycomyces,* using classical action spectroscopy, showed that several photoreceptors control this response (Schäfer *et al.,* 1983b; Galland, 1983). Action spectroscopy should, therefore, be model-related action spectroscopy (called analytical-action spectroscopy) as suggested by Hartmann and Cohen-Unser (1972). The aim of this method is not only an estimation of the absorption cross section but also a kinetic analysis of the photoreceptor system and its transduction chain based on reaction models. The estimation of the absorption cross section from fluence response curves is not only a problem in the case of a photochromic pigment such as phytochrome, but also if the active form of the pigment is not stable. A fast, dark reversion of the active form is assumed for the photoreceptor cryptochrome (Lipson and Presti, 1980).

Mathematical modeling should follow the general principles pointed out by Fukshansky and Mohr (1980) and Fukshansky and Schäfer (1983) and be based, in the case of photomorphogenesis, mainly on *in vivo* spectroscopy of phytochrome. In the case of analytical action spectroscopy, the first analysis was made by Hartmann and Cohnen-Unser (1972). A detailed discussion of this problem, based on new experimental results, should include a mathematical analysis of several models and would, therefore, exceed the scope of this book. Discussion here is limited to a summary of the general problems involved in action spectroscopy and the approaches used for analytical action spectroscopy under induction and HIR conditions. Plant photomorphogenetic response types can be divided into induction responses and so-called high irradiance responses (HIR) (see Chapter 4). Induction responses are effected by a brief (i.e., minutes) red or blue light irradiation, which can be reversed by subsequent irradiation with far-red light. Induction responses require relatively low light energies and obey approximately the Bunsen–Roscoe reciprocity law. The HIR responses usually require prolonged (i.e., hours) irradiation and do not exhibit reciprocity (Mancinelli and Rabino, 1978).

II. OPTICAL PROBLEMS IN ACTION SPECTROSCOPY

Hartmann and Cohnen-Unser (1972) and Hartmann (1977) have described in great detail the problems involved in obtaining variable fluence rates and in measuring and calculating the effective photon fluence rate applied. The treatment here is therefore restricted to a short discussion of these problems.

A. Variation of photon-fluence rates

For correct action spectroscopy, the fluence must be varied by several orders of magnitude whilst the fluence rate remains constant for the irradiation period. In addition, the temperature should be the same under all experimental conditions.

To minimise interference by stray light, the light sources should be separated from the room where the seedlings are grown. This room should be temperature controlled ($\pm 0.2 - 0.5\,^{\circ}C$). At very high fluence rates, an additional temperature-controlled water bath is necessary for the sample to avoid significant heating by irradiation. Further details are given in Chapter 3.

The fluence can be varied by changing the irradiation time and/or fluence rate. Varying irradiation time is desirable only if reciprocity holds, i.e., if the response is a function only of the product of the photon-fluence rate and time. If reciprocity does not hold, the fluence should be varied by changing the fluence rate, which can be achieved by changing the power input of the irradiance source, by altering the distance between irradiation source and the object, or by inserting neutral-density glass filters in the optical pathway. In the first and third cases, it is necessary to determine (using a spectroradiometer) whether the spectral distribution of the light has changed as a result of a change of the colour temperature of the irradiation source or by the fact that the transmission of light by neutral-density filters is wavelength dependent. Furthermore, changes in the optical system (e.g., several neutral density filters) may alter the degree of polarisation, which must be kept constant. Although altering the distance between the light source and the irradiated object is a convenient method for changing fluence rates, this method is not suitable for altering the fluence rate by more than two orders of magnitude.

B. Radiation measurements

The general principles of radiation measurements are discussed in Chapter 4, so discussion is restricted here to specific problems encountered in action spectroscopy. Although knowledge of the fluence-rate gradient in the sample is required, it is possible only to measure the fluence rate of the radiation field with a cosine-corrected or spherically integrating measuring system. The cosine-corrected measurement ignores back reflection, but this can be measured using the spherically integrating system. In the latter case, the fact that the radiation field is distorted by the plants placed in the radiation field is usually ignored. These influences depend strongly on the background reflection and the plant material. These problems are greater if polychromatic radiation is used because light attenuation by the plants can be strongly wavelength dependent, thereby resulting in large differences in the spectral photon distribution between incoming and back-reflected light. Careful measurements, using a spectroradiometer, are necessary to correct for these distortions. With monochromatic radiation, the choice of method must be based on the conditions used, i.e., back reflection, plant material, leaf and/or cotyledon position, response localisation, etc. The problems are discussed in detail in Chapter 4.

C. Photon-fluence rate gradients in plant tissues

Plant tissues are normally highly scattering, nonhomogeneous samples. The internal photon-fluence rate gradient is dependent on at least three factors. These are the attenuation of radiation by screening, the absorption inhomogeneities (sieve effect), and the attenuation of radiation by scattering.

The effect of self-screening by the photoreceptor pigment can normally be excluded in photomorphogenesis, although it must be kept in mind, as pointed out by Hartmann (1977), that the local pigment concentrations can be very high. The apparent absorbance caused by phytochrome, including amplification by scattering (Butler, 1962), of an etiolated cotyledon is always less than 10^{-2} and therefore negligible.

Because biological objects are optically complex, it is necessary to estimate an effective photon-fluence rate, i.e., the average photon-fluence rate in tissue with homogeneous pigment distribution. The general distribution of phytochrome or cryptochrome (functional or nonfunctional) is not known. Thus even with knowledge of the photon-fluence rate gradients, only rough estimations of the effective photon-fluence rate can be made by assuming a homogeneous pigment distribution. This is useful only if the responses of whole organs, and not those of specialised cells, are analysed (see anthocyanin in mustard, for example, Steinitz and Bergfeld, 1977).

The influence of screening, sieve-effect, and scattering on photon-fluence rate gradients and measured absorption spectra is described in Chapter 6.

D. Effect of screening pigments on the shape of photon-fluence rate response curves

(1) Static effects
The previous section discussed the influence of light attenuation and scattering on the internal photon-fluence rate gradient. If these optical disturbances do not change during the period of irradiation and do not depend on the photon-fluence rate, they will be referred to as static effects. All other

effects will be described as dynamic. All of these effects are wavelength dependent and will, therefore, lead to a wavelength-dependent alteration of the effective photon-fluence rate:

$$E_{eff}(\lambda) = f(\lambda) \cdot E(\lambda,0) \qquad \text{Eq. (1)}$$

where $E(\lambda,0)$ is the incident photon fluence rate at wavelength λ, $E_{eff}(\lambda)$ is the effective photon fluence rate at wavelength λ, and $f(\lambda)$ is the correction factor.

Therefore, in a situation where the response, or relative response, is plotted as a function of the logarithm of $E(\lambda,0)$ (the so-called logarithmic photon-fluence rate response curves) the static optical disturbances can only lead to a parallel shift of the photon-fluence rate response curves. The modification of the applied photon-fluence rate $E(\lambda,0)$ by a factor $f(\lambda)$, which is only wavelength dependent, will result in an additive component of

$$\log E_{eff}(\lambda) = \log[f(\lambda) \cdot E(\lambda,0)] = \log f(\lambda) + \log E(\lambda,0) \qquad \text{Eq. (2)}$$

Therefore, optical effects can never result in a change of the slope of the semilogarithmically plotted-photon fluence rate response curves, irrespective of the functional relationship between response and effective photon-fluence rate. If, on the other hand, the response is plotted as a linear function of the photon-fluence rate, even static screening will lead to alterations of the slope of the curves. This is the main justification for the frequent use of logarithmic fluence rate response curves. Furthermore, biological responses often obey the Weber–Fechners law [Section III. B, Eq. (9)], which will lead to linear fluence rate response curves if the response is plotted versus the logarithm of the fluence rate.

For most of the well-tested cases in photomorphogenesis it has been found that the fluence rate response curves are not parallel if the response is plotted as a function of the logarithm of the fluence rate. The discussion above clearly shows that a wavelength dependence of the slope of semilogarithmic fluence rate response curves can only be due to a dynamic (fluence-rate dependent) screening (Johnson, 1980) and/or that a photoreceptor system (i.e., photochromic pigment or two or more nonadditive-acting photoreceptors) controls the response.

(2) Dynamic effects

Under conditions of prolonged irradiation (i.e., those normally used to analyse the high irradiance responses), stationary optical behaviour of the biological object cannot be expected. In studies of etiolated tissue chlorophyll accumulation, which is dependent on both wavelength and photon-fluence rate, will alter the optical properties of the system. The observable changes in chlorophyll content and chloroplast size will lead to wavelength-dependent and photon-fluence-rate-dependent changes in scattering and sieve effect.

The observed redistributions of phytochrome (Mackenzie *et al.,* 1975) and the wavelength- and photon-fluence-rate dependence of Ptot destruction (Schäfer *et al.,* 1975, 1976) will also lead to a change of the effective photon-fluence rate:

$$E_{\text{eff}}(\lambda, t) = f[\lambda, t, E(\lambda, 0)] \cdot E(\lambda, 0) \qquad \text{Eq. (3)}$$

Both a static and a dynamic effect is observed in the phytochrome-mediated inhibition of mesocotyl growth of *Avena* seedlings after 24 h irradiation (Schäfer *et al.,* 1982). At low fluence rates, a strong but static screen by protochlorophyll exists because this photoresponse is saturated before significant protochlorophyll pool changes are measureable. The contribution of protochlorophyll in the primary leaf as a screen for the inhibition of mesocotyl growth was tested by comparison of two characteristic photon-fluence rate response curves for irradiation from the top, one from the side and one from the bottom with parallel light beams. In all cases where Response (λ_1)/Response (λ_2) is dependent on the incident direction of the light, screening has to be considered. In this study on inhibition of *Avena* mesocotyl growth, the ratio of the apparent photoconversion cross section S_{685}/S_{658} changed from 7.3 (irradiation from above) to 1.7 and 1.1 (irradiation from the side or from below, respectively). Screening did not alter the slope of the curves.

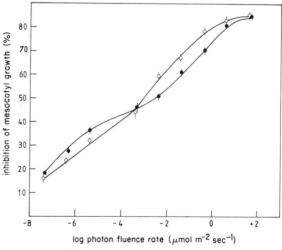

Fig. 1. Light-induced inhibition of mesocotyl growth in 54-h dark-grown *Avena sativa* L. seedlings. The inhibition was measured after 5 min at 658 nm (O) or 685-nm (●) light of various fluence rates followed by a 24-h dark period (Schäfer *et al.,* 1982).

A simple test for screening can be described, using the inhibition of meso-cotyl growth in 54-h dark-grown *Avena* seedlings (Fig. 1). Irrespective of the screen, the response will depend only on photoequilibrium (φ_λ) at high fluence rates if φ_λ has been established. When protochlorophyll acts as a screen, the effective fluence rate at 657 nm can be much less than at 685 nm, resulting in a lower effectiveness of 685 nm as compared with 650 nm at low fluence rates. However, at higher fluence rates, the response curves cross over because $\varphi_{657} > \varphi_{685}$. At high fluence rates, strong screening takes place and results in a change in the slope of the fluence rate response curves (Schäfer *et al.,* 1982).

III. ACTION SPECTRA UNDER INDUCTION CONDITIONS

Action spectra in photomorphogenesis can be divided into two groups: those under induction conditions where the law of reciprocity is valid, and those under continuous light conditions where the response displays fluence-rate dependency. For simplicity, we start the discussion with action spectra under induction conditions and choose the problem of photoconversion as the simplest example.

A. The problem of photoconversion

The simplest situation will be achieved if an action spectrum of photocon-version or photoexcitation is performed. In this case only pigment parame-ters are measured, excluding the metabolic signal transduction chain. It is necessary to start with a model analysis. *In vivo* measurements of photore-versibility under various conditions (Schäfer and Schmidt, 1974; Schmidt *et al.,* 1973) and measurements using immuno-recognition of the protein mol-ecule (Mackenzie *et al.,* 1975; Hunt and Pratt, 1980) are all consistent with a simple phytochrome model as shown in Fig. 2.

It should be pointed out that we ignore the observation that the rate of Pr

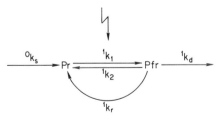

Fig. 2. Model of the dynamics of the photochrome system: ${}^0k_s$ = rate constant of Pr synthesis, ${}^1k_1$, ${}^1k_2$ = rate constants of the light actions Pr $\rightarrow$ Pfr and Pfr $\rightarrow$ Pr, ${}^1k_d$ = rate constant of Pfr destruction, ${}^1k_r$ = rate constant of Pfr $\rightarrow$ Pr dark reversion.

synthesis seems to be under phytochrome control in some monocotyledonous seedlings (Duke *et al.*, 1977; Gorton and Briggs, 1981; Gottmann
and Schäfer, 1982, 1983).

Both the phytochrome dark reactions and irradiation time influence photoconversion. Under induction conditions the irradiation time t is usually
chosen so that $^{0}k_s$, $^{1}k_d$, and $^{1}k_r$ are very small compared to $k_1 + k_2$. The
resulting simpler model for photoconversion of phytochrome (Pr $\underset{k_2}{\overset{k_1}{\rightleftharpoons}}$ Pfr)
will therefore neglect all dark reactions. The solution of the dynamic of Pr
and Pfr is now very simple.

$$Pr(t) = [Pr(0) - Pr(\infty)]e^{-(k_1+k_2)t} + Pr(\infty)$$
$$Pfr(t) = [Pfr(0) - Pfr(\infty)]e^{-(k_1+k_2)t} + Pfr(\infty) \qquad \text{Eq. (4)}$$

where Pr,fr(t) = phytochrome (Pr or Pfr) content at time t, Pr,fr(0) = phytochrome (Pr or Pfr) content at time 0, and Pr,fr(∞) = phytochrome (Pr or
Pfr) content at infinite time.

The terms k_1 and k_2 are the product of the effective photon-fluence
rate and the photoconversion cross section. Schematic curves are shown
in Fig. 3. To obtain linear curves, one has to plot log [Pfr(t) − Pfr(∞)]/
[Pfr(0) − Pfr(∞)] versus effective fluence (Fig. 4).

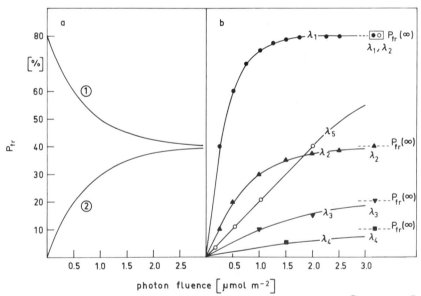

Fig. 3. (a) Schematic plot of phytochrome photoconversion kinetics after a red ① or far-red ②
light pretreatment. (b) Phytochrome photoconversion kinetics after a far-red light pretreatment [Pfr(0) ≈ 0], using various wavelengths establishing Pfr(∞) values between 10 and 80%.
In all cases, Pfr is plotted as a function of photon fluence.

$$\log\frac{\text{Pfr}(t) - \text{Pfr}(\infty)}{\text{Pfr}(0) - \text{Pfr}(\infty)} = -(\sigma_r + \sigma_{fr}) \cdot E_{eff}(\lambda,0) \cdot t = -(\sigma_r + \sigma_{fr}) \cdot F_{eff}(\lambda,0)$$

Eq. (5)

where σ_r = photoconversion cross section Pr $\rightarrow$ Pfr, σ_{fr} = photoconversion cross section Pfr $\rightarrow$ Pr, and $F_{eff} = E_{eff} \cdot t$.

Equation (4) shows that under the chosen conditions the response [Pr(t),Pfr(t)] is a function of the product $(\sigma_r + \sigma_{fr}) \cdot E_{eff}(\lambda,o) \cdot t = (\sigma_r + \sigma_{fr}) \cdot F_{eff}(\lambda,0)$, and therefore reciprocity should hold. An experimental indication that reciprocity does not hold would indicate that the simple assumptions are not valid.

Based on Eq. (5), $(\sigma_r + \sigma_{fr})$ can be obtained if Pfr(0), Pfr(t), Pfr(∞) and $E_{eff}(\lambda,0) \cdot t = F_{eff}(\lambda,0)$ are measured. Because

$$\text{Pfr}(\infty) = (\sigma_{fr}/\sigma_r + \sigma_{fr})$$

Eq. (6)

both absolute cross sections can be obtained. The main problem for *in vivo* spectroscopy is the measurement of $F_{eff}(\lambda,0)$ (see Section II. C). For *in vitro* measurements, Butler *et al.* (1964) solved this problem by using a special cuvette which had a very large area and a small sample thickness during actinic irradiation and a large thickness and small area during Pfr,Pr measurements. By ignoring background reflection, $F_{eff}(\lambda,0) \simeq F(\lambda,0)$.

B. Construction of action spectra

The generally accepted method of constructing action spectra (based on the theory of classical action spectroscopy) is indicated in Fig. 4. It is necessary to know for each wavelength what fluence is required to obtain a standard response (0.4 in Fig. 4). The reciprocal of these values is then plotted as a function of the wavelength [see Eq. (7)].

The calculation of the action spectrum depends strongly on the way the fluence response curves are presented. For photoconversion kinetics it is obvious, by comparing Figs. 3a and 4, that only on the basis of normalised fluence response curves (i.e., [response (F_λ) – response(∞)]/ [response(0) – response(∞)] plotted versus F_λ) will meaningful action spectra be obtained.

The situation becomes even more complex if the fluence response curves are not monotonous functions of the fluence—i.e., the same response is obtained for one wavelength at two or three different fluences. In these cases normal reciprocity does not hold for the total fluence range tested. Therefore, action spectra are constructed normally only for the threshold response by extrapolating the fluence response curves to response zero. This extrapolation should be described carefully in each case. The best way is to extrapolate the linear part of the logarithmic fluence rate response curve back to

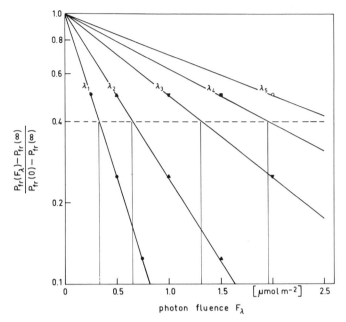

Fig. 4. Replot of the photoconversion kinetics shown in Fig. 3a. $[Pfr(F_\lambda) - Pfr(\infty)]/$ $[Pfr(0) - Pfr(\infty)]$ is plotted semilogarithmically as a function of photon fluence.

zero. It is obvious that this is useful only if the deviations of the measured values from the extrapolated curve are not too large.

In the relationship between incident light quanta and the response, not only the characteristics of the photoreceptor but also those of the transduction chain are involved. To overcome this problem, an extrapolation to a zero response level is useful (cf. Shropshire, 1972). In a special case like phototropism, a simultaneous irradiation with a test wavelength from one side and a reference wavelength from the other side allows an adjustment of the fluence rate of the test wavelength to obtain a zero response (cf. Shropshire, 1972). In all other cases, the absolute responsivity of the system should be checked from time to time because the measurement of an action spectrum is often very time consuming and changes in absolute responsivity have to be expected. This check can be done by using a test light which produces about a 50% saturation level.

Responsivity to both a reference light and a dark control treatment should be measured. It must be emphasised that in action spectroscopy the only parameters that are to be varied are the number of photons and the energy of the photons reaching the photoreceptor. Therefore, all other parameters, such as temperature, humidity, pH, water supply, experimental material, and pretreatments should be constant. This is not always possible because the measurement of an action spectrum can require long experimental periods and large ranges of fluence rates are used. The differences in tissue temperature at different fluence rates can be substantial. Temperature control of the irradiated tissue is very important and at high fluences a water bath may be necessary as well as temperature control of the growth room. Even when taking this precaution there will still be differences in the tissue temperatures, and measurements of this parameter would be very useful.

Phytochrome presents serious problems for the construction of action spectra because it is a photochromic pigment, i.e., a pigment existing in two forms which are interconvertible by light. If we have a simpler system of a photoreceptor existing only in one stable form, "classical" action spectroscopy can be applied (Schäfer *et al.*, 1983a).

$$P \xrightarrow{k} P^* \searrow R$$

where $k = \sigma \cdot E_{\lambda,\sigma}$, conversion (excitation) cross section, and R = response.

Based on the laws of Grotthus–Draper and Einstein–Stark we obtain:

If
$$R_{\lambda_1}(E_{\lambda_1}) = R_{\lambda_2}(E_{\lambda_2})$$
$$\Rightarrow \sigma_{\lambda_1} \cdot E_{\lambda_1} = \sigma_{\lambda_2} \cdot E_{\lambda_2}$$

and therefore
$$\sigma_{\lambda_1}/\sigma_{\lambda_2} = (1/E_{\lambda_1}){:}(1/E_{\lambda_2}) \qquad \text{Eq. (7)}$$

This is the basic equation of classical action spectroscopy and will be valid irrespective of whether reciprocity holds or changes in sensitivity occur during the irradiation period. A further consequence of Eq. (7) is that logarithmic fluence rate response curves will be parallel (Schäfer *et al.*, 1983a).

Some comments should be made with respect to general problems using action spectroscopy As shown in the case of phytochrome photoconversion kinetics, a plot of $\log\{[R(F_\lambda) - R(\infty)]/[R(0) - R(\infty)]\}$ versus F_λ was linear and useful in estimating photoconversion cross sections. If R were proportional to P^*, the following equation would be valid:

$$R(t) = \alpha \cdot P(t = 0) \cdot (1 - e^{-\sigma_\lambda E_\lambda \cdot t}) \qquad \text{Eq. (8)}$$

where α = proportionality factor.

In this case, and only in this case, a plot of $\log\{[R(t = \infty) - R(t)]/[R(\infty)]\}$ versus $E_\lambda \cdot t$ gives a straight line with a slope σ_λ (see Hartmann and Cohnen-Unser, 1972). If R is not proportional to P^*, a plot of R versus $\log (E \cdot t)$ gives parallel curves and allows one to estimate the relative values.

Another problem is: why is the response proportional to the logarithm of photon fluence (rate)? This is the so-called Weber–Fechner law, which is valid for a large number of physiological and psychophysical responses:

$$R = \alpha \cdot \log(N_\lambda \cdot t) \qquad\qquad \text{Eq. (9)}$$

where α = proportionality factor.

This relationship does not characterise the photoreceptor dynamics but the dynamics of the transduction chain, including adaptation phenomena (i.e., changes in sensitivity depending on time and/or irradiation treatment). A logarithmic amplifier between P* and R must be postulated because Eq. (9) is also observed at very low fluence rates, where the amount of the photoreceptor remains virtually constant. Classical action spectroscopy will provide no information about this postulated logarithmic amplifier. This complexity increases when dealing with a photochromic pigment like phytochrome. As was shown in Figs. 3 and 4, it is primarily values for $k_1 + k_2$, which are obtained by measuring photoconversion kinetics. The induction spectrum extrapolated for a *zero response level* should reflect k_1 and, therefore, the Pr absorption spectrum (Fig. 5). At high response levels, i.e., high fluence rates and established photoequilibria, the action spectrum should reflect the wavelength dependence of the photoequilibrium $\varphi_\lambda = k_1/(k_1 + k_2)$.

In the case of the action spectrum for the reversion of the red-light induced effect, a coincidence with k_2 can be observed only at a wavelength where the Pr → Pfr reaction (k_1) is negligible, i.e., for $\lambda > 700$ nm.

General rules for the construction of an action spectrum are as follows.

1. Keep all other parameters than light as constant as possible, especially temperature.

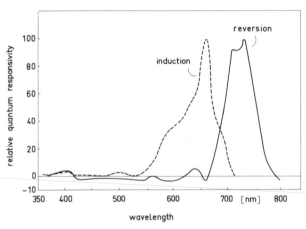

Fig. 5. Action spectra for induction and reversion of the light-dependent opening of the plumular hook of the bean seedling. (After Withrow *et al.*, 1957.)

2. Check for responsivity of the system by either using a null balance method or by measuring dark controls and the responsiveness to a test wavelength.

3. Test for reciprocity. Only if reciprocity holds may fluence be changed by changing the irradiation time. If reciprocity does not hold, fluence should be changed by altering the fluence rate. In this case, either the distance between lamp and sample should be varied or neutral-density glasses can be inserted. Measurements of the transmission of the glasses is absolutely necessary because the spectral quality may be altered.

4. Normally the fluence has to be varied over several orders of magnitude and the response should be plotted as a function of the logarithm of the photon fluence. If a thermopile is used for light measurements instead of a quantum meter, it is most convenient to convert the measurements directly to mol m^{-2} to be able to plot photon-fluence instead of energy-fluence response curves. The energy content of the photons must be taken into account for the construction of the action spectrum:

$$\frac{N_{\lambda_{max}}(J\ m^{-2})}{N_\lambda(J\ m^{-2})} \cdot \frac{\lambda}{\lambda_{max}}$$

must be plotted as a function of λ for a constant response level.

5. To obtain an action spectrum, $N_{\lambda_{max}}/N_\lambda$ is plotted as a function of wavelength for a constant response level, where $N_{\lambda_{max}}$ is the most effective wavelength. If the logarithmic fluence response curves are not parallel, the action spectrum should be given for several response levels.

6. The interpretation of an action spectrum, especially if based on non-parallel logarithmic fluence response curves, must always take into account possible optical artefacts before new photoreceptors are postulated. Furthermore, kinetic properties of the photoreceptors have to be taken into account, and therefore the interpretation will be often model-related (analytical) action spectroscopy.

IV. ACTION SPECTRA UNDER CONTINUOUS IRRADIATION CONDITIONS

A. General principles

Action spectra under continuous irradiation conditions are normally more complex than under induction conditions because not only the light and dark reactions but also the dynamics of the transducing system have to be considered.

Figure 6 shows an action spectrum for the high-energy photoinhibition of

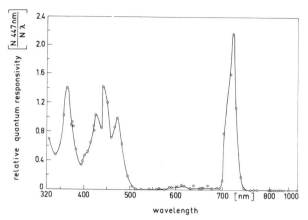

Fig. 6. Action spectrum for photoinhibition of hypocotyl lengthening in lettuce seedlings (*Lactuca sativa* L., cv. Grand Rapids) by continuous light. The action spectrum was measured for the period 54–72 h after sowing. (After Hartmann, 1967.)

lettuce hypocotyl growth.   The seedlings were continuously irradiated for 18 h.   This action spectrum shows no similarity to any known pigment absorption spectrum in the far-red waveband.   Nevertheless, there is strong evidence that phytochrome is the functional pigment in the far-red–red region of the spectrum.   This has been shown by using two different techniques.   In the first, Hartmann used dichromatic irradiation to show that the far-red peak of action for controlling the inhibition of the growth of the previously dark-grown *Lactuca sativa* hypocotyl may be due to an *optimal* effectiveness of about 0.03 = [Pfr]/[Ptot] in this system (Hartmann, 1966). A brief theoretical background to this approach is given in Section IV. B. The same technique has been used since this pioneer work to show the involvement of phytochrome in several continuous light responses (cf. Chapter 2).   In the second technique, Mancinelli and Rabino (1978) demonstrated that, for the control of anthocyanin accumulation in mustard and red cabbage seedlings, the effect of continuous far-red light could be substituted for by pulsed far-red light if the total number of photons was the same and the intervening dark periods were shorter than 1 min.   Reversion experiments showed clearly that this pulsed-light response is under the control of phytochrome.

B. Problems involved in action spectroscopy under prolonged irradiation

Because knowledge of the dynamics of cryptochrome is very poor, discussion will again be limited to phytochrome.   Three obvious main problems are involved.   First, the total amount of phytochrome (Ptot) will not be constant under prolonged irradiation. Second, the optical properties of the plant material will usually not be constant during the experiment. Third, the responsiveness of the system may change during the experiment.

Under prolonged irradiation, normally 10–24 h, the dark reactions of phytochrome cannot be ignored (cf. Frankland, 1972). Because it has been shown that the rate of synthesis of Pr is not affected by light (Schäfer *et al.,* 1972; Schäfer, 1978; Jabben *et al.,* 1980) and the rate of disappearance of phytochrome is dependent on both wavelength and photon-fluence rate (Schäfer *et al.,* 1975, 1976), the amount of Ptot should depend on wavelength and photon-fluence rate. This has been shown to be true for continuous far-red light in cotyledons of *Sinapis alba,* even under steady-state conditions (Fig. 7). On the other hand, it has been shown that in monocotyledonous seedlings the rate of phytochrome synthesis in the dark depends on the light pretreatment and is probably under phytochrome control (Duke *et al.,* 1977; Gorton and Briggs, 1981; Gottmann and Schäfer, 1982, 1983). Whether the rate of synthesis under continuous irradiation (Smith, 1981) in dicotyledonous seedlings is also under phytochrome control has yet to be tested. Furthermore, these parameters and the photoconversion rates, and therefore Ptot, depend on the age of the seedling (Schäfer, 1978; Schäfer and Mohr, 1980). The expected strong effects of dynamic screening under prolonged irradiation have been described in Section II. B.

Changes in responsiveness are usually observed if the phytochrome sys-

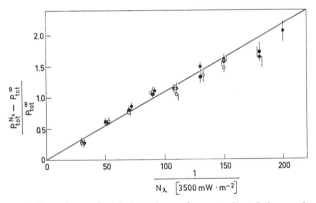

Fig. 7. Fluence rate dependence of total phytochrome in mustard cotyledons under continuous far-red light. To test whether steady-state levels have been reached, i.e., a balance between *de novo* synthesis of Pr and Pfr destruction, three dark–light treatments were chosen starting from almost zero or far below or above the Ptot steady-state level. Ptot$^{N_\lambda}$ = total amount of spectrophotometrically detectable phytochrome under steady-state conditions at the fluence rate N_λ; Ptot$^\infty$ = total amount of spectrophotometrically detectable phytochrome under steady-state conditions at infinitely high fluence rates. (After Schäfer and Mohr, 1974.)

tem is coupled to an endogenous rhythm. Mohr *et al.* (1979) showed, in the case of light-controlled anthocyanin accumulation in *Sinapis alba* seedlings, that even the effect of an inductive light pulse is affected by a light-pulse pretreatment or a continuous light pretreatment. Dark-grown *Sinapis alba* hypocotyls exhibit almost no sensitivity to inductive light pulses; this contrasts with the high sensitivity of light-grown seedlings to inductive light pulses (Beggs *et al.,* 1981). It was shown that this sensitivity change is induced by a classical HIR (Beggs *et al.,* 1981). These observations indicate a complex dependency of the responsiveness of the system on the age of the seedlings and on the preirradiations. Changes in responsivity are also to be expected during prolonged irradiation treatments.

Whether photomorphogenetic responses to blue light can be understood on the basis of phytochrome alone, or whether the action of a separate blue light receptor has to be postulated, can be tested using a method based on action spectroscopy. The interaction between a pigment and light can be described for a simple pigment ($P \rightarrow P^*$) by one rate constant and for a photochromic pigment ($A \leftrightharpoons B$) by two rate constants. Therefore, the photochromic pigment, phytochrome, cannot distinguish between light of different spectral distributions if the number of photoconversions per unit time $Pr \rightarrow Pfr$ and $Pfr \rightarrow Pr$ is the same (Gammermann and Fukshansky, 1971; Schäfer, 1981; Schäfer *et al.,* 1983a). This is equivalent to the statement that the photomorphogenetic output of the phytochrome system can only depend on k_1/k_2 and $k_1 + k_2$, where $k_1 = E(\lambda)\epsilon_{r\lambda}\phi_{r\lambda}$, $k_2 = E(\lambda)\epsilon_{fr\lambda}\phi_{fr\lambda}$, $E(\lambda)$ = spectral photon fluence rate of the light between λ_1 and λ_2, $\epsilon_{r,fr}$ = extinction coefficients, and $\phi_{r,fr}$ = quantum yields of Pr,Pfr at wavelength λ.

To test for the involvement of a blue-light absorbing photoreceptor it is necessary to estimate k_1 and k_2 by measuring photoconversion kinetics of phytochrome *in vivo* for the total spectrum. Using photon-fluence rate response curves, responsiveness can then be plotted as a function of k_1/k_2 at a constant value of $k_1 + k_2$. If all points of the spectrum are on one curve, it can be assumed that this response is only under phytochrome control. A significant deviation of the blue part from the red–far-red curve indicates the involvement of another photoreceptor (Fig. 8) (see also Chapter 2).

These considerations are the basis for the use of dichromatic irradiation to test for the involvement of phytochrome in photomorphogenesis (Hartmann, 1966, 1977) as described in section IV. A. Keeping the photon-fluence rate constant at 760 nm and varying that at 660 nm will result in a change in k_1/k_2 and $k_1 + k_2$ for the phytochrome system. In this case, the dichromatic curve is almost equivalent to a series of monochromatic irradiations between 760 and 660 nm at a photon fluence rate used for the 760-nm light.

The main implicit assumption of this method should not be ignored; this is

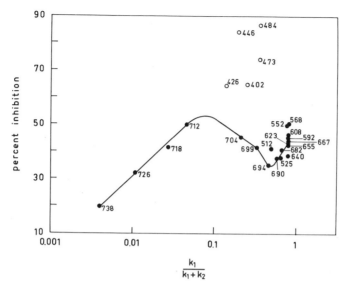

Fig. 8. Plot of percent inhibition of hypocotyl growth against $k_1/(k_1 + k_2)$ as measured by *in vivo* spectroscopy. 54-h dark grown *Sinapis alba* seedlings were irradiated for 24 h with monochromatic light. (After Beggs, 1980, and including data of Holmes and Schäfer, 1981.)

that the measured k_1/k_2 and $k_1 + k_2$ should be the same (or very close) to the physiologically active values of k_1/k_2 and $k_1 + k_2$. Using photoconversion kinetics, total phytochrome is measured and, therefore, average values of k_1/k_2 and $k_1 + k_2$ are obtained. If only the phytochrome of special cells (or cell compartments) controls the measured response, then it is necessary to measure the photoconversion kinetics of this phytochrome. This is almost impossible, at least *in vivo*.

One further difficulty with respect to action spectroscopy is the fact that for green plants the continuous-light response action spectra become simpler again if the interference of photosynthesis with the photomorphogenetic response is small (see hypocotyl growth in young green *Sinapis* seedlings, Beggs *et al.,* 1980). In principle, a spectrum similar to Pr absorption is obtained even for continuous irradiation (Fig. 9). A significant shift of the action maxima compared to the Pr absorption spectrum is obtained for green plants. This is probably due to chlorophyll screening as shown by the parallel shift of the photon-fluence rate response curves if green plants and herbicide-treated plants (Norflurazon, or SAN 9789) are compared (Fig. 10). Why an induction-type action spectrum is obtained even under continuous irradiation is not clear. It indicates that the Pr → Pfr photoconversion is rate limiting under these conditions, although a fast competing dark reaction is not known in light-grown plants.

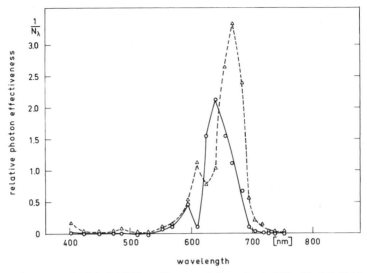

Fig. 9. Action spectra for light inhibition of hypocotyl elongation in 54-h-old white-light-grown *Sinapis alba* L. seedlings. The curves are plotted for 50% inhibition and are taken from photon-fluence rate response curves similar to those shown in Fig. 10. O—O, Hoaglands solution; △---△, 5 × 10⁻⁶ M SAN 9789 (Norflurazon) solution. (After Beggs *et al.*, 1980.)

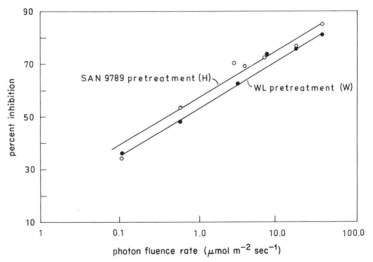

Fig. 10. Fluence-rate response curves for inhibition of hypocotyl elongation in 54-h-old white-light-grown *Sinapis alba* L. seedlings. The curves are for screened = white-light-grown material with high chlorophyll levels (●) and unscreened = SAN 9789 (Norflurazon)-treated material with almost no detectble amounts of chlorophyll (O). Monochromatic irradiation at 655 nm was used in this experiment. (After Beggs *et al.*, 1980.)

It should be mentioned that action spectra in green plants are normally more complex because of the involvement of photosynthesis. An action spectrum in light-grown seedlings without the contribution of photosynthesis, i.e., at constant photosynthetically active irradiation, has not been described.

V. FUTURE AIMS

As has already been pointed out, analytical action spectroscopy should be based on models, and the outcome of this approach depends primarily on the validity of the model used. The modeling is based on *in vivo* spectroscopy and, therefore, introduces the kinetic behaviour of the "bulk phytochrome." Until it has been proved that the parameters of the "bulk" phytochrome are different from the physiological action of phytochrome, only the measured values may be used.

The goals should be to improve knowledge and understanding of phytochrome and other photoreceptors not only in etiolated but also in light-grown seedlings, including a characterisation of the dynamics of the system under prolonged continuous irradiation. The use of the herbicide Norflurazon might be very helpful for this analysis (Jabben *et al.,* 1980), because with the absence of chlorophyll it is possible to analyse the dynamics of the "bulk" phytochrome system under prolonged irradiation.

Further detailed analysis of dichromatic fluence-rate response curves and of cyclic light – dark treatments enables the comparison of the kinetic parameters measured by *in vivo* spectroscopy and those deduced from action spectroscopy.

To analyse the transduction chain of phytochrome, the first step is to measure the loss of reversibility under cyclic irradiation conditions. A major problem appears to be the difficulty of analysing the phytochrome-induced changes of the responsiveness of the system toward phytochrome. Only if these interactions are not too strong, and if the optical behaviour of the system is known, can analytical action spectroscopy provide an understanding of the kinetics of the functional phytochrome and its transduction chain(s).

REFERENCES

Beggs, C. J. (1980). Dissertation, University of Freiburg, Freiburg, West Germany.
Beggs, C. J., Holmes, M. G., Jabben, M. and Schäfer, E. (1980). *Plant Physiol.* **66,** 615–618.
Beggs, C. J., Geile, W., Holmes, M. G., Jabben, M., Jose, A. M. and Schäfer, E. (1981). *Planta* **151,** 135–140.

128   EBERHARD SCHÄFER AND LEONID FUKSHANSKY

Butler, W. L. (1962). *J. Opt. Soc. Am.* **52**, 292–299.
Butler, W. L., Hendricks, S. B. and Siegelman, H. W. (1964). *Photochem. Photobiol.* **3**, 521–528.
de Fabo, E. (1980). *In* "The Blue Light Syndrome" (H. Senger, ed.), pp. 187–197. Springer-Verlag, Berlin and New York.
Duke, S. O., Naylor, A. W. and Wickliff, J. L. (1977). *Physiol. Plant.* **40**, 59–68.
Engelmann, T. W. (1882). *Bot. Ztg.* **40**, 419–426.
Frankland, B. (1972). *In* "Phytochrome" (K. Mitrakos and W. Shropshire, Jr., eds.), pp. 195–225. Academic Press, New York.
Fukshansky, L. and Mohr, H. (1980). *In* "Photoreceptors and Plant Development" (J. de Greef, ed.), pp. 135–144. Antwerpen Univ. Press, Antwerp.
Fukshansky, L. and Schäfer, E. (1983). *Encycl. Plant Physiol., New Ser.* **16** (W. Shropshire Jr., and H. Mohr, eds.) pp. 69–95. Springer-Verlag, Berlin and New York.
Galland, P. (1983). *Photochem. Photobiol.* **37**, 221–228.
Gammermann, A. Ya. and Fukshansky, L. (1971). *Fiziol. Rast. (Moscow)* **18**, 114–118.
Gorton, H. L. and Briggs, W. R. (1981). *Plant, Cell Environ.* **4**, 449–454.
Gottmann, K. and Schäfer, E. (1982). *Photochem. Photobiol.* **35**, 521–525.
Gottmann, K. and Schäfer, E. (1983). *Planta* **157**, 392–400.
Hartmann, K. M. (1966). *Photochem. Photobiol.* **5**, 349–366.
Hartmann, K. M. (1967). *Z. Naturforsch., B: Anorg. Chem., Org. Chem., Biochem., Biophys., Biol.* **22B**, 1172–1175.
Hartmann, K. M. (1977). *In* "Biophysik" (W. Hoppe, W. Lohmann, H. Mark and H. Ziegler, eds.), pp. 197–222. Springer-Verlag, Berlin and New York.
Hartmann, K. M. and Cohnen-Unser, I. (1972). *Ber. Dtsch. Bot. Ges.* **85**, 481–551.
Holmes, M. G. and Schäfer, E. (1981). *Planta* **153**, 267–272.
Hunt, R. and Pratt, L. H. (1980). *Plant, Cell Environ.* **3**, 91–95.
Jabben, M., Heim, B. and Schäfer, E. (1980). *In* "Photoreceptors and Plant Development" (J. de Greef, ed.), pp. 145–158. Antwerpen Univ. Press, Antwerp.
Johnson, C. B. (1980). *Plant, Cell Environ.* **3**, 45–51.
Lipson, E. D. and Presti, D. (1980). *Photochem. Photobiol.* **32**, 383–392.
McCree, K. J. (1976). *In* "Light and Plant Development" (H. Smith, ed.), pp. 461–465. Butterworth, London.
Mackenzie, J. M., Jr., Coleman, R. A., Briggs, W. R. and Pratt, L. H. (1975). *Proc. Natl. Acad. Sci. U.S.A.* **72**, 799–803.
Mancinelli, A. L. and Rabino, I. (1978). *Bot. Rev.* **44**, 129–180.
Mohr, H., Drumm, H., Schmidt, R. and Steinitz, B. (1979). *Planta* **146**, 369–376.
Ng, Y. L., Thiemann, K. V. and Gordon, S. A. (1964). *Arch. Biochem. Biophys.* **107**, 550–558.
Schäfer, E. (1978). *Photochem. Photobiol.* **27**, 775–580.
Schäfer, E. (1981). *In* "PLants and Daylight Spectrum" (H. Smith, ed.) pp. 461–480. Academic Press, London.
Schäfer, E. and Mohr, H. (1974). *J. Math. Biol.* **1**, 9–15.
Schäfer, E. and Mohr, H. (1980). *Photochem. Photobiol.* **31**, 495–500.
Schäfer, E. and Schmidt, W. (1974). *Planta* **116**, 257–266.
Schäfer, E., Marchal, B. and Marmé, D. (1972). *Photochem. Photobiol.* **15**, 457–464.
Schäfer, E., Lassig, T.-U. and Schopfer, P. (1975). *Photochem. Photobiol.* **22**, 193–202.
Schäfer, E., Lassig, T.-U. and Schopfer, P. (1976). *Photochem. Photobiol.* **24**, 567–572.
Schäfer, E., Lassig, T.-U. and Schopfer, P. (1982). *Planta* **154**, 231–240.
Schäfer, E., Fukshansky, L. and Shropshire, W., Jr. (1983a). *Encycl. Plant Physiol., New Ser.* **16**, (W. Shropshire Jr., and H. Mohr, eds.) pp. 39–68. Springer-Verlag, Berlin and New York.

Schäfer, E., Heim, G. and Löser, G. (1983b). *Ber. Dtsch. Bot. Ges.* **96**, 497–509.
Schmidt, W., Marmé, D., Quail, P. and Schäfer, E. (1973). *Planta* **111**, 329–336.
Senger, H., ed. (1980). "The Blue Light Syndrome." Springer-Verlag, Berlin and New York.
Shropshire, W., Jr. (1972). *In* "Phytochrome" (K. Mitrakos and W. Shropshire, Jr., eds.), pp. 161–182. Academic Press, New York.
Shropshire, W., Jr. (1980). *In* "The Blue Light Syndrome" (H. Senger, ed.), pp. 172–186. Springer-Verlag, Berlin and New York.
Smith, H. (1981). *Nature (London)* **293**, 163–165.
Song, P. S. (1980). *In* "The Blue Light Syndrome" (H. Senger, ed.), pp. 157–171. Springer-Verlag, Berlin and New York.
Steinitz, B. and Bergfeld, R. (1977). *Planta* **133**, 229–235.
Vanderhoef, L. N., Quail, P. H. and Briggs, W. R. (1979). *Plant Physiol.* **63**, 1062–1967.
Withrow, R. B., Klein, W. H. and Elstad, V. (1957). *Plant Physiol.* **32**, 453–462.

6

In Vivo *Spectrophotometry*

J. Gross, M. Seyfried, L. Fukshansky, and E. Schäfer

I. INTRODUCTION

The main difficulties which are encountered when measuring light absorption in living plant material arise from the inhomogeneities of the samples. Very often plant tissues are highly scattering, in many cases they are highly anisotropic (e.g., light pipe effects of some stem tissues, see Mandoli and Briggs, 1982a,b) and the pigments are distributed unevenly within the tissue as well as in the cells constituting the tissue (e.g., chloroplast pigments). Further complications arise from high contents of "shielding pigments" with respect to the content of the pigment to be determined. Also, changes in light scattering or geometrical properties with time (e.g., because of swelling or sedimentation of algal cells), fluorescence of some of the substances in the tissue (e.g., chlorophyll), and light-dependent changes in the absorption characteristics of some of the pigments (e.g., xanthophylls and phytochrome) cause serious problems for quantitative determinations. Therefore, certain considerations must be taken into account in designing spectrophotometers for *in vivo* spectrophotometry. In general, the instruments should be capable of measuring small differences in light absorbance in inhomogeneous, highly scattering samples with strong wavelength-dependent background absorption. They should provide a facility for actinic-light irradiation of the sample, a shield for the light detector against fluorescence from the sample, and a controlled-temperature environment. Normally, only measurements of relative absorbance changes are of interest, and direct measurements of absorbance by the sample are not possible. In the first part of this chapter we shall concentrate on instrumentation for measurements of absorbance. In the second part, instrumentation and theoretical approaches for measuring absorbance, reflectance, and transmittance of biological material will be described, together with interpretation of the data.

TECHNIQUES IN PHOTOMORPHOGENESIS
0-12-652990-6

A. Measurement of absorbance in scattering material

The detector should collect most, or at least a representative fraction, of the light scattered by the sample. This can be achieved by placing the sample as close as possible to the light-sensitive surface of the detector. The greater the sensitive area and the closer the sample is placed to this surface, the closer is the approach to a collection of radiation from an angle of 2πsr from the sample. It should be noted that there are also negative aspects to using large detector areas. For example, with photomultipliers the dark current increases with increasing sensitive area, and photomultipliers are temperature dependent; the dark current is exponentially dependent on the cathode temperature. It is important that the sizes and positioning of the sample and detector be chosen to allow that part of the transmitted or scattered light from the sample that contains the wanted information to reach the detector.

The optical geometry can be improved by placing an opal glass between the sample and the detector. The opal glass scatters all impinging light isotropically, thereby causing a representative fraction of it to be collected even on a relatively small-faced photomultiplier. The main disadvantage of this technique is that most of the measurable signal is lost. This, in turn, leads to a lower signal-to-noise ratio. If this is compensated for by increasing the intensity of the measuring beam, care must be taken to avoid unwanted light effects on the sample, such as phototransformation or bleaching of pigments.

The scattered light from the sample can also be collected with a light pipe or hollow conus with a silvered inner surface. One can thereby collect or even concentrate the radiation on a relatively small detector area. The use of an Ulbricht sphere to collect all transmitted or reflected light will be discussed in Section III.

B. Sample geometry and actinic irradiation

Although the fluence rate of the measuring beam must be low enough to avoid phototransformation within the sample, the signal-to-noise ratio of the photodetector decreases with decreasing fluence rate. Therefore, large sample areas must be used to ensure sufficient light flux at low density. A compromise must be found for the sample thickness. With increasing sample thickness the absorption signal increases while the available light flux decreases.

The geometry of the sample is of special importance for the determination of photochromic pigments. The actinic radiation used to transform the pigment must be sufficient to produce saturating transformation throughout the sample volume. This makes it necessary to keep the path length of the actinic light in the sample short.   Spectrophotometers for measurements of photochromic pigments must allow for actinic irradiation either parallel or perpendicular to the measuring beam direction.   The sample compartment must therefore be large enough for the installation of mirrors and/or have side windows.

II. GENERAL PURPOSE SPECTROPHOTOMETERS

A. The "classical" design: monochromatic light on the sample

(1) Single beam, single wavelength spectrophotometers
This type of spectrometer is the simplest and cheapest.   Its use in *in vivo* measurements is very limited because of the lack of an optical balance or reference cell.   Thus it should be used for fixed-wavelength measurements only.   The samples may not change their geometrical and scattering properties in the course of time.   (For determination of small absorbance changes it is desirable that one can "prebalance" the instrument changing the monochromator slit width or by using an optical attenuator.)   In addition, the stability of the lamp and drift compensation of the amplifier section of the instrument must be extremely good.   These shortcomings are avoided in most of the other types of spectrophotometers because there is a direct compensation of these parameters between the reference and measuring signals.

(2) Dual beam, single wavelength spectrophotometers
In dual beam, single wavelength spectrophotometers, two separate cells are used for the measurement.   They allow for the direct comparison of two samples (e.g., light or chemically treated versus untreated) and can be used to scan difference spectra of two samples.   The quality of the spectra depends on the geometrical similarity of both light paths and samples.   If two separate light detectors are used for the measuring and reference beams, the similarity of their spectral response is important.   For difference spectra, instruments using the same detector for both beams are preferable.   The limitations for the determination of absorbance spectra *in vivo* arise from the difficulties in preparing geometrically equal samples, so this type of instrument is normally used for time- or treatment-dependent measurements at a fixed wavelength.   All of the considerations made about scattering samples apply to these instruments with the additional complication of two separate measuring beams.

Dual beam, single wavelength (split beam) spectrophotometers have the advantage of internal compensation for lamp fluctuations but are very sensitive to changes of the geometrical properties of the samples with respect to each other. It should be pointed out that the measurement of small absorption changes with these instruments should be made with equal light fluxes on both detectors. This reduces the problem of measuring small changes on a high background to measuring them on a zero background and, therefore, minimizing the influence of electrical instabilities of the amplifer section on the measured signal. Therefore, it should be possible to balance the measuring beam intensities by means of continuously variable optical attenuators.

(3) Dual wavelength spectrophotometers

Dual wavelength spectrophotometers are the most versatile and are adaptable to different types of measurements. The four possible arrangements of sample(s) and detector(s) are shown in Fig. 1. In most of the commercially available instruments, measurements can be made with either one or two cuvettes. Because of the equal spectral response on both measuring beams, arrangement (b) would appear to be preferable to (a), but with careful selection of the detectors, (a) can also provide satisfactory measurements. Of the one-cuvette arrangements, (d) is superior to (c) if highly inhomogeneous samples are to be measured. This is because in (c) (which has two noncoincident measuring beams) both optical paths may differ in their geometry, whereas in (d) both share the same and, thereby, identical path.

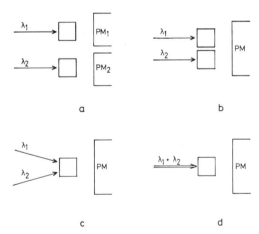

Fig. 1. Four possible arrangements of sample(s) and photomultiplier(s) in dual wavelength photometers. (a) Two samples and two photomultipliers, (b) two samples and one photomultiplier, (c) one sample and two noncoincident measuring beams, and (d) one sample and coincident (alternatingly chopped) measuring beams.

Not all of the potential uses of dual wavelength spectrophotometers can be discussed here, but special emphasis is placed on the detection of small absorbance changes and the measurement of photochromic pigments.

In the two-sample arrangements, one of the samples is treated (e.g., by an actinic-light irradiation) and the other stays untreated. Geometrical changes of the samples with respect to each other are not compensated. In the one-sample arrangements, the change of the absorbance due to the treatment is measured. The difficulties discussed with the dual beam, single wavelength spectrophotometers also apply to the two-sample measurements here. The most important advantages of the dual wavelength instruments for *in vivo* measurements appear if they are used with both beams coinciding. If the two wavelengths are close together, the geometrical fluctuations of the sample are compensated. Therefore, very stable measurements can be made even with unstable samples such as suspensions, in which the particles tend to sediment. The choice of the two wavelengths used is a compromise between small wavelength differences for optimal suppression of geometry changes and a large wavelength difference to provide the highest sensitivity. For photochromic pigments, typical pairs may consist of one absorption maximum versus an isobestic point or the absorption maxima of the two forms of the pigment.

B. The reverse optics design: polychromatic light on the sample

Two modern spectrophotometer designs are shown in Fig. 2. Whereas in "classical" instruments the sample is illuminated with monochromatic light and all light transmitted by the sample is detected (Fig. 2a), the reverse optics is based on polychromatic illumination of the sample and spectral analysis of the transmitted light. In this example (Fig. 2b), the detector consists of a linear array of detectors which measure the light at several hundred wavelengths simultaneously, thereby giving immediate information about the whole absorption spectrum of the sample. With highly scattering samples, the most important difference between the two arrangements is that there is no imaging of the light source through the sample with the classical spectrophotometer, whereas the reverse-optics system images the light source through the sample onto the monochromator entrance slit. This results in the loss of all scattering light since only the directly transmitted light can enter the monochromator and be measured. With clear or moderately scattering samples having high quality plane parallel surfaces, the advantages of the reverse-optics design are apparent. But even *in vitro* measurements of samples such as membrane vesicle suspensions can be made with great sensitivity and accuracy. The radiation load for the sample is low because the measurement of the whole spectrum takes only a few hundred millisec-

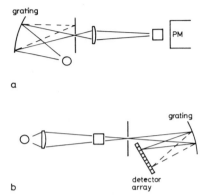

Fig. 2. The "classical" and reverse optics for spectrophotometers. (a) In the classical spectrophotometer the sample is irradiated with monochromatic light. (b) In the reverse-optics spectrophotometer the sample is irradiated with polychromatic light. The light transmitted by the sample is spectrally analysed by a multichannel detector.

onds. This type of instrument cannot be used with samples which have uneven surfaces (e.g., pieces of tissue) but the speed with which it provides spectral information can be very useful for following relatively fast spectral changes in the majority of biological samples.

C. Comparison of commercial spectrophotometers

A comparative measurement of an algal suspension (*Chlorella,* about 10^5 cells/ml) absorption spectrum with different photometers is shown in Fig. 3. The most striking difference between the spectra obtained by using instruments with a short distance (Fig. 3a) or a light pipe between sample and detector and those with a relatively long distance (3b) is that a high, nearly wavelength-independent apparent "background absorbance" is observed in the latter.

These data show that the directly transmitted light contains most of the information. If the fluence rate of this light is not sufficient to obtain an acceptably high signal-to-noise ratio, also scattered light from the sample should be collected on the detector. This has the disadvantage that in most cases the effective optical path length of the light is unknown and that the data become strongly dependent on the detector geometry. Whenever possible, measurements should be made with directly transmitted light only in order to get device-independent data.

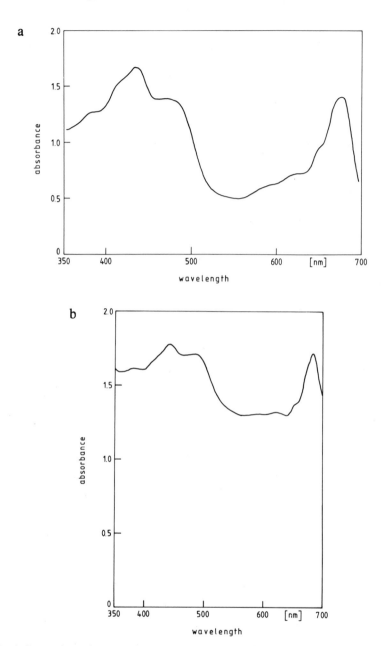

Fig. 3. Comparison of absorption spectra obtained with different commercially available spectrophotometers. In all cases the same sample, a cell suspension of *Chlorella* (about 10⁵ cells/ml) has been used. (a) Shimadzu: short distance between sample and photomultiplier, (b) Kontron Uvicon 820: large distance between sample and photomultiplier, (c) Beckmann DU-8: medium distance between sample and photomultiplier, (d) Hewlett – Packard HP8450: spectrophotometer in reverse-optics design. (Fig. continued on next page.)

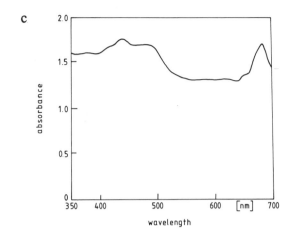

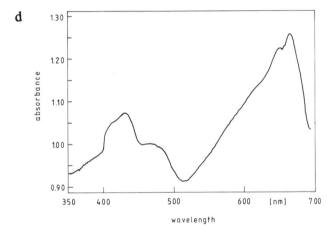

Fig. 3. (Continued)

D. A specialized spectrophotometer for photochromic pigment determinations

The dual wavelength, differential spectrophotometer ("ratiospect") used for the determination of phytochrome or other photochromic pigments is based on the original designs of G. S. Birth, W. L. Butler, B. Chance, and K. H. Norris. The most extensive modifications to the original design were made by L. H. Pratt; a simplified scheme of the optical system is shown in Fig. 4. Two measuring beams of nearly parallel light are derived from a tungsten – iodine incandescent lamp (12 V, 25 W). Both beams pass optical attenuators, one with fixed, the other with continuously variable attenuation. The

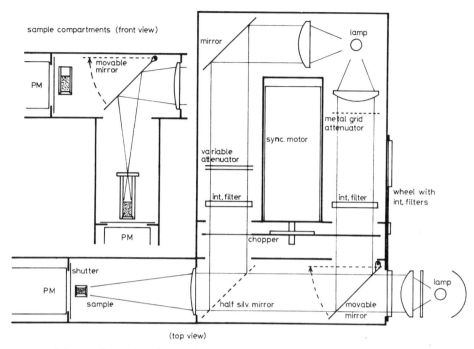

Fig. 4. Schematic drawing of the optical system of the dual-wavelength spectrophotometer for photochromic pigment determinations. The measuring light is derived from the lamp in the top right corner, the actinic light from the lamp in the lower right.

variable attenuator consists of two polarization filters which can be rotated in opposite directions; this provides homogeneous attenuation over the whole area of the beam.   The fixed attenuator is a metal grid and is placed as close as possible to the condenser lens.   The attenuated beams then pass interference filter monochromators.   After this, both beams are blocked alternately by a rotating chopper blade with three 60°-sector holes.   The chopper rotates at 1500 rpm, giving a frequency of 75 Hz for the chopped light.   This frequency has been chosen for minimum electrical interference with noise from 50-Hz line frequency.   The chopper is driven by a synchronous motor operated directly from the mains.   Both beams are then made coincident by a system of one fully reflecting and one half-silvered mirror. The joined beams are then focused on the sample.   Two sample holders are provided, one for a horizontal, the other for a vertical light path.   The light can be directed to either of the two samples by means of a movable mirror. The horizontal light path is made for standard 1-cm cuvettes.   The vertical light path allows the use of path lengths of up to 4.5 cm.   The sample is followed immediately by a shutter which protects the photomultiplier from high fluence rates. The cathodes of the photomultiplier are placed as close as possible to the shutter.

The path of the actinic light coincides with one of the measuring light beams behind the chopper. The light source is a high-output tungsten–iodine incandescent lamp (24 V, 150 W). The light passes a heat-absorbing filter, a condenser lens, and an interference filter. Four interference filters are mounted on a wheel. The appropriate filter is selected by turning the wheel with an electric motor. For actinic irradiation, the mirror, which normally deflects the measuring beam, is turned by 45°, using an electro-magnet. The mirror blocks the measuring beam and allows the actinic light to pass to the half-silvered mirror, where about one-half of its energy is lost by reflection to the spectrophotometer casing, and the other half of the actinic light passes to the sample.

The major advantage of this design is that the light paths for both measuring wavelengths are nearly identical because both beams are derived from the same light source, pass the same chopper with a 180° phase difference, and enter from the same direction and with the same divergence. A nearly homogeneous distribution of the light can be reached on the surface of the sample for the measuring as well as for the actinic light. Most of the light transmitted and scattered by the sample can be received by the photomulti-plier cathode because it is placed close behind the sample. The optical balance by the variable light attenuator gives high sensitivity for small differential absorption changes, even with poor electrical stability of the signal amplifiers.

The design has two shortcomings, both of which are related to the light path for the actinic radiation. Because the actinic light has to pass the half-silvered mirror, about one-half of its intensity is lost. This can be overcome by using high-power lamps and normally will not cause difficulties. A more critical point is the high precision needed for positioning the movable mirror when changing from measuring to actinic light. The mechanical precision at this point influences the accuracy and repeatability of the measurement. The 45° stop for the mirror has to be made of hardened materials and with large surfaces getting into contact in order to reduce deformation by the repeated impact by the carrier of the mirror. A better method is to use glass plates instead of half-transmitting mirrors. This has the advantage of increasing the flux of the actinic light and the mirror does not have to be moved, but the disadvantage is that the light flux of the measuring beams is decreased by a factor of about 5. This problem can be overcome by running the measuring lamp at a higher output.

The electronic circuitry for amplification and processing of the signal consists mostly of commercially available units (Fig. 5). The primary signal

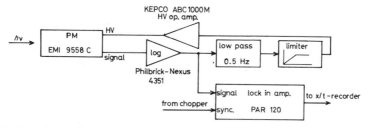

Fig. 5. Schematic drawing of the signal processing and the high-voltage generation electronic circuitry for the dual-wavelength spectrophotometer.

is the anode current of the photomultiplier which is translated into voltage by a logarithmic amplifier. The very low frequencies of this voltage are filtered out and are used to regulate the photomultiplier cathode voltage in a feedback loop. This results in a constant mean current from the photomultiplier and thus eliminates the effect of changing light attenuation through the sample occurring at both measuring wavelengths and variation in lamp output. The absorbance-difference signal is extracted from the logarithmic amplifier output by a lock-in amplifier which is synchronized by a 75-Hz reference signal from the chopper. The resulting signal is displayed on a x/t recorder.

The spectrophotometer is controlled by a timer unit which enables automated measurements with different measuring and actinic light periods, and different wavelengths. To do this, all mechanical elements, such as the shutters for the photomultipliers, the movable mirrors, and the filter wheel, must be equipped with sensor switches which signal the position of the elements to the control circuits.

III. ANALYSIS OF TRANSMITTANCE, REFLECTANCE, ABSORBANCE, AND LIGHT FLUX GRADIENTS

A. Solutions

Since the problems of scattering and heterogeneities do not exist in clear solutions, all of the instruments described in Section II can be used for the measurement of transmittance. Only surface reflection must be accounted for when determining absorbance from the transmittance value. This can be calculated directly if the refractive index of the solution is known, or derived from an additional measurement of the transmittance of a reference cuvette. Errors which may occur are due only to instrument artefacts (e.g., stray-light problems, too large bandwidth, too high scanning speed, etc.). The experimentally verified Beer–Lambert–Bouguer absorption law states that:

$$T = e^{-\alpha\rho L}; \qquad A = -\ln T = \alpha\rho L \qquad\qquad \text{Eq. (1)}$$

where T is transmittance, A is absorbance, α is molar absorptivity, ρ is concentration of the pigment, and L is the thickness of the object, and applies to homogeneous objects and only takes absorption into account.

B. Suspensions

In diluted suspensions, the scattering may be low enough to use the instruments described in Section II. However, sample heterogeneities cannot be ignored because absorption by suspended particles (sieve effect) will disturb transmission measurements.

Sieve effect

The sieve effect is caused by heterogeneous pigment distribution in the plane perpendicular to the direction of light propagation; this results in different parts of the beam cross section intersecting different amounts of the pigment. From the simplified scheme in Fig. 6, it can be seen that this leads to changes in measured absorbance which are not connected with any real changes in the optical properties of the pigment. These changes depend on the molar absorptivity and, therefore, on the wavelength λ. It can be seen in Fig. 6 that Beer's law [Eq. (1)] cannot be applied to a whole beam; it is applicable only within a subarea where the homogeneity of pigment distribution can be assumed (in Fig. 6, this is the column with fractional area γ). In biological tissues, the pigment is randomly distributed with respect to different parts of the incident light. If the light beam is considered to consist of thin elementary rays, the heterogeneous spatial distribution of the pigment will cause a distribution of elementary rays differently attenuated depending on the amount of pigment intersected. The transmittance (and absorbance) is obtained by summing up over all the elementary rays, whereas for a single ray, Beer's law is applicable, and the amount x of pigment crossed by a ray is a random variable with a density probability $f(L,x)$:

$$T = \frac{I_t}{I_0} = \frac{\displaystyle\int I_0 f(L,x)e^{-x\alpha}dx}{I_0} = \int f(L,x)e^{-\alpha x}dx \qquad \text{Eq. (2)}$$

Here I_0 and I_t are incident and transmitted fluence rates, respectively, and L is the thickness of the sample.

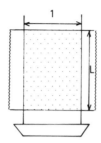 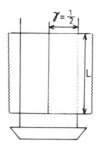

Fig. 6. Simplified presentation of sieve-effect origin. The condensation of pigment within an area having a small fractional cross section leads to changes in the measured transmittance T and absorbance A. (a) Object with pigment homogeneously distributed over the whole irradiated area has transmission $T_0 = e^{-\alpha \rho_0 L}$ (α = molar absorptivity, ρ_0 = spatial density, L = thickness, beam cross section is made to equal 1 and absorbance $A_0 = -\ln T_0 = \alpha \rho_0 L$. (b) Object with the same amount of pigment heterogeneously distributed (all pigment is gathered within a column having fractional area $\gamma = \frac{1}{2}$ of the beam cross-section) has transmittance $T = (1 - \gamma) + \gamma e^{-\rho \alpha_0 L} = (1 - \gamma) + \gamma e^{-1/\gamma \rho_0 \alpha_0 L} = \frac{1}{2}(1 + e^{-2A_0})$ and absorbance $A = -\ln T = -\ln \frac{1}{2}(1 + e^{-2A_0})$ (here $\rho = 1/\gamma \rho_0$ is the new spatial density). In the case $T_0 = 0.8$ and $A_0 = \ln 0.8 = 0.223$, one receives $T = 0.820$; $A = 0.498$; $\beta = (A_0 - A)/A_0 = 11.2\%$. In the case $T_0 = 0.5$ and $A_0 = \ln 0.5 = 0.693$, one receives $T = 0.625$; $A = 0.470$; $\beta = (A_0 - A)/A_0 = 32.2\%$. β is defined as the sieve-effect factor.

Obviously, for the same amount of pigment m with the same molar absorptivity α and fixed L, Eq. (2) can give a difference value of T, depending on the spatial distribution of the pigment [i.e., on $f(x,L)$]. This is why the universal characterization of a pigment must be related to a fixed spatial distribution and, naturally, the homogeneous distribution is generally chosen. However, in a tissue the heterogeneous spatial distribution of a pigment occurs because the pigment is part of the biological structure and reflects real conditions under which the pigment functions. Therefore, one is interested in absorption spectra which correspond to the situation described in Eq. (2). To correct absorption spectra for changes caused by spatial heterogeneities of the pigment, the sieve-effect factor β is applied. This is defined as the relative change of absorbance between homogeneous and heterogeneous spatial patterns:

$$\beta = \frac{A_0 - A}{A_0} \qquad \text{Eq. (3)}$$

Here A_0 is from Eq. (1) and A, from Eq. (2).

Analogously, the sieve effect between two different heterogeneous spatial patterns is defined as:

$$\beta_{ij} = \frac{A_j - A_i}{A_j} \qquad \text{Eq. (3a)}$$

where A_j and A_i are from Eq. (2) with functions f_i and f_j ($f_i \neq f_j$), respectively.

Properties of the sieve-effect factor β, correlations between β and α (i.e., λ), dependence of β on parameters of pigment spatial distribution, and object thickness must be known in order to choose the optimal conditions for measurements and to estimate the influence of possible artefacts. These properties, which are analysed by using the theory of sieve effect, depend predominantly on the properties of the function f. The theory of sieve effect consists of two parts (Fukshansky, 1982). The first covers general theory, i.e., statements which can be made in the absence of any information about f and are valid for any f. The second part covers the statistical theory, i.e., more detailed statements which can be made when the type of f (and, perhaps, even parameters of f) is known on the basis of probabilistic reasoning. Examples of general theory statements are that any heterogeneity in pigment distribution can only diminish the absorbance of a corresponding homogeneous object (Fukshansky, 1978) and that the sieve effect is independent of object thickness L (Fukshansky, 1982).

The simplest situation to which the statistical theory can be applied is a suspension of identical particles (Fig. 7a). Here, the function f can be evaluated on the basis of the following three simplifying postulates. First, the particles are assumed to be identical cubes (with edge d) with a parallel orientation to the direction of light propagation. Second, the object is divided along the direction of light propagation into $n = d^{-2}$ cuboids (columns) each having a length L and cross section d^2, and there are only n fixed positions available for a particle in a plane perpendicular to the direction of light propagation. Also the cross section of a particle must coincide with that of a column. Third, the object is divided by planes perpendicular to the direction of light propagation into $m = L/d$ layers having thickness d. There are only m fixed positions available for a particle along the direction of light propagation. Each particle must be placed within a layer.

If all three postulates are applied, the idealised suspension is described as a number of regularly oriented cubes of volume d^3 randomly placed in a spatial lattice having cubical compartments of the same volume (Fig. 7b). This corresponds to the idealisation made by Duysens (1956) who first introduced a probabilistic approach to the sieve effect (discrete cubic particle model). If the third postulate is removed, the more realistic semicontinuous cubic particle model (Fukshansky, 1978) occurs (Fig. 7c).

Statistical theory can be used to elucidate changes in the sieve effect promoted by various transformations of the spatial pattern, such as subdivision of particles into smaller particles with density remaining constant, subdivision of initial particles into particles containing less pigment with size d remaining constant, or increase in size d with a simultaneous decrease in density ρ and preservation of the amount of pigment in a particle.

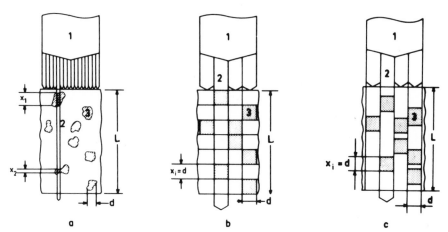

Fig. 7. The real (a) and idealized (b) and (c) suspension traversed by the beam of light. 1 = beam; 2 = elementary ray; 3 = particle; x_1, x_2, and x_i = path lengths of a ray inside a particle (in a real suspension x_i are random quantities); d = linear size of a particle; L = thickness of the vessel; (b) and (c) correspond to discrete and semicontinuous schemes, respectively.

Spatial distribution of pigments in tissues is usually more complicated than those in suspensions because of a layered structure containing different cell types in different layers, the absence of pigment in some cells, vacuolisation, and sequestering.   The theory for such objects deals predominantly with hierarchical models: absorbing particles are gathered into specific regions in clusters which are randomly distributed within the object, while particles are randomly distributed within a cluster.   The results of application of this theory are shown in Fukshansky (1982).

C. Plant tissues

Plant tissues are much more complex than suspensions, and both scattering and absorption statistics make the interpretation of spectrophotometrical measurements very complex.   The instrumentation described in Section II will give information only on a photon-flux ratio I/I_0.   The same ratio can be caused by high scattering and low absorbance or low scattering and high absorbance.   Therefore, absorption a can only be measured by measuring the reflectance R and the transmittance T: $(1 = a + R + T)$.

The interpretation of the data is normally based on the assumption of pigment distribution and a theory of the light gradient within the sample.

The simplest and most commonly used theory to describe the optical properties of intensely scattering material is the Kubelka–Munk theory

(Kubelka, 1948; see also Kortüm, 1969; Fukshansky and Kazarinova, 1980).

(1) Macrohomogeneous material

(a) Kubelka–Munk theory
In the idealised scheme underlying the Kubelka–Munk theory (Fig. 8), the radiation within the object is considered to consist of two diffuse light fluxes, I and J, which propagate in opposite directions (in Fig. 8, I is directed from above as incident flux I_0, J is directed from below, and the object is presented as an infinite strip with parallel sides). Photons may be absorbed with the unknown probability k (phenomenological absorption coefficient), which is the same for both fluxes.

Scattering manifests itself in the transfer of photons from I to J and vice versa, and occurs with the unknown probability s (phenomenological coefficient of scattering) and is the same for both fluxes. Specular reflection is not included in this approach.

This scheme leads to the following linear system of differential equations:

$$\frac{dI}{d\xi} = -(k+s)I + sJ$$

$$-\frac{dJ}{d\xi} = -(k+s)J + sI \qquad\qquad \text{Eq. (4)}$$

where ξ is the spatial coordinate coinciding to I_0; the fluence rates $I(\xi,\lambda)$ and $J(\xi,\lambda)$ are functions of ξ and the wavelength λ.

The two boundary conditions necessary to solve Eq. (4) can be provided by any two out of four values of I and J measured at the boundaries of the object (i.e., when $\xi = 0$, $\xi = L$, where L is the thickness of the object):

$$I(0,\lambda) = I_0(\lambda); \qquad J(0,\lambda) = R(\lambda)I_0(\lambda) \qquad \text{Eq. (4a)}$$
$$I(L,\lambda) = T(\lambda)I_0(\lambda); \qquad J(L,\lambda) = 0 \qquad \text{Eq. (4b)}$$

Here

$$R(\lambda) = \frac{J(0,\lambda)}{I_0(\lambda)}, \qquad T(\lambda) = \frac{I(L,\lambda)}{I_0(\lambda)}$$

are reflectance and transmittance of the object at wavelength λ, respectively. In the case of a reflecting background (backing) behind the object, part of the transmitted radiation will be returned to the object and the second expression from Eq. (4b) will be changed correspondingly.

$$I(L,\lambda) = T(\lambda)I_0(\lambda); \qquad J(L,\lambda) = R_g(\lambda)T(\lambda)I_0(\lambda) \qquad \text{Eq. (4c)}$$

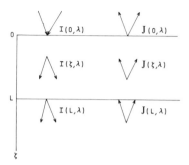

Fig. 8. Graphical presentation of the idealization underlying the Kubelka–Munk theory. ξ = spatial coordinate, L = thickness of the object, λ = wavelength, I and J = photon fluxes propagating in opposite directions.

where $R_g(\lambda)$ is the reflectance of the backing (when taken together with the object in question).

System (4) with boundary conditions in Eq. (4a) has the solution:

$$I = \left[\frac{R-a}{b} \sinh(sb\xi) + \cosh(sb\xi) \right]$$

$$J = I_0 \left[\frac{a(R-1)}{b} \sinh(bs\xi) + R \cosh(bs\xi) \right] \qquad \text{Eq. (5)}$$

where auxiliary constants a and b are:

$$a(\lambda) = \frac{k(\lambda) + s(\lambda)}{s(\lambda)}; \qquad b(\lambda) = \sqrt{a^2(\lambda) - 1} \qquad \text{Eq. (6)}$$

and sinh and cosh are hyperbolical sine and cosine.

When Eq. (4b) (object without backing or with a completely absorbing "black backing") or Eq. (4c) (reflective backing) are substituted in Eq. (5), this system and Eq. (6) can be solved with respect to k and s.

For some values of parameters s and b, the hyperbolical fluence rate gradients of Eq. (5) will be very close to exponential curves (Hartmann and Cohnen-Unser, 1972).

Since the Kubelka–Munk theory considers two processes (absorption and scattering), two independently measured parameters are required (T and R).

If R and T have been measured for a sample of thickness L, we obtain:

$$s = \frac{1}{Lb} \text{Arsinh} \left(\frac{bR}{T} \right)$$

$$k = s(a - 1) \qquad \text{Eq. (7)}$$

where

$$a = \frac{1 + R^2 + T^2}{2R}, \qquad b = \sqrt{a^2 - 1}$$

or, if s and k are known

$$R = \frac{1}{a + b \coth(bsL)}$$

$$T = \frac{b}{a \sinh(bsL) + b \cosh(bsL)} \qquad \text{Eq. (8)}$$

where coth, sinh, and cosh are the hyperbolic cotangent, sine, and cosine, respectively, and the expression for the fluence rate gradient is:

$$I(x) = I_0 T \left[\frac{a + 1}{b} \sinh(bsL) + \cosh(bsL) \right] \qquad \text{Eq. (9)}$$

The correlation between phenomenological Kubelka–Munk coefficients k and s and effective absorption and scattering coefficients of the absorbing and scattering medium σ_a and σ_s can be expressed approximately (Brinkworth, 1971) as:

$$k \approx 2\sigma_a \qquad s \approx 2\sigma_s$$

The important difference between scattering and nonscattering samples, which is seen in the differences between the Kubelka–Munk theory and Beer's law, is concerned with the additivity of absorbance as a function of thickness. In a nonscattering sample with thickness $L = L_1 + L_2$, the absorbance of the whole sample is the sum of the absorbances of its parts (having thicknesses L_1 and L_2). Correspondingly, the fluence rate at the boundary of part 1 (having thickness L_1) will be equal to the fluence rate within the intact sample at depth L_1. This is not true for scattering samples where the fluence rate at a certain depth is formed from the interaction between all the layers which constitute the sample, including those which are cut off when transmittance of a part of the sample is measured.

(b) Reflectance and transmittance measurements

Application of the Kubelka–Munk theory to biological samples requires knowledge of the diffuse reflectance and transmittance values of the sample.

(i) Measurements. Total diffuse transmittance and reflectance are usually determined with an integrating sphere, which is a hollow sphere with a diffusely reflective white inner coating, (usually $BaSO_4$ or MgO). It can be used for measurements in the UV and visible range. The principal arrangement for transmittance measurements is shown in Fig. 9a. For this mode of

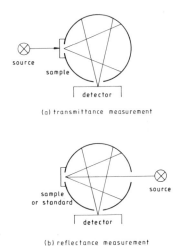

(a) transmittance measurement

(b) reflectance measurement

Fig. 9. (a) Integrating sphere for transmittance measurements (schematically). (b) Integrating sphere for reflectance measurements.

operation the sphere has two ports, one to accept the sample (sample port) and the other to detect the emitted radiation (detector port). Remittance measurements require an additional entrance aperture in the sphere for irradiation (Fig. 9b). The transmitted or reflected light is distributed by multiple reflections over the inner surface of the sphere, part of it reaching the detector through the measuring port. A baffle prevents light from the sample hitting the detector directly. The sphere accepts light transmitted or reflected over an angle of 2π sr. In an alternate arrangement, the sample is arranged in the centre of the sphere; here the light coming from the sample is measured over 4π sr; i.e., all light not absorbed is assessed.

Whereas these experimental layouts are based on specular illumination and diffuse measurement [usually termed $R_{0/d}$, which indicates perpendicular ($0°$) illumination and diffuse measurement], possibilities also exist for diffuse illumination and either diffuse (integral) or specular measurement. For diffuse illumination, a second sphere is needed (Duntley, 1942); the light source is positioned inside the sphere and the sample is shielded from direct illumination by a baffle.

Calibration is carried out either with a blank sample port for transmission or against a remission standard with known reflectance, usually a $BaSO_4$ tablet freshly pressed from high-purity powder. The technically simpler type of integrating sphere is as shown in Fig. 9b, where the sample is substituted for the reference (substitution type analogous to a single beam, single wavelength mode). With a comparison-type integrating sphere, the standard remains at an additional port (reference port) during measurement. The comparison is made by alternately switching the measuring beam from sample to reference and back. This is analogous to a double beam, single wavelength spectrophotometer (Section II. A. 1). The comparison type is preferable since the substitution of a high reflectance standard by a (generally) low reflectance sample, both of which form part of the sphere wall, changes the overall reflection of the sphere wall, thus causing a bias in the reflectance values. Fortunately, this bias can be estimated easily and corrected for, if the reflectances of the inner surfaces of the sphere and sample are known. If the sphere surface is large compared to the sample surface, the error and other deviations from the "ideal sphere" (Jacquez and Kuppenheim, 1955) are small. Since the signal decreases with increasing sphere size, optimal sphere diameter is determined by the sample size.

The scanning of a diffuse transmission or remission spectrum is done point by point for most sphere types. The monochromator is usually arranged in the irradiation beam, i.e., between the light source and the sphere. The geometry $T_{d/0}$ suggests a spectral analysis of the transmitted light, but mirrors or light guides provide for monochromatic illumination inside the sphere if desired. Light guides can be used to achieve measuring geometries not originally supplied by a commercial integrating sphere (see also Chapter 3). Nowadays light guides have any desired diameter and acceptance angles (which are identical to the angles of output) from 120° to 20°. Small acceptance angles are very appropriate for skew illumination since they give a distinct light spot, which is otherwise achieved only by a complicated optical system or a laser beam.

In some cases it is desirable to measure only the diffuse part of the transmited or reflected light; to do this, the sphere can be equipped with a light trap to eliminate the specularly reflected or transmitted light. The design of spheres for specific tasks is described by Egan and Hilgeman (1975). Some commercially available instruments are described in Kortüm (1969) and the different geometries are discussed in Stenius (1954). Most modern spectrophotometers have scattered-light accessories, some of which also provide for reflectance measurements. Very often the opal-glass method, if properly applied, gives satisfactory results. Shibata (1957) suggests an opal-glass method, by which diffuse reflectance spectra are obtained without an integrating sphere.

(ii) Practical considerations. If reflectance and transmittance measurements of plant tissue are for application of the Kubelka–Munk theory, special attention must be paid to the experimental conditions. The prolonged measuring time necessary with a simple integrating sphere often necessitates the use of a cuvette. For flat samples, such as leaves or leaf

sections, the cuvette should also be flat to ensure well-defined irradiation of the sample. Versatile cuvettes of this type are easily prepared by enclosing the sample between two cover slips, joined by wax or glazier's putty. Water is injected afterwards with a syringe. Another method is to prepare flat steel rings with a broad rim and a very smooth surface. Wet cover slips stick to the surface, thus enclosing the sample and an appropriate amount of water in the inner part (Fig. 10). These cuvettes provide constant optical properties of the sample for at least 2 h. Whether or not the sample is measured in a cuvette, the Kubelka–Munk theory may not be applied straightforwardly with the measured transmittance and reflectance values. As pointed out in Section III. C. 1. a, the Kubelka–Munk theory does not apply if specular reflection is present and is restricted to macrohomogeneous samples with the measuring light incident diffusely or alternatively at 60°. Biological tissues show specular reflection owing to the differences in refractive index between the tissue and the surrounding medium; they are often heterogeneous owing to a layered structure (e.g., leaves or skin). Diffuse incidence is usually difficult to achieve; even the 60° requirement is normally not fulfilled because the incident beam is bent according to Snell's law to smaller angles.

The specular reflection problem can be accounted for mathematically. A general approach to this type of extension of the Kubelka–Munk theory (i.e., the inclusion of specular reflection) is described by Reichman (1973). In this theory, a set of four independent measurements is needed to find the parameters s and k of the Kubelka–Munk theory. A typical set could be two pairs of reflectance and transmittance measurements at two different angles of incidence. This approach is mathematically complex and is currently inadequate for biological material of nonuniform shape. A simplified ap-

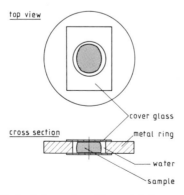

Fig. 10. "Adhesion" cuvette for flat samples. The metal ring must be flat and very smooth (polished).

proach (Seyfried *et al.,* 1983) is applicable if a flat object of high scattering power and known refractive index is to be dealt with. In this case, the light inside the sample may be considered to be almost perfectly diffuse, and the specular reflection of the light traveling toward the sample boundaries from outside (irradiation) or inside (scattered light) may be computed in a straightforward manner. For collimated incidence at angle α, reflectance is calculated from the Fresnel formula:

$$R(\alpha) = \frac{1}{2}\left(\frac{\sin^2(\alpha - \beta)}{\sin^2(\alpha + \beta)} + \frac{\tan^2(\alpha - \beta)}{\tan^2(\alpha + \beta)}\right) \qquad \text{Eq. (10)}$$

where the angle β is derived from:

$$\frac{\sin \alpha}{\sin \beta} = n = \frac{n_0}{n_1} \qquad \text{Eq. (10a)}$$

For perpendicular incidence we have:

$$R(0) = \frac{(n_1 - n_0)^2}{(n_1 + n_0)^2} \qquad \text{Eq. (10b)}$$

For diffuse incidence, a formula given by Walsh (see Kortüm, 1969) applies if the light comes from the side with lower refractive index n_0

$$R_{\text{diff}} = 2 \int_0^{\pi/2} \sin \alpha \cos \alpha \, R(\alpha,n) \, d\alpha \qquad \text{Eq. (11)}$$

These values are tabulated in Kortüm (1969) and Foster and Stearns (1971). If the diffuse light comes from the side with higher refractive index n_1 (i.e., the light scattered in the sample), the reflectance is given by

$$R_{n_1 \to n_0} = 1 - \frac{T_{n_0 \to n_1}}{n^2} \qquad \text{Eq. (12)}$$

For crown glass ($n_1 = 1.5$) in air, $R_{1.5 \to 1} = 0.596$.

In Section III. C. 1. a it is stated that the absorbances of two scattering layers are not simply additive as is the case for clear solutions. The "fusion" of two layers is achieved by a method originally conceived by Stokes (1860). The details of this fusing procedure are shown in Fig. 11 [Eqs. (13) and (14)]. A number of interreflections contribute to the total transmittance and reflectance; the contributions form a geometric progression. Here the reflecting boundaries are considered as "pseudo" layers of zero thickness and absorption and known reflectance. The fusing procedure applied to the sample layer and a reflecting boundary and again applied to these two fused layers and the second reflecting boundary gives formulae for the total reflec-

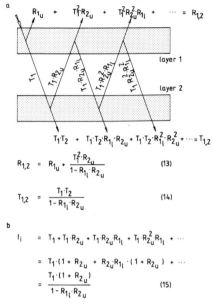

Fig. 11. (a) "Fusion" of two layers of scattering material. The layers are separated for clarity only; in reality there is no discontinuity (gap). R and T mean reflectance and transmittance, the subscripts 1 and 2 designate the layer, the subscripts u and l indicate whether the reflectance comes from irradiation of the upper or lower surface of a layer. Series of subscripts are to be read from the left hand side (irradiation side). So $R_{1,2}$ is the reflectance of the fused layers 1 and 2 upon irradiation in layer l. (b) Calculation of light flux I_i between two layers of scattering material. Both layers are measured separately. Compare cleavage method in Section III. C. 1.

tances and transmittance of the sample. When the sample is measured in a cuvette, the same formulae apply, but the two boundary layers are now complex, i.e., themselves fused from different layers. The measured reflectances and transmittance are now available as a system of three equations, two for the reflectances from either side, and one for the transmittance, which is independent of orientation (Kubelka, 1954). All three formulae are functions of R_u, R_l, and T, the reflectance from the upper or lower surface and transmittance values of the object without specular reflection.

Unfortunately, these values all appear in all three formulae and cannot be separated. Therefore, they can be computed only by numerical computation, as functions F_1, F_2, and F_3, each of which should give zero for the correct values of T, R_u and R_l. This representation facilitates the application of standard numerical methods, e.g., a generalized Newton approximation.

$$F_1 = \frac{T_c' T_c T}{D} - T_{csc}$$

$$F_2 = T_c' T_c R_u + \frac{T_c' T_c R_c T^2}{D} - (R_{cs_uc} - R_c')(1 - R_c R_u)$$

$$F_3 = T_c' T c R_l + \frac{T_c' T_c R_c T^2}{D} - (R_{cs_lc} - R_c')(1 - R_c R_l)$$

where we put

$$D = 1 - R_c(R_u + R_l - R_c R_u R_l + T^2 R_c)$$

for abbreviation.

Primes indicate specular reflection (collimated incidence), the subscript c stands for cover and means either the reflecting boundary or the complex boundary layer. The subscripts u and l represent upper and lower faces, series of subscripts are to be read from the left (illuminated side), e.g., R_{cs_lc} is the reflectance of the sample s in the cuvette (two boundary layers) upon irradiation from the lower side. The application of this approach is restricted to cases where the light fluxes leaving the object are highly diffuse. In most cases only two to seven iterations are needed to achieve $|F| < 10^{-8}$, where $|F| = \max(|F_1|, |F_2|, |F_3|)$.

(c) Estimation of light gradients

The methods discussed up to now provide some means to estimate fluence rate gradients in scattering tissues. It should be noted that backscattering can cause the light flux in the uppermost layer of the tissue to be considerably higher (up to a factor of three) than the incident light flux.

(i) From Kubelka–Munk theory. The simplest approach would be to measure the true (i.e., unbiased by surface reflection) reflectance and transmittance values and apply the Kubelka–Munk theory [Eq. (7), compare Section III. C. 1. a]. The approach is valid only if the object is macrohomogeneous, i.e., if the reflectances from both orientations are equal (minimum requirement!). Where this is not true, an approximation procedure may be applied if the sample can be considered bilayered. This is the case for most biological samples (leaves, skin, etc.). Kubelka–Munk coefficients are calculated using the transmittance and each reflectance. These coefficients are attributed to the top or lower layer; then we compute the reflectances and transmittance according to Eq. (8), and the layers are fused using Eqs. (13) and (14). These procedures yield reflectances and transmittance of the total object, which are generally very close to the measured values. The scattering and absorption coefficients of the layers may then be adjusted slightly to achieve better approximation of the transmittance and reflectances. The thickness of the two layers must also be known, and this is derived from microscope sections.

The approximated Kubelka–Munk constants are now available for calculation of the incoming light fluxes on each of the layers according to Eqs. (8), (13), and (14) for any illumination conditions (including background light) and for calculation of light gradients according to Eq. (9) in each layer (Fig. 11a). These calculations require a computer; the programme is easily extended to include more layers and specular reflection, thus approximating real objects in air. Details of the computation of light gradients and examples of light gradients are described by Seyfried and Fukshansky (1983).

(ii) From the cleavage method. This approach is suggested in Fig. 11b. If the optical properties (reflectances and transmittance) of two arbitrarily obtained parallel sections of an object have been measured, the fluence rate at the plane of the sectioning is evaluated by Eq. (15). The optical properties of the whole object may be measured and calculated as a control (Seyfried, 1984). This approach is applicable only if the sections still obey the requirement that the light transmitted and reflected be highly diffuse. If this is not the case, the reflectances and transmittance of the whole object and the thicker section may be used to compute the light intensity in the section plane. In this case, no control is available. After randomly cleaving a series of samples at different depths, the light gradient is available as a mean curve through the points calculated from the measurements.

(iii) From phytochrome photoconversion measurements. Light gradients can be determined using the light-measuring properties of phytochrome. The method is applicable only for relatively thick objects of high phytochrome content. Experiments of this type have been carried out in our laboratory on the cotyledons of *Cucurbita pepo;* these cotyledons satisfy the above conditions.

The procedure is as follows. Excised etiolated cotyledons are placed on a black metal plate cooled to 0°C and irradiated, e.g., with red light at nonsaturating light fluxes for phytochrome photoconversion from Pr to Pfr. The cotyledons are then cut into slices in planes perpendicular to the irradiation (Fig. 12). Each slice is assayed for phytochrome (in a ratiospect, see Section II. d) to determine the Pfr/Ptot gradient with increasing penetration depth. For different irradiation times at constant fluence, the photoconversion in the top layer, or even better, the Pfr/Ptot gradient extrapolated to the irradiated surface, gives a reference system that allows calculation of light gradients relative to the light flux in the top layer. Fairly precise knowledge of the light flux in this top layer can be derived from the method described in the previous subdivision.

It is then possible to calculate true *in vivo* cross sections of phytochrome. If only the light flux in the top layer is used, the light flux gradients can be

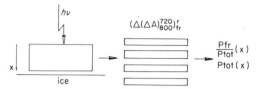

Fig. 12. Method for measuring the spatially dependent phytochrome photoconversion in a (scattering and absorbing) cotyledon.

recalculated on the basis of the Pfr/Ptot gradient.   Alternatively, the light gradients as calculated from a method of the previous sections may be taken for granted and thus (possibly) spatially dependent spectroscopic properties of phytochrome *in vivo* can be evaluated.

This approach is also applicable with other photoconvertible pigments, e.g., protochlorophyll.   If slices of varying thickness are assayed, the relative pigment content apparently increases with layer thickness.   This is an effect of increased scattering due to more scattering material, i.e., an increased path length of the measuring light (Butler, 1962).   For constant layer thickness, the photoconversion readings yield the spatially dependent pigment distribution.   Low pigment concentration does not lead to problems in easily reproducible samples.   If n slices taken from the same depth are piled up, the signal increases more than n-fold, again owing to increased scattering.

(iv)   With photoacoustic spectroscopy. Using photoacoustic spectroscopy (PAS), an evaluation of light gradients is also possible.   This method is based on the fact that absorbed light energy is dissipated as heat and may be recorded by sensitive microphones if the incident beam is chopped.   The chopping frequency determines the analytical depth of the method; an absorption profile of the sample is readily obtained therefrom.   PAS complements other methods of absorption determination since only the dissipated absorbed energy (e.g., energy not used in photosynthesis) is measured in PAS.   For details and literature see Balasubramanian and Rao (1981) or Bults *et al.* (1982).

REFERENCES

Balasubramanian, D. and Rao, C. M. (1981).   *Photochem. Photobiol.* **34,** 749–752.
Brinkworth, B. J. (1971).   *J. Phys. D* **4,** 1105–1106.
Bults, G., Horwitz, B. A., Malkin, S. and Cahan, D. (1982).   *Biochem. Biophys. Acta* **679,** 452–465.
Butler, W. L. (1962).   *J. Opt. Soc. Am.* **52,** 292–299.
Duntley, S. Q. (1942).   *J. Opt. Soc. Am.* **32,** 61–70.
Duysens, L. N. M. (1956).   *Biochem. Biophys. Acta* **19,** 1–12.

Egan, W. B. and Hilgeman, T. (1975). *Appl. Opt.* **14,** 1137–1142.

Foster, W. H. and Stearns, E. I. (1971). *J. Opt. Soc. Am.* **61,** 60–62.

Fukshansky, L. (1978). *J. Math. Biol.* **6,** 177–196.

Fukshansky, L. (1982). *In* "Plants and the Daylight Spectrum" (H. Smith, ed.), pp. 21–40. Academic Press, New York.

Fukshansky, L. and Kazarinova, N. (1980). *J. Opt. Soc. Am.* **70,** 1101–1111.

Hartmann, K. M. and Cohnen-Unser, I. (1972). *Ber. Dtsch. Bot. Ges.* **85,** 481–555.

Jacquez, J. A. and Kuppenheim, H. F. (1955). *J. Opt. Soc. Am.* **45,** 460–470.

Kortüm, G. (1969). "Reflectance Spectroscopy: Principles, Methods, Applications." Springer-Verlag, New York.

Kubelka, P. (1948). *J. Opt. Soc. Am.* **38,** 448–456.

Kubelka, P. (1954). *J. Opt. Soc. Am.* **44,** 330–335.

Mandoli, D. F. and Briggs, W. R. (1982a). *Plant, Cell Environ.* **5,** 137–145.

Mandoli, D. F. and Briggs, W. R. (1982b). *Proc. Natl. Acad. Sci. U.S.A.* **79,** 2902–2906.

Reichman, J. (1973). *Appl. Opt.* **12,** 1811–1815.

Seyfried, M. (1983). Ph.D. Thesis, University of Freiburg, Freiburg, West Germany.

Seyfried, M. and Fukshansky, L. (1983). *Appl. Opt.* **22,** 1402–1408.

Seyfried, M., Fukshansky, L. and Schäfer, E. (1983). *Appl. Opt.* **22,** 492–496.

Shibata, K. (1975). *J. Opt. Soc. Am.* **47,** 172–175.

Stenius, A. S. (1955). *J. Opt. Soc. Am.* **45,** 727–732.

Stokes, G. G. (1860–1862). *Proc. R. Soc. London* **11,** 545–557.

7

Determination of Phytochrome Parameters from Radiation Measurements

P. M. Hayward

Although the spectrophotometric determination of parameters relating to phytochrome has reached a high level of sophistication and sensitivity (Chapter 5), it is still not possible to use this method with tissues containing chlorophyll. Thus for most light-grown plants (with the exceptions of naturally achlorophyllous tissue and herbicide-treated tissue, the use of which might be objected to on the grounds that the material is "atypical"), it is usually necessary to use an indirect method to calculate the state of phytochrome within the plant. Such calculations are based on measurements of the light environment in which the plant is growing and on established knowledge about the action of light on phytochrome, usually derived from spectrophotometric measurements (and hence making the results again dependent on etiolated tissue).

I. THE PHYTOCHROME SYSTEM

Several light (photochemical) and dark (thermal) reactions interact to determine the concentrations of Pr* and Pfr at any given instant. However, it is obviously only possible to account for the former in the type of calculation outlined above since the dark reactions are independent of the incident light flux. If for the present we consider only the light reactions, the phytochrome system is reduced to the simple interconversion $Pr \rightleftharpoons Pfr$. Both these reac-

* Abbreviations: Pr (Pfr), form of phytochrome absorbing red (far-red) light; $[Pr_\infty]$ ($[Pfr_\infty]$), relative amount of Pr (Pfr) at photoequilibrium; Ptot = Pr + Pfr.

TECHNIQUES IN PHOTOMORPHOGENESIS
0-12-652990-6

tions appear to be first order (Butler *et al.*, 1964a), the rate in each direction (R_1, R_2) therefore being a function of phytochrome concentration ([Pr],[Pfr]) and a rate constant (k_1, k_2). This may be represented in classical systems analysis notation (Forrester, 1961) as

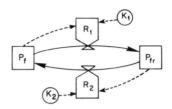

where $R_1 = \dfrac{d[Pr]}{dt} = [Pr]k_1$ and $R_2 = \dfrac{d[Pfr]}{dt} = [Pfr]k_2$

The photoconversions may thus be described completely in terms of two parameters, k_1 and k_2. These are both functions of the light environment and, as we shall see later, may be calculated from measurable quantities. However, it is often more useful to describe the system in terms of another pair of parameters; the relative amount and flux of Pfr.

The relative amount of Pfr after prolonged (i.e., saturating) irradiation is the "photoequilibrium" φ:

$$\varphi = \frac{[Pfr_\infty]}{[Pr_\infty] + [Pfr_\infty]} = \frac{R_2/k_2}{R_1/k_1 + R_2/k_2} = \frac{R_2 k_1}{R_1 k_2 + R_2 k_1}$$

But at equilibrium, $R_1 = R_2$

Therefore $$\varphi = \frac{k_1}{k_1 + k_2}$$

The flux of Pfr is usually measured as the rate of photoconversion, i.e., the cycling rate (v, or sometimes θ) where $v = k_1 + k_2$.

Since $k = (\ln 2)/t_{1/2}$ (where $t_{1/2}$ is the duration of the irradiation required to reach half the photoequilibrium), it is relatively easy to find v by successive spectrophotometric determinations of the proportion of Pfr present.

The formulae presented above make no allowance for the selective attenuation of light before it reaches a phytochrome molecule. If the results of the calculations are intended to represent the state of phytochrome in a green plant, where the incident light is absorbed by pigments other than phytochrome (primarily chlorophyll), then the values must be corrected accordingly, although this is not easy since the structure of a leaf is complex and the pigments are not distributed uniformly. However, Holmes and Fukshansky (1979) calculated the effect of natural light and canopy light on the photoequilibrium within a leaf. They showed that under various conditions different gradients of φ would be set up across the thickness of the leaf (Fig. 1). Most authors choose to ignore such effects, making the arbitrary assumption that phytochrome is not appreciably screened by other pigments. This is probably reasonable if comparative measurements are being made. (For further discussion see Chapter 6.)

The true state of phytochrome *in vivo* is further complicated by the presence of thermal (nonphotochemical) reactions as mentioned above. If measurements were made at 0°C, the nonphotochemical reactions could be ignored since they virtually stop at this temperature. However, at room temperature, photoequilibria do not usually exist in the form described. The relative amount of Pfr under these conditions is partly determined by the various dark reactions, although in some circumstances they may still not be significant since dark reactions are much slower in green material than in dark-grown plants. For example, the half-life of Pfr destruction in light-grown tissue is approximately 8 h as compared to 30–40 min in dark-grown tissue (Heim *et al.,* 1981).

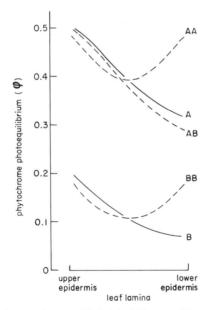

Fig. 1. Predicted phytochrome photoequilibria (φ) within a model green leaf (after Holmes and Fukshansky, 1979). Leaves were irradiated as follows: (A) Natural daylight on upper epidermis only (average $\varphi = 0.40$); (B) canopy light on upper epidermis only (average $\varphi = 0.12$); (AA) natural daylight on upper and lower epidermis (average $\varphi = 0.43$); (AB) natural daylight on upper epidermis, canopy light on lower epidermis (average $\varphi = 0.38$) (BB) canopy light on upper and lower epidermis (average $\varphi = 0.13$).

In particular we should note that when dark reactions are significant, [Ptot] can no longer be considered to remain constant. However, an equilibrium will again be reached, although R_1 need no longer equal R_2, and a distinction must therefore be made between "photoequilibrium" due solely to light reactions [i.e., $\varphi = k_1/(k_1 + k_2)$] and a second type of equilibrium, the "photostationary state" (Hartmann, 1977), which includes the effects of dark reactions (i.e., $\psi = [\text{Pfr}]/[\text{Ptot}]$). A similar situation exists with the photoconversion rate at room temperature. This was originally defined for $R_1 = R_2$, but with dark reactions included, the two rates of photoconversion may not be equal, even at equilibrium.

Under prolonged irradiation and at room temperature, both the parameters φ and ν therefore lose their meaning in the sense in which they were first defined, as measures of the amount and flux of Pfr. If we need to know the state of phytochrome in these conditions (i.e., ψ rather than φ) then the effect of the dark reactions must be included in a model of the whole system. On the other hand, φ and ν can still be used as characteristics of the light environment with respect to their effect on the light reactions of phytochrome. The pair of parameters can be used in this way to describe completely the light environment and are equivalent to the pair k_1 and k_2 for this purpose

$$k_1 = \nu\varphi$$
$$k_2 = \nu(1 - \varphi)$$

II. DETERMINATION OF THE PHOTOEQUILIBRIUM

We have seen that $\varphi = k_1/(k_1 + k_2)$. Now the rate constants k_1 and k_2 are the product of the incident photon flux and the probability that a photon will cause the appropriate transformation. The latter is known as the "photoconversion cross section" (or "relative quantum efficiency" as used by Pratt and Briggs, 1966) and is itself determined by two factors, the extinction coefficient and the quantum yield (i.e., the number of molecules affected as a proportion of the number of quanta absorbed). Thus:

$$k_1 = I_\lambda \sigma_{r\lambda} \quad \text{and} \quad k_2 = I_\lambda \sigma_{fr\lambda}$$

where I_λ = incident photon flux at wavelength λ (mol sec^{-1} m^{-2}), $\sigma_{r\lambda}$ $(\sigma_{fr\lambda})$ = photoconversion cross section at wavelength λ for Pr $\rightarrow$ Pfr (Pfr $\rightarrow$ Pr) (m^2 mol^{-1}); and

$$\sigma_{r\lambda} = \epsilon_{r\lambda}\Phi_r \quad \text{and} \quad \sigma_{fr\lambda} = \epsilon_{fr\lambda}\Phi_{fr}$$

where $\epsilon_{r\lambda}$ $(\epsilon_{fr\lambda})$ = extinction coefficient of Pr (Pfr) at wavelength λ (m^2 mol^{-1}), and Φ_r (Φ_{fr}) = quantum yield for Pr (Pfr) (dimensionless). The quantum yield is assumed to be independent of wavelength. Were it not, it would be indicative of more than one chromophoric absorption centre. (In fact, the value of Φ in the UV is different from that in the visible part of the spectrum (Pratt and Butler, 1970).

A. Monochromatic light

When the entire incident photon flux is at single wavelength,

$$k_1 = I_\lambda \epsilon_{r\lambda} \Phi_r \quad \text{and} \quad k_2 = I_\lambda \epsilon_{fr\lambda} \Phi_{fr}$$

and the photoequilibrium becomes

$$\varphi = \frac{I_\lambda \epsilon_{r\lambda} \Phi_r}{I_\lambda \epsilon_{r\lambda} \Phi_r + I_\lambda \epsilon_{fr\lambda} \Phi_{fr}} = \frac{1}{1 + \dfrac{\epsilon_{fr\lambda}}{\epsilon_{r\lambda}} \dfrac{\Phi_{fr}}{\Phi_r}}$$

Relative values of ϵ and Φ may therefore be used, which, as we shall see later, makes the calculation considerably simpler. As we would expect, at photoequilibrium the result is independent of incident photon flux. This means that we may construct a spectrum showing the photoequilibrium set up at all wavelengths of monochromatic light (Hartmann, 1966; Pratt, 1978) as shown in Fig. 2.

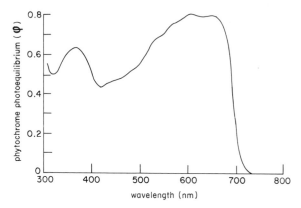

Fig. 2. Spectrum of phytochrome photoequilibrium (φ) produced with monochromatic light. Calculated from the absorption spectra of Butler *et al.* (1964b) with maximum $\varphi = 0.80$ and $\Phi_r/\Phi_{fr} = 1.23$. (After Hartmann, 1966.)

B. Polychromatic light

The same relationship as for monochromatic light still holds, except that now the terms in the equation must be integrated across the appropriate part of the spectrum (λ_1 to λ_2)

$$\varphi = \frac{\int_{\lambda_1}^{\lambda_2} (I_\lambda \epsilon_{r\lambda}) d\lambda \; \Phi_r}{\int_{\lambda_1}^{\lambda_2} (I_\lambda \epsilon_{r\lambda}) d\lambda \; \Phi_r + \int_{\lambda_1}^{\lambda_2} (I_\lambda \epsilon_{fr\lambda}) d\lambda \; \Phi_{fr}}$$

$$= \frac{1}{1 + \dfrac{\int_{\lambda_1}^{\lambda_2} (I_\lambda \epsilon_{fr\lambda}) d\lambda \; \Phi_{fr}}{\int_{\lambda_1}^{\lambda_2} (I_\lambda \epsilon_{r\lambda}) d\lambda \; \Phi_r}}$$

Again, only relative values of ϵ and Φ (and I) are required. The integrations will usually be carried out by a summation of n discrete intervals so that $\int_{\lambda_1}^{\lambda_2} (I_\lambda \epsilon_\lambda) \, d\lambda$ becomes $\sum_{i=1}^{n} (I_j \epsilon_j) \dfrac{\lambda_2 - \lambda_1}{n}$, where $j = \lambda_1 + \left[(i - \frac{1}{2}) \dfrac{\lambda_2 - \lambda_1}{n} \right]$.

By calculating j in this manner, the values of I and ϵ are taken from the middle of each of the n intervals (Fig. 3). However, in practice, n will be large and, as pointed out by Schäfer (1981), the error introduced by simply taking readings at the side of each interval will be small. Many spectroradiometers give values of I for discrete wavelength intervals anyway and it is most straightforward simply to use these in the calculation. The wavelength interval used is then simply $\dfrac{\lambda_2 - \lambda_1}{n}$. Given that the number of intervals for both summations is the same, the equation then becomes

$$\varphi = \frac{1}{1 + \dfrac{\displaystyle\sum_{i=\lambda_1}^{\lambda_2} (I_i \epsilon_{fri}) \; \Phi_{fr}}{\displaystyle\sum_{i=\lambda_1}^{\lambda_2} (I_i \epsilon_{ri}) \; \Phi_r}}$$

C. Use of the ratio of red:far-red light

In phytochrome studies, the light environment is often described in terms of the ratio of the light fluxes at the absorption maxima of Pr and Pfr. This quantity, the red:far-red ratio (R:FR) has been called ζ (Monteith, 1976;

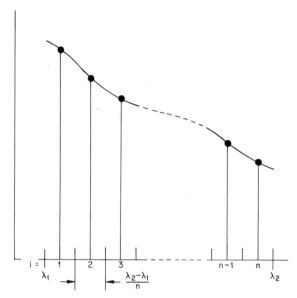

Fig. 3. Method of integration by summing the values at the midpoint of discrete intervals. The wavelength values required (j) are found from the equation $j = \lambda_1 + (i - \frac{1}{2}) \dfrac{\lambda_2 - \lambda_1}{n}$ for i = 1 to n:

$$i = 1, \quad j = \lambda_1 + \frac{1}{2} \frac{\lambda_2 - \lambda_1}{n}$$

$$i = 2, \quad j = \lambda_1 + 1\frac{1}{2} \frac{\lambda_2 - \lambda_1}{n}$$

$$\vdots$$

$$i = n, \quad j = \lambda_1 + (n - \frac{1}{2}) \frac{\lambda_2 - \lambda_1}{n} = \lambda_2 - \frac{1}{2} \frac{\lambda_2 - \lambda_1}{n}$$

See text for further details.

Holmes and Smith, 1977) and may be defined as

$$\zeta = \frac{\displaystyle\int_{\lambda=655}^{665} (I_\lambda)\, d\lambda}{\displaystyle\int_{\lambda=725}^{735} (I_\lambda)\, d\lambda}$$

It is possible to calculate reasonably accurate values of φ based on this ratio so long as there are no large peaks elsewhere in the spectrum, especially in the blue region where phytochrome has a minor absorption band. This may not be a very serious problem since blue light is some fifty times less effective at bringing about phototranformations than red light (Jabben *et al.*, 1982) and, as they conclude, "the contribution of blue light to phytochrome-mediated responses is negligible ($<3\%$) in the natural environment."

By using ζ, we are effectively assuming that the light is "dichromatic" at wavelengths of 660 and 730 nm and, by knowing the extinction coefficients of Pr and Pfr at these wavelengths, we may calculate φ on the basis of R:FR

$$\varphi = \cfrac{1}{1 + \cfrac{I_{660}\epsilon_{fr660} + I_{730}\epsilon_{fr730}}{I_{660}\epsilon_{r660} + I_{730}\epsilon_{r730}}\cfrac{\Phi_{fr}}{\Phi_r}}$$

$$= \cfrac{1}{1 + \cfrac{\zeta\epsilon_{fr660} + \epsilon_{fr730}}{\zeta\epsilon_{r660} + \epsilon_{r730}}\cfrac{\Phi_{fr}}{\Phi_r}}$$

If we assume that $\dfrac{\Phi_{fr}}{\Phi_r} = 1.0$ (Pratt, 1975) and that $\epsilon_{r730} \approx 0$, then the equation simplifies to

$$\varphi = \cfrac{1}{1 + \cfrac{\zeta\epsilon_{fr660} + \epsilon_{fr730}}{\zeta\epsilon_{r660}}}$$

$$= \cfrac{\cfrac{\epsilon_{r660}}{\epsilon_{r660} + \epsilon_{fr660}}}{1 + \cfrac{\epsilon_{fr730}}{\epsilon_{r660} + \epsilon_{fr660}}\cfrac{1}{\zeta}}$$

and using the values of ϵ from Pratt (1978, Fig. 7), which are corrected so that $[Pfr_\infty]_{660} = 0.75$ (see below) then

$$\varphi = \cfrac{0.75}{1 + \cfrac{0.45}{\zeta}}$$

The hyperbolic relationship thus expected between φ and R:FR was demonstrated empirically by Smith and Holmes (1977) who plotted spectrophotometrically measured values of φ against the value of ζ produced by a variety of broad-band actinic light sources. This has been used as a "calibration" curve to estimate the photoequilibrium from the measured R:FR ratios of natural broad-band sources. Its use was limited at low values of R:FR because of the steepness of the curve (Smith, 1982) although a mathemati-

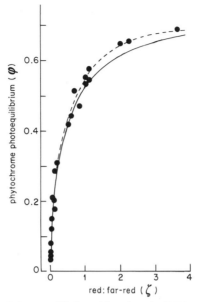

Fig. 4. Measured values of photoequilibrium (φ) against the R:FR ratio for broad-band light sources (Smith and Holmes, 1977.) Broken line, rectangular hyperbola fitted to measured data; solid line, predicted curve using extinction coefficients of phytochrome from Pratt (1978) and assuming $\Phi_r/\Phi_{fr} = 1.0$ and $\epsilon_{r730} = 0.0$. All data calculated on the assumption that maximum $\varphi = 0.75$.

cally fitted function should enable all values of R:FR to be converted to estimated φ. The hyperbolic curve fitted to the results of Smith and Holmes (1977) by a "least squares" fit, gave the equation

$$\varphi = \frac{0.75}{1 + \dfrac{0.35}{\zeta}}$$

which may be compared to that calculated above (Fig. 4).

III. DETERMINATION OF REACTION RATES

If we wish to calculate the rate constants of the two light reactions k_1 and k_2 and hence the rate of photoconversion v we must know the absolute values of ϵ, Φ, and I rather than their relative values. Once these have been found by experiment (described later), then the rate constants may be calculated. As we have already seen, for a single wavelength of light

$$k_1 = I_\lambda \sigma_{r\lambda} \quad \text{and} \quad k_2 = I_\lambda \sigma_{fr\lambda}$$

and integrating for polychromatic light

$$k_1 = \int_{\lambda_1}^{\lambda_2} (I_\lambda \epsilon_{r\lambda}) d\lambda \; \Phi_r \quad \text{and} \quad k_2 = \int_{\lambda_1}^{\lambda_2} (I_\lambda \epsilon_{fr\lambda}) d\lambda \; \Phi_{fr}$$

Given the values of Φ_r and Φ_{fr} together with those for ϵ_r and ϵ_{fr} at each wavelength, we may calculate k_1 and k_2. The rate of photoconversion is then given by $k_1 + k_2$.

IV. OBTAINING VALUES FOR PARAMETERS

For the calculations described, we need information about the light environment and about the effect of light on phytochrome. These are best considered separately.

A. Incident light flux

Light measurement in general is dealt with in Chapter 4 and will not be considered here except for some points specific to our calculations.

For monochromatic light of known wavelength, any instrument capable of measuring energy or photon flux at that wavelength will suffice. Thus a quantum meter or a radiometer would often be sufficient for measuring the light transmitted by a narrow band interference filter since the light is practically all of a single, known wavelength.

In polychromatic light, measurements at just two wavelengths in the red and far-red regions of the spectrum may be sufficient, as we have seen. Again these could be measured with a quantum meter or radiometer, using narrow band filters. In general, however, spectroradiometric data for the whole spectrum are required. If the measurements are of energy flux, they should be converted to photon flux (since it is quanta rather than energy that are perceived) according to the equation

$$E = \frac{hcN_A}{\lambda}$$

where E = energy per mole of photons (J mol^{-1}), h = Planck's constant ($= 6.626 \times 10^{-34}$ J sec), c = speed of light ($= 2.998 \times 10^8$ m sec^{-1}), N_A = Avogadro's number ($= 6.022 \times 10^{23}$ mol^{-1}), and λ = wavelength (m). Therefore, 1 μmol sec^{-1} = $8.359 \times 10^{-3} \times \lambda$W ($\lambda$ now measured in nm).

B. Photoconversion cross section

If only the relative values of σ_r and σ_{fr} are required (for calculating photo-equilibria) the ratio may be obtained from action spectra such as those of Butler et al. (1964a) or Pratt and Briggs (1966). However, it is also possible to measure ϵ_r/ϵ_{fr} and Φ_r/Φ_{fr} separately, and this is rather easier to do since Φ_r/Φ_{fr} is independent of wavelength and $\epsilon_r/\epsilon_{fr} = A_r/A_{fr}$, where A is absorbance† (Hartmann, 1966).

Relative absorption spectra may be measured by irradiating a sample with either red (660 nm) or far-red (730 nm) actinic light before measuring the absorbance at each wavelength across the spectrum. Several such absorption spectra have been published (Butler et al., 1964b; Mumford and Jenner, 1966; Butler, 1973; Gardner and Briggs, 1974; Yamamoto and Smith, 1981a,b; Roux et al., 1982) but most have tended to use partly degraded preparations, their molecular weights ranging from 60,000 to 120,000. The preparation of apparently undegraded phytochrome (MW 124,000) has only recently been accomplished (Bolton and Quail, 1981, 1982) and its absorption spectra measured (Vierstra and Quail, 1983a).

The two spectra thus obtained with red and far-red actinic light overlap (Fig. 5) and cannot, therefore, be assumed to be those of pure Pfr and Pr, respectively.

The absorption of Pr in far-red light is low and consequently so is the photoequilibrium (φ_{730}). For many purposes the difference from unity is often ignored although Schäfer et al. (1972) showed that the amount of Pr changes with time and that for *Sinapis alba* (mustard) seedlings, φ_{730} may be as high as 0.14 after 36 h in continuous broad-band far-red, dropping to 0.075 between 72 and 120 h.

Because of the considerable absorption by Pfr in the red part of the spectrum, the photoequilibrium in red light (φ_{665}) is much less than 1.0, and the difference must be taken into consideration. The values of φ_{665} as measured over the last 20 years are shown in Table I together with the values of the other parameters we shall be considering shortly. Since most of these values were obtained from more or less degraded samples, we shall assume that the results of Vierstra and Quail (1983b) obtained with 124,000 daltons phytochrome are the most reliable. On this basis, we may take φ_{665} to be 0.862 ± 0.005. However, it must be remembered that measurements must of necessity be made *in vitro*, which may change the molecular environment and with it the *in vivo* absorbance spectra (Butler, 1973), as demonstrated by Yamamoto and Smith (1981a).

---

† Absorbance (= optical density or extinction) = $\log (I_0/I)$. The other measure of attenuation is absorptance = $(I_0 - I)/I_0$, where I_0 = incident flux and I = transmitted flux. Absorption is commonly used as an inclusive term for absorbance and absorptance.

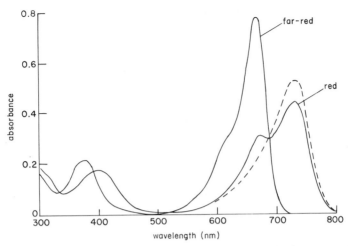

Fig. 5. Absorption spectra of 124-kD Avena (oat) phytochrome. Solid lines, absorption after irradiation with red or far-red light; broken line, absolute absorption spectrum of Pfr calculated on the basis of $\varphi = 0.86$ in red light. (After Vierstra and Quail, 1983a.)

To correct for the incomplete transformation, we may use the fact that absorbances are additive so that at any given wavelength

$$A_{730} = A_{Pr}[Pr_\infty]_{730} + A_{Pfr}[Pfr_\infty]_{730}$$

and
$$A_{665} = A_r[Pr_\infty]_{665} + A_{Pfr}[Pfr_\infty]_{665}$$

where A_{730} (A_{665}) = measured absorbance after 660 nm (730 nm) actinic irradiation, and A_{Pr} (A_{Pfr}) = absorbance of sample if all phytochrome converted to Pr (Pfr). Solving these equations simultaneously gives the values required for our calculations

$$A_{Pr} = \frac{A_{665}[Pr_\infty]_{730} - A_{730}[Pr_\infty]_{665}}{[Pfr_\infty]_{665} + [Pr_\infty]_{730} - 1} \approx \frac{A_{660} - 0.14\,A_{730}}{0.86}$$

$$A_{Pfr} = \frac{A_{730}[Pfr_\infty]_{665} - A_{665}[Pfr_\infty]_{730}}{[Pfr_\infty]_{665} + [Pr_\infty]_{730} - 1} \approx A_{730}$$

The other ratio required in the calculation, Φ_r/Φ_{fr}, may also be calculated [see Butler (1973) for method]. The value of 1.76 ± 0.10 from Vierstra and Quail (1983b) (Table I) is probably the most reliable, although it should be noted that the value may vary between species (Yamamoto and Smith, 1981b).

The calculation of reaction rates requires a knowledge of the absolute values of ϵ and Φ. The values of ϵ for undegraded phytochrome was found to be 7.3 and 4.9×10^7 cm^2 mol^{-1} at the absorption maximum of Pr and Pfr,

Table I. Extinction coefficients (ϵ), quantum yields (Φ) and photostationary state in saturating red light ([Pfr$_\infty$]) for various phytochrome preparations at specified wavelengths (λ)

Source	MW[a] (kd)	$\epsilon_{r(\lambda)}$ ($\times 10^7$ cm² mol⁻¹)	$\epsilon_{fr(\lambda)}$	Φ_r/Φ_{fr}	Φ_r	Φ_{fr}	[Pfr$_\infty$]$_{(\lambda)}$	Reference
Oat	60	>1.6(665)	>1.0(725)	1.5	—	—	0.81(665)	Butler et al. (1964a)
Oat	60	—	—	1.23	—	—	—	Hartmann (1966)[b]
Oat	60	7.6(665)	4.6	—	—	—	—	Mumford and Jenner (1966)
Corn	60	—	—	1.0	—	—	0.8(660)	Pratt and Briggs (1966)
Rye	42	1.64(665)	0.88	—	—	—	—	Correll et al. (1968)
Rye	120	7(665)	4	—	—	—	—	Tobin and Briggs (1973)
Rye	120	—	—	—	0.28	0.20	—	Gardner and Briggs (1974)[c,d]
Oat	60	—	—	—	0.21	0.15	0.81(665)	Pratt (1975)[c]
Oat	120	—	—	—	0.17	0.17	0.75(665)	Pratt (1975)
Rye	120	—	—	—	0.17	0.17	0.76(665)	Pratt (1975)
Pea	120	—	—	—	0.17	0.17	0.77(665)	Pratt (1975)
Rye	120	—	—	1.53	—	—	0.84(665)	Yamamoto and Smith (1981b)
Pea	120	—	—	1.06	—	—	0.80(665)	Yamamoto and Smith (1981b)
Oat	60	8.2(667)	—	—	—	—	—	Roux et al. (1982)
Oat	120	10.2(667)	4.35(724)	—	—	—	—	Roux et al. (1982)
Oat	30% 118, 70% 124	12.1(668)	—	1.43	0.15	0.11	0.83(648)	Litts et al. (1983)
Oat	50% 118, 50% 114	7.5(666)	4.7(730)	0.98	0.11	0.12	0.791(665)	Vierstra and Quail (1983b)
Oat	124	7.3(666)	4.9(730)	1.76	0.17	0.10	0.862(665)	Vierstra and Quail (1983b)

[a] 60- and 120-kD species are often referred to as "small" and "large" phytochrome, respectively. Most "large" phytochrome is probably in reality a mixture of 114–118-kd species.

[b] Based on [Pfr$_\infty$] = 0.81.

[c] Based on extinction coefficients of Tobin and Briggs (1973).

[d] Later corrected by Pratt (1975) to give $\Phi_r = \Phi_{fr} = 0.17$.

respectively (Table I).   From these values, the absorption spectra can be quantitatively calibrated to give ϵ_r and ϵ_{fr} at all wavelengths.

By measuring experimentally the rate constant with known photon flux, and with a knowledge of the appropriate extinction coefficient, the absolute value of Φ_r and Φ_{fr} may easily be determined

$$\Phi_r = \frac{k_{1\lambda}}{I_\lambda \epsilon_{r\lambda}} \quad \text{and} \quad \Phi_{fr} = \frac{k_{2\lambda}}{I_\lambda \epsilon_{fr\lambda}}$$

Vierstra and Quail (1983b) found values of 0.17 ± 0.02 and 0.10 ± 0.02, respectively.

V. LIGHT EQUIVALENCE

All light conditions which produce the same values of k_1 and k_2 may be considered to be equivalent with respect to phytochrome, although, of course, they may produce very different effects with, for example, chlorophyll and photosynthesis.   Having ascertained the rate constants produced by a natural, wide-band light environment, it might then be useful sometimes to be able to define the wavelength and photon flux of a monochromatic light which is equivalent to it (Fukshansky et al., 1981; Schäfer, 1981).   The wavelength of the monochromatic light will be that which produces the same photoequilibrium.   As we have already seen, this is independent of photon flux and we therefore simply look for that wavelength which satisfies

$$\frac{\epsilon_{fr\lambda}}{\epsilon_{r\lambda}} = \left(\frac{1}{\varphi} - 1 \right) \frac{\Phi_{fr}}{\Phi_r}$$

When φ is greater than about 0.4, there will be more than one wavelength which satisfies the condition (Fig. 2).   Having found the wavelength which produces the same photoequilibrium, we may find the appropriate photon flux from

$$I = \frac{k_1 + k_2}{\epsilon_{r\lambda}\Phi_r + \epsilon_{fr\lambda}\Phi_{fr}}$$

where k_1 and k_2 are the rate constants determined from the broad-band spectrum.

REFERENCES

Bolton, G. W. and Quail, P. H. (1981).   *Plant Physiol.* **67**, Suppl., 130.
Bolton, G. W. and Quail, P. H. (1982).   *Planta* **155**, 212–217.

Butler, W. L. (1973). *In* "Phytochrome" (K. Mitrakos and W. Shropshire, eds.), pp. 185–192. Academic Press, London.

Butler, W. L., Hendricks, S. B. and Siegelman, H. W. (1964a). *Photochem. Photobiol.* **3**, 521–528.

Butler, W. L., Siegelman, H. W. and Millar, C. O. (1964b). *Biochemistry* **3**, 851–857.

Correll, D. L., Steers, E., Jr., Towe, K. M. and Shropshire, W., Jr. (1968). *Biochem. Biophys. Acta* **168**, 46–57.

Forrester, J. W. (1961). "Industrial Dynamics." MIT Press, Cambridge, Massachusetts.

Fukshansky, L., Beggs, C. J., Holmes, M. G., Jabben, M. and Schäfer, E. (1981). *Proc. Eur. Symp. Light Mediated Plant Dev.*, 9.14.

Gardner, G. and Briggs, W. R. (1974). *Photochem. Photobiol.* **19**, 367–377.

Hartmann, K. M. (1966). *Photochem. Photobiol.* **5**, 349–366.

Hartmann, K. M. (1977). *In* "Biophysik" (W. Hoppe, W. Lohmann, H. Markl and H. Ziegler, eds.), pp. 197–222. Springer-Verlag, Berlin and New York.

Heim, B., Jabben, M. and Schäfer, E. (1981). *Photochem. Photobiol.* **34**, 89–93.

Holmes, M. G. and Fukshansky, L. (1979). *Plant, Cell Environ.* **2**, 59–65.

Holmes, M. G. and Smith, H. (1977). *Photochem. Photobiol.* **25**, 533–538.

Jabben, M., Beggs, C. and Schäfer, E. (1982). *Photochem. Photobiol.* **35**, 709–712.

Litts, J. C., Kelly, J. M. and Lagarias, J. C. (1983). *J. Biol. Chem.* **258**, 11025–11031.

Monteith, J. L. (1976). *In* "Light and Plant Development" (H. Smith, ed.), pp. 447–460. Butterworth, London.

Mumford, F. E. and Jenner, E. L. (1966). *Biochemistry* **5**, 3657–3662.

Pratt, L. H. (1975). *Photochem. Photobiol.* **22**, 33–36.

Pratt, L. H. (1978). *Photochem. Photobiol.* **27**, 81–105.

Pratt, L. H. and Briggs, W. R. (1966). *Plant Physiol.* **51**, 467–474.

Pratt, L. H. and Butler, W. L. (1970). *Photochem. Photobiol.* **11**, 503–509.

Roux, S. J, McEntire, K. and Brown, W. E. (1982). *Photochem. Photobiol.* **35**, 537–543.

Schäfer, E. (1981). *In* "Plants and the Daylight Spectrum" (H. Smith, ed.), pp. 461–480. Academic Press, London.

Schäfer, E., Marchal, B. and Marmé, D. (1972). *Photochem. Photobiol.* **15**, 457–464.

Smith, H. (1982). *Annu. Rev. Plant Physiol.* **33**, 481–518.

Smith, H. and Holmes, M. G. (1977). *Photochem. Photobiol.* **25**, 546–550.

Tobin, E. M. and Briggs, W. R. (1973). *Photochem. Photobiol.* **18**, 487–495.

Vierstra, R. D. and Quail, P. H. (1983a). *Biochemistry* **22**, 2498–2505.

Vierstra, R. D. and Quail, P. H. (1983b). *Plant Physiol.* **72**, 264–267.

Yamamoto, K. T. and Smith, W. S., Jr. (1981a). *Plant Cell Physiol.* **22**, 1149–1158.

Yamamoto, K. T. and Smith, W. S., Jr. (1981b). *Plant Cell Physiol.* **22**, 1159–1164.

8

Phytochrome Purification *

Lee H. Pratt

I. INTRODUCTION

Phytochrome isolation and purification methods are derived from the initial extraction and *in vitro* spectrophotometric detection of phytochrome by Butler *et al.* (1959), who identified phytochrome as a chromoprotein, and from its initial partial purification by Siegelman and Firer (1964), who introduced the use of brushite chromatography for this application. Mumford and Jenner (1966), who started with etiolated oat shoots, first reported the purification of phytochrome to homogeneity, although in a monomeric, 60,000-dalton (d) proteolytically degraded form. Shortly thereafter, Correll *et al.* (1968a) reported purification of phytochrome to homogeneity from etiolated rye shoots, obtaining tetramers and hexamers of a 42,000-d monomer (Correll *et al.*, 1968b). Evidently, either oat and rye phytochromes were quite different or there were methodological problems of a then unknown nature.

A. Proteolysis as a probelm

The two most serious problems in the isolation and purification of phytochrome are (1) the low concentrations of phytochrome generally found in plant tissues and (2) the marked susceptibility of phytochrome to proteolysis. The first problem is easily dealt with by utilisation of etiolated shoots for

---

* Abbreviations: bistrispropane, 1,3-bis[tris(hydroxymethyl)methylamino]propane; DEAE, diethylaminoethyl; EDTA, ethylenediaminetetraacetic acid; MOPS, morpholinopropanesulphonic acid; PAGE, polyacrylamide gel electrophoresis; PMSF, phenylmethylsulphonyl fluoride; SAR, specific absorbance ratio, A_{667}/A_{280} with phytochrome as Pr; SDS, sodium dodecyl sulphate; tris, tris(hydroxymethyl)aminomethane.

TECHNIQUES IN PHOTOMORPHOGENESIS
0-12-652990-6

phytochrome extraction since, at least in the case of oats, they contain about 50- to 100-fold more phytochrome per unit extractable protein than do light-grown shoots (Jabben and Deitzer, 1978; Hunt and Pratt, 1979a). The second problem, which led to the observations of Mumford and Jenner (1966) and probably accounted for the small monomer size reported by Correll *et al.* (1968b), has been more difficult to solve.

This more difficult problem of proteolysis was dealt with first when Gardner *et al.* (1971) demonstrated that phytochrome with approximately 120,000-d monomers could be extracted from both oats and rye. They further documented that both oat and rye phytochrome *in vitro* underwent proteolytic degradation to approximately 60,000-d polypeptides, with the proteases responsible being derived from the plant tissue itself (for discussions, see Briggs and Rice, 1972; Briggs *et al.*, 1972). For more than a decade thereafter it was often assumed (e.g., see Pratt, 1978) that the approximately 120,000-d polypeptide represented the native form of the molecule. Most of what we know about this pigment from investigations *in vitro* is derived from the study of this large form of phytochrome or its still photoreversible, 60,000-d degradation product (Pratt, 1982). Recently, however, Vierstra and Quail (1982a) have documented that even this larger form of the molecule has lost some 50 or so amino acids, an observation confirmed by Kerscher and Nowitzki (1982). It now appears safe to conclude that for oat phytochrome, at least, the native molecule has a monomer size of 124,000 d (Vierstra and Quail, 1982a).

In order to facilitate unambiguous and unbiased reference to these different sizes of phytochrome, the following terminology will be used. The photoreversible, 60,000 d, monomeric chromopeptide originally characterized by Mumford and Jenner (1966) will be referred to as 60-kd phytochrome. The larger, but still partially degraded form of the pigment, originally characterised by Rice and Briggs (1973), will be termed 120-kd phytochrome. Phytochrome of about 120 kd exists in solution as a 240-kd dimer (Hunt and Pratt, 1980). The still larger and presumably native form will be referred to as 124-kd phytochrome. The reader must be warned that these sizes refer to those obtained with oat phytochrome and may not reflect precise monomer estimates for phytochrome from other species. In addition, this initial proteolytic cleavage from 124 to about 120 kd (more precisely to a mixture of 118- and 114-kd polypeptides) has only been thoroughly described for phytochrome obtained from oat shoots (Vierstra and Quail, 1982a,b, 1983), although similar if not identical cleavage patterns have also been observed for phytochrome from other species (Vierstra *et al.*, 1984. The reader must therefore be further cautioned that our present understanding of phytochrome *in vitro* and problems associated with its purification in a native state are now in a state of rapid flux.

It is apparent that both 60-kd and 120-kd phytochrome are artifactual forms of the molecule. Emphasis must therefore be given to working with the 124-kd species. Two protocols for purification of 124-kd phytochrome are available at the time of this writing (Vierstra and Quail, 1983; Litts *et al.,* 1983). The protocol developed by Vierstra and Quail will be described in this chapter although it is obviously too early to determine which, if either, of these methods will eventually find the most widespread application. Additionally, purification of 120-kd phytochrome will be presented here for two reasons. (1) The procedures used for its purification have already found wide application in several laboratories and thus one can have confidence in them. (2) It will be necessary for at least the next several years, during which time 124-kd phytochrome will be more fully characterised, to work also with the 120-kd degradation product in order to compare new information with that previously obtained. Only by such comparative work can one be certain whether reported differences may be ascribed to different monomer sizes of this chromoprotein or to other factors.

B. Units

Several expressions have been used to describe phytochrome quantity and purity, sometimes leading to difficulties when comparing data reported by different laboratories. In particular, the use of arbitrary units, which precludes rigorous comparison of data, should be avoided. The following expressions are recommended. One unit of phytochrome is that quantity, which when dissolved in 1.0 ml, has an absorbance in the Pr form of 1.0 at 667 nm for a 1-cm light path. For 120-kd oat phytochrome, 1 unit is approximately 1.2 mg as determined by quantitative amino acid analysis (Roux *et al.,* 1982). A comparable determination for 124-kd oat phytochrome indicates that 1 unit is about 1.0 mg (Litts *et al.,* 1983). The simplest estimate of purity (see Section IV below) is the specific absorbance ratio (SAR), which is the ratio of A_{667}/A_{280} with the pigment as Pr. SAR is analogous to specific activity as used to estimate the purity of an enzyme. Since visible extinction of phytochrome is readily changed by a variety of conditions, however, these spectral measurements must be used with caution (for discussions, see Briggs and Rice, 1972; Pratt, 1983).

Other units for expression of phytochrome quantities may be converted to that used here if sufficient information is given (see Pratt, 1983, for discussion). For example, based upon absorbance spectra for highly purified phytochrome, 1 unit/ml would be expected to give a $\Delta\Delta A$ of about 1 or 0.5 for a 1-cm light path when measured at 667 versus 724 nm or 724 versus 800 nm, respectively. Except for assay of crude or minimally purified plant extracts, in which case phytochrome quantities are low and interfering pigments may be present, it is generally easiest to make single-wavelength measurements rather than dual-wavelength photoreversibility measurements.

II. PURIFICATION METHODS

It is sometimes argued that phytochrome purification is a difficult task. The opposite perspective will be presented here. Neither highly specialised instrumentation and laboratory facilities nor unusual dedication is required. The methods described here may be used, without assistance, by a single investigator under relatively comfortable working conditions and without the need to work longer than 8–10 h in a single day.

A. Laboratory requirements

The only special laboratory requirement is the need to work under green lighting that is not appreciably absorbed by phytochrome. Generally, the fluence rate of green light used in most laboratories in which phytochrome-mediated morphogenesis is being investigated is unnecessarily low. Sylvania 40-W green fluorescent bulbs wrapped with one layer each of Roscolene No. 874 (medium green) and Roscolene No. 877 (medium blue-green) filters (Rosco Laboratories Inc., Port Chester, New York 10573) (Fig. 1) provide enough light to work comfortably if one lamp is used per 5 m² of laboratory area and if the lamps are uniformly distributed on the ceiling. Such an illumination level does not result in measurable phytochrome photoconversion even after several hours.

B. Brushite and hydroxyapatite preparation

Add 4 l of 1.0 M $CaCl_2$ to an approximately 70-l plastic container. While stirring continuously with a paddle, add 20 l each of 1.0 M $CaCl_2$ and 1.0 M K_2HPO_4 at equal flow rates of about 4 ml/min (Siegelman *et al.,* 1965). When addition is complete, remove the paddle and allow the contents of the container to settle and remain undisturbed for 1 week. At the end of the week, decant the supernatant and wash the brushite crystals five times with distilled water by first mixing with water, then permitting the crystals to settle, and finally decanting the supernatant. Store the washed brushite at about 4°C, at which temperature it may be kept for at least 1 year. Smaller quantities of brushite may be prepared by scaling down the above procedure and reducing the flow rate proportionately.

To prepare the brushite for use, mix overnight at room temperature

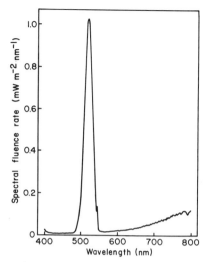

Fig. 1. Spectral fluence rate of a green fluorescent lamp (Sylvania F40G) covered with one layer each of Roscolene No. 874 (medium green) and Roscolene No. 877 (medium blue-green) filters (Rosco Laboratories Inc., Port Chester, New York 10573) measured at a distance of 1.2 m from the lamp. Below 480 nm and above 550 nm the curve is at baseline (i.e., zero output) level, which is not a flat line with the instrument used. (Measurement courtesy of K. H. Norris.)

3920 ml of H_2O, 314 g of K_2HPO_4, and 2000 ml of a 1:1 mixture of settled brushite in H_2O. Different quantities of brushite may be "charged" in this way by adjusting all quantities in proportion. For immediate use, this charged brushite is poured into a column and equilibrated with 3 l or more of 15 mM potassium phosphate, 14 mM 2-mercaptoethanol, pH 7.8. If the charged brushite is not to be used immediately, wash it two to four times by resuspension in 15 mM potassium phosphate, pH 7.8, followed by settling and decanting of the supernatant. This charged and washed brushite may be kept for several months at 4°C. For use, pour the brushite into a column and equilibrate with two column volumes of 15 mM potassium phosphate, 14 mM 2-mercaptoethanol, pH 7.8. Approximately 1 l of settled bed volume is used for each 2 kg monocotyledonous shoots extracted or for each 1 kg of dicotyledonous tissue extracted.

To make hydroxyapatite (Siegelman *et al.,* 1965), first prepare (but do not charge) a known quantity of brushite as described above. A 4-mol batch is convenient and may be made by reducing the starting quantities of 1.0 M $CaCl_2$ and 1.0 M K_2HPO_4 to 20% of those given above, and by reducing the flow rate from 4 to 0.8 ml/min. From a 45% (w/v) KOH stock solution, add 4 mol (equal to the number of mol of brushite prepared) of KOH while mixing well. Allow to stand for 3 days at room temperature. The pH should remain above 11.7. Wash five times with water as described above. The product is ready for use and may be kept at 4°C for several years. Do not use a magnetic stirring bar during preparation since this causes breakage of the crystals, which results in low flow rates.

C. Tissue

The most abundant source of phytochrome is etiolated tissue, which is therefore preferentially used for extraction. It is most convenient to freeze the tissue after harvest. Tissue may be stored for at least 1 year at −20°C. While 124-kd oat phytochrome may be purified from frozen oat shoots, Kerscher and Nowitzki (1982) report that freezing and thawing of plant tissues liberates higher protease levels. Use of freshly harvested tissue may therefore be advisable when purifying 124-kd phytochrome.

D. Purification of 120-kd phytochrome

Details of the methods used for conventional purification of phytochrome will be given per kg of tissue extracted. The protocol may be adjusted as desired for different quantities of tissue. Extraction of 4 kg at one time normally represents a practical upper limit because as additional tissue is extracted and additional phytochrome is obtained, this gain in phytochrome is substantially negated by proteolysis to 60 kd of aleady extracted pigment. The procedures described here have been developed from earlier protocols (Hopkins and Butler, 1970; Rice et al., 1973; Pratt, 1973). Throughout, maintain phytochrome as close to 0°C as possible, work as rapidly as possible to minimise proteolysis, and maintain phytochrome in the red-absorbing form.

(1) Extraction

Extract 1 kg of frozen tissue in a 4-l Waring blender with 630 ml of H_2O (room temperature), 70 ml of 0.5 M tris (pH adjusted to 8.5 at 25°C), 10 ml of 2-mercaptoethanol, and a small amount of any antifoaming agent. Tissue is added to the blender rapidly at first and then slowly while operating at low speed to prevent freezing of the extraction mixture while at the same time keeping it near the freezing point. When addition of tissue is complete, extract at top speed for 60 sec.

Rapidly filter the extract through 3 layers of cheesecloth. Depending on the tissue extracted, the pH of the filtered extract should be between 7.4 and 7.8. If necessary, add an appropriate quantity of 2.5 M tris to the extract buffer before tissue homogenization. Do not permit the pH to fall below 7.0 at any time.

Add 0.5 ml of $3M$ $CaCl_2$/100 ml of the filtered extract. Since this addition decreases the pH, add 2.5 M tris to bring the pH back to 7.8. Centrifuge the extract for 15 min at maximum speed (about 30,000 g) in a 6×250-ml rotor in a refrigerated centrifuge. Discard the pellet.

For each 100 ml of supernatant add 1.0 ml of 0.5 M K_3PO_4 (pH increases slightly) and then 1.0 ml of 0.5 M EDTA, pH 7.8 (pH decreases slightly). Adjust pH to 7.8 with 2.5 M tris if it is much below this value.

(2) Brushite chromatography

Add the prepared extract to an equilibrated brushite column (see Section II. B.). A 130-mm diameter column operated at a flow rate of 3 l/h, is convenient when the bed volume is 1 l or more. After the extract has entered the column, wash with 15 mM potassium phosphate, 14 mM 2-mercaptoethanol, pH 7.8. When the column eluate exhibits negligible A_{280} (or more conveniently when it has turned from yellow to colourless), elute with 60 mM potassium phosphate, 14 mM 2-mercaptoethanol, pH 7.8. After a volume of 60 mM buffer equal to 75% of the bed volume has passed through the column, begin to monitor fractions either by photoreversibility assay (see Chapter 6) or absorbance at 667 nm. Collect fractions that are about 10% of the bed volume in size and pool those containing phytochrome.

(3) Ammonium sulphate fractionation

Precipitate phytochrome by rapid addition of 20 g/100 ml of finely powdered $(NH_4)_2SO_4$. Monitor the pH during $(NH_4)_2SO_4$ addition and keep it near 7.8 with 2.5 M tris. Stir for about 10 min and ensure that the $(NH_4)_2SO_4$ is completely dissolved.

Centrifuge for 15 min at 15,000 g. Discard the supernatant since it contains 60-kd phytochrome, which is not precipitated by the relatively low $(NH_4)_2SO_4$ concentration used here (Correll and Edwards, 1970). Dissolve the pellet in 15 ml of 5 mM bistrispropane, 10 mM KCl, pH 7.4 at 4°C.

Store the dissolved pellet overnight (or for a few days if desired) at −20°C. Elapsed time to this stage of purification should be about 7–12 h.

(4) DEAE chromatography

Prepare a DEAE–cellulose column in advance. Pour sufficient DEAE–cellulose slurry (mEq/g) in water into a 24-mm-diameter column to form, at a flow rate of 200 ml/h, a settled bed volume of 70 mm/kg tissue extracted. Wash the column sequentially with 400 ml of saturated NaCl, 400 ml of

0.1 N KOH, and 500 ml H_2O. Equilibrate the DEAE–cellulose with 5 mM bistrispropane, 0.1 M KCl, pH 7.4 at 4°C. Equilibrium is reached when the pH of the eluate is stable at 7.4. Finally, equilibrate with 5 mM bistrispropane, 10 mM KCl, pH 7.4 at 4°C. When the pH of the eluate is stable and at 7.4, the column is ready for use.

Thaw the frozen sample and clarify by centrifugation for 15 min at 40,000 g. Rapidly remove residual $(NH_4)_2SO_4$, which prevents binding of phytochrome to DEAE, by passage of the sample at 300 ml/h through a 24 $\times$ 550 mm Sephadex G-25 column that has previously been equilibrated with 5 mM bistrispropane, 10 mM KCl, pH 7.4 at 4°C. Avoid dialysis, which is too slow and would thus permit proteolysis of phytochrome.

Apply the desalted sample to the DEAE–cellulose column, which is operated throughout at a flow rate of 200 ml/h. After sample addition is complete, wash the column with 1 column volume of 5 mM bistrispropane, 10 mM KCl, pH 7.4 at 4°C. Elute phytochrome with a nonlinear gradient of KCl. The gradient is formed by connection of the inlet of the column to a closed reservoir containing 250 ml of 5 mM bistrispropane, 10 mM KCl, pH 7.4 at 4°C. This closed reservoir is in turn connected to an open reservoir containing 500 ml of 5 mM bistrispropane, 300 mM KCl, pH 7.4 at 4°C. Degraded, 60-kd phytochrome elutes at a lower ionic strength than does 120-kd phytochrome (Correll and Edwards, 1970). Hence, if the elution profile indicates two populations of phytochrome eluting from the DEAE column, pool only the population that elutes last.

Concentrate the phytochrome-containing pool from the DEAE column by addition of an equal volume of saturated $(NH_4)_2SO_4$ (pH adjusted to 7.8 with tris). Collect the precipitate by centrifugation and dissolve in sufficient 0.1 M sodium phosphate, pH 7.8, to give a phytochrome concentration near 1.5 unit/ml. Clarify by centrifugation.

(5) Gel filtration chromatography

Chromatograph the clarified, redissolved pellet through a suitable gel filtration column. A 35 $\times$ 800-mm Sephadex G-200 or Sephacryl S-300 column works well for this purpose. Although Bio-Gel P-300 had also been used successfully for filtration of 120-kd phytochrome (Pratt, 1973), newly manufactured Bio-Gel P-300 has different characteristics such that it is no longer suitable for this application (L. H. Pratt, unpublished observations).

The sample should be added to the gel exclusion column at the end of the second day. The flow rate of the column should be adjusted so that phytochrome is eluted early on the third day, about 12–18 h after addition to the column. The 120-kd phytochrome elutes from G-200 at a volume about 20% greater than the void volume.

Concentrate the phytochrome pool by addition of an equal volume of saturated $(NH_4)_2SO_4$, pH 7.8. Collect the precipitate by centrifugation, dissolve it in 0.1 M sodium phosphate, pH 7.8, to give a phytochrome concentration of about 1 unit/ml, and clarify by centrifugation.

Store the clarified sample at or below $-70°C$.

(6) Immunoaffinity chromatography

The only special requirement for immunopurification of phytochrome is a supply of specific antibodies to phytochrome (see Chapter 9). The immunopurification protocol is circular in that the final product (purified phytochrome) is needed for its initiation (Fig. 2). The protocol described here (Hunt and Pratt, 1979b; Cordonnier and Pratt, 1982) assumes that conventionally purified phytochrome and specific antiserum to phytochrome are available. Once the procedure has been established, however, immunopurified phytochrome may be used in place of conventionally purified phytochrome.

(a) Phytochrome immobilization

Conventionally purified 120-kd phytochrome is coupled to CNBr-activated agarose (e.g., Pharmacia 74301 or Sigma C-1942). This phytochrome need

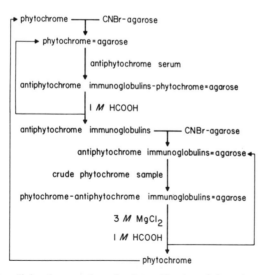

Fig. 2. Flow chart outlining the procedures for the purification of phytochrome by an immunoaffinity method. Thick arrows follow the steps needed to purify antiphytochrome immunoglobulins and phytochrome while the thin arrows indicate how various components may be utilised repeatedly. Covalent bonds are indicated by double lines, immunochemical bonds by single lines.

not be completely purified if the antiserum to phytochrome, which is used in the following step, is monospecific.

Add solid NaCl to a phytochrome-containing solution in 0.1 M sodium phosphate, pH 7.8, to make it 0.5 M in NaCl. Adjust the pH to 8.3 with 1 N NaOH. For each 1 mg of protein in the phytochrome sample add 120 mg dry weight of CNBr-activated agarose that has been previously washed extensively with 1 mM HCl. Mix gently overnight at 4°C by rotating end over end. Collect the beads by centrifugation for 2 min at 1500 g and discard the supernatant (after checking for nonimmobilized phytochrome if desired). For each ml of swollen agarose, add 4 ml of 0.1 M monoethanolamine, adjusted to pH 9.0 with HCl, to block remaining reactive sites on the agarose. Mix overnight at 4°C as above and then transfer the agarose to a plastic syringe that has a volume about five to ten times greater than the volume of the hydrated agarose. The syringe is then treated as a small chromatographic column. Wash the agarose, which now contains coupled phytochrome, with about 10 column volumes of 25 mM MOPS–tris, 5 mM EDTA, pH 7.8 (prepared by mixing 25 mM MOPS, 5 mM EDTA with 25 mM tris, 5 mM EDTA). Store the column away from white light at 4°C in 25 mM MOPS–tris, 5 mM EDTA, pH 7.8, 0.02% NaN$_3$. With care, this immobilized phytochrome may be used an indefinite number of times.

(b) Immunopurification and immobilization of antiphytochrome immunoglobulins

It is not feasible to couple to agarose crude antiserum to phytochrome because the antibodies to phytochrome represent at best about 1–2% of the protein in the serum. Coupling of crude serum would result in an immunoaffinity column too large for practical use. Hence, it is necessary to purify the antiphytochrome immunoglobulins from the serum.

Rinse the column containing immobilized phytochrome with 25 mM MOPS–tris, 5 mM EDTA, pH 7.8, to remove the NaN$_3$. Add enough antiserum to phytochrome almost to fill the syringe. Stopper the syringe and gently mix the serum with the agarose for about 1 h. Permit the agarose to settle in the syringe, drain and collect the serum, and then wash the agarose with 1 M NaCl, 10 mM MOPS–tris, pH 7.8, until the A_{280}^{1cm} of the eluate is below 0.03 (50–100 ml are needed for a column containing 10 ml of agarose). Equilibrate the agarose with 25 mM MOPS–tris, 5 mM EDTA, pH 7.8. Attach the outlet of the syringe to the inlet of a Sephadex G-25 column (bed volume about 10–20 times that of the agarose bed volume), which is operated with a minimal volume of liquid above the G-25. The G-25 should be equilibrated with 25 mM MOPS–tris, 5 mM EDTA, pH 7.8.

Layer over the agarose 1 column volume of 1 N HCOOH. Percolate the HCOOH into the agarose and follow immediately with 25 mM MOPS–tris,

5 mM EDTA, pH 7.8.   The HCOOH breaks the bonds between the immobilized phytochrome and the antiphytochrome immunoglobulins.   The immunoglobulins then enter the G-25 column within which they are rapidly
separated from the HCOOH to minimise denaturation.   Monitor the eluate
of the G-25 column at 280 nm and collect the immunoglobulins, which will
elute at the void volume.

Precipitate the immunoglobulins by addition of an equal volume of saturated $(NH_4)_2SO_4$, pH 7.8.   Collect the precipitate by centrifugation, dissolve
in 0.1 M sodium phosphate, pH 7.8, to give an immunoglobulin concentration of about 1 mg/ml, and clarify by centrifugation.   Store at $-20°C$.
Immunoglobulin concentration is determined by assuming $E_{280}^{1\%, 1\,cm} = 13.6$
(Sober, 1970).

The once-adsorbed serum that was collected earlier may be readsorbed as
above since it is unlikely that all of the antibodies to phytochrome were
removed.   If necessary, the serum may be concentrated by precipitating with
half-saturation $(NH_4)_2SO_4$ and redissolving the collected precipitate in
0.1 M sodium phosphate, pH 7.8.

Purified antiphytochrome immunoglobulins are then coupled to CNBr-
activated agarose and stored in a syringe exactly as described above for
coupling and storage of phytochrome.

For storage, wash the agarose column containing immobilized phytochrome and the G-25 column with 25 mM MOPS–tris, 5 mM EDTA, pH
7.8, 0.02% NaN_3.

(c) Immunopurification of phytochrome

Phytochrome may be immunopurified from a completely crude extract or
from an extract at any stage of purification.   Use of brushite-purified phytochrome represents the best compromise between maximum yield and
minimum contamination.   To utilize brushite-purified phytochrome, follow the conventional purification protocol given above up to and including
the brushite chromatography step and the initial 20 g/100 ml $(NH_4)_2SO_4$
fractionation.   Dissolve the collected $(NH_4)_2SO_4$ pellet in 0.1 M sodium
phosphate, pH 7.8, rather than the bistrispropane buffer.   Clarify the redissolved pellet by centrifugation.

Add the phytochrome-containing solution (about 0.5 units of phytochrome per mg of coupled immunoglobulins are satisfactory although the
ratio may vary over a wide range) to the immobilized immunoglobulins.
Mix gently for 15–30 min or percolate the solution slowly through the
agarose.   Elute the phytochrome-containing solution, which may be saved
and readsorbed later if desired.

Wash the agarose with 1 M NaCl, 10 mM MOPS–tris, pH 7.8, until the
$A_{280}^{1\,cm}$ of the eluant is less than 0.03 (for 10 ml of agarose, 50–100 ml is

required). Then equilibrate the agarose with 25 mM MOPS–tris, 5 mM EDTA, pH 7.8, and connect the outlet to a G-25 column as described above for immunoglobulin purification. Elute phytochrome from the column with 3 M MgCl$_2$ (titrated at 2°C to pH 7.8 with solid tris) as described above for immunoglobulin elution with 1 N HCOOH. Phytochrome also elutes from the G-25 column at its void volume. Phytochrome elution is monitored at 667 nm.

Concentrate the eluted phytochrome by addition of one-half its volume of saturated (NH$_4$)$_2$SO$_4$, pH 7.8, which precipitates 120-kd phytochrome and leaves 60-kd phytochrome in solution (Correll and Edwards, 1970). Collect the precipitate by centrifugation, dissolve it in 0.1 M sodium phosphate, 1 mM EDTA, pH 7.8, to give a phytochrome concentration of near 1 unit/ml, and clarify the redissolved sample by centrifugation. Store the purified phytochrome, which should be undenatured, at −70°C or below.

If desired, add the original phytochrome-containing solution back to the column to adsorb phytochrome a second time (Cordonnier and Pratt, 1982). Repeat the wash and 3 M MgCl$_2$-elution procedures just described. Continue to adsorb and elute phytochrome with 3M MgCl$_2$ until less than 10% of the initial phytochrome quantity remains.

Wash the agarose and G-25 to remove MgCl$_2$. Elute the agarose exactly as described above except that 1 N HCOOH is used rather than 3 M MgCl$_2$. Monitor the eluant of the G-25 column at 280 nm since phytochrome will have been irreversibly denatured and partially bleached at 667 nm by the brief 1 N HCOOH exposure. Concentrate phytochrome by precipitation with (NH$_4$)$_2$SO$_4$ as described above for the MgCl$_2$-eluted phytochrome. Do not attempt to clarify the redissolved phytochrome as much of it, having been denatured, will no longer go into solution. Although denatured, this phytochrome is nevertheless as pure as that eluted by MgCl$_2$ and may therefore be used for a variety of purposes, including amino acid analysis and injection into rabbits for raising more antibodies.

Regenerate the agarose containing immobilised immunoglobulins by first performing mock elutions with a second 1 N HCOOH treatment and then with a 3 M MgCl$_2$ treatment. No attempt is made to recover any phytochrome that might be eluted. The agarose and G-25 are finally washed with 25 mM MOPS–tris, 5 mM EDTA, pH 7.8, 0.02% NaN$_3$, for storage. This agarose column, which contains immobilized immunoglobulins, may also be used an indefinite number of times.

(7) Comments

Phytochrome purified by the methods described above is the partially degraded, 120-kd species. It must be emphasized that without working quickly and maintaining phytochrome-containing samples near 0°C, it is not possible to obtain this large form from many plant tissues. With appropriate caution, however, it is possible to use either the conventional (brushite, DEAE–cellulose, gel filtration) or the immunoaffinity procedure to obtain 120-kd phytochrome from a variety of species (Pratt, 1973; Hunt and Pratt, 1979b; Cordonnier and Pratt, 1982). The procedures thus have general applicability. Typical results to be expected from these two procedures when purifying oat phytochrome are given in Tables I and II.

Phytochrome produced by the conventional procedure is typically less than 50% pure. It is suitable for a wide variety of applications but must be used with the knowledge that such preparations still contain sufficient endoprotease activity that the pigment will likely degrade further to the 60-kd size over a period of many hours to a few days, even at 0°C. By contrast, immunopurified phytochrome approaches 100% purity and is generally free of protease activity. It can therefore be used for applications that require long incubations. As noted, while the $MgCl_2$-eluted phytochrome appears

Table I. Flow chart for conventional purification of 120-kd phytochrome from 4 kg of etiolated oat shoots.

Day	Fraction	Volume (ml)	Phytochrome (units)	SAR	Yield (%)
1	Crude extract after clarification	5360	ND[a]	ND	—
1	Brushite pool before $(NH_4)_2SO_4$ fractionation	850	78[b]	0.024	100
2	After $(NH_4)_2SO_4$ fractionation and immediately prior to G-25	90	57	0.065	73
2	DEAE–cellulose pool I[c]	50	13.8	0.20	17.6
2	After $(NH_4)_2SO_4$ concentration	5	11.3	0.21	14.5
3	Sephadex G-200 pool (centre cut)	34	5.6	0.38	7.2
	(side cut)	26	2.6	0.25	3.4
3	After $(NH_4)_2SO_4$ concentration				
	(centre cut)	5.0	4.1	0.37	5.3
	(side cut)	2.5	1.8	0.27	2.3
	DEAE–cellulose pool II[c]	91	15.0	0.10	19.2
	After $(NH_4)_2SO_4$ concentration, Sephadex G-200 chromatography, and final $(NH_4)_2SO_4$ concentration	4.6	5.3	0.19	6.8

[a] ND = not determined.

[b] Value estimated from photoreversibility measurements.

[c] Phytochrome eluting from the DEAE–cellulose column was divided into two pools, a centre fraction labelled I and the combined side fractions labelled II. Pool II was frozen and chromatographed later through the Sephadex G-200 column with the results indicated. Total yield following Sephadex G-200 chromatography of both pools was 11.2 units of 120-kd phytochrome (yield = 14.4%). Data are from P. H. Quail and L. H. Pratt.

Table II. Flow chart for the purification of 120-kd phytochrome from 2 kg of etiolated oat shoots by the immunoaffinity method.

Fraction	Volume (ml)	Phytochrome (units)	SAR	Yield (%)
Following brushite chromatography and (NH$_4$)$_2$SO$_4$ fractionation	39	24	0.051	100
MgCl$_2$ pools after (NH$_4$)$_2$SO$_4$ concentration	5.0	6.4	0.77	27
HCOOH pool after (NH$_4$)$_2$SO$_4$ concentration	1.0	11[a]	ND[b]	46

[a] Value estimated by A_{280} because phytochrome was bleached in red spectral region by HCOOH exposure.
[b] ND = not determined because phytochrome was bleached in red region. The immunoaffinity column contained 52 mg of rabbit antiphytochrome immunoglobulins. The brushite sample was adsorbed and eluted twice by 3 M MgCl$_2$ prior to HCOOH elution.

undamaged, the HCOOH-eluted phytochrome is denatured. Nevertheless, it is as pure as that eluted by MgCl$_2$. If the starting phytochrome sample is adsorbed repeatedly as described, the immunopurification yield can be quite high (a total of 73% in Table II).

When purifying phytochrome from material other than oats, modifications of the above procedures may be helpful. Addition of insoluble polyvinylpyrrolidone to the extraction medium does not interfere with the early stages of phytochrome purification and can help minimise the effects of phenolics when they are abundant. If clogging of the brushite column is a problem, addition of a layer of Miracloth to the three layers of cheesecloth during the initial filtration and/or filtration of the crude extract through celite in a large Buchner funnel after centrifugation will improve the flow rate. If difficulty is experienced with phytochrome passing through the brushite column without binding, prior fractionation of the crude extract with (NH$_4$)$_2$SO$_4$ and/or use of 5 mM rather than 15 mM potassium phosphate for initial equilibration of the brushite column should help. If the brushite is initially equilibrated with 5 mM potassium phosphate and then washed with 5 mM potassium phosphate after sample addition, the column should be washed with 15 mM potassium phosphate prior to elution of phytochrome with 60 mM potassium phosphate.

E. Purification of 124-kd phytochrome

Methods for the purification of this apparently native form of phytochrome are only now being developed. The method chosen for presentation here, which was developed by R. D. Vierstra and P. H. Quail (1983), is still being refined.† Application to tissues other than etiolated oat shoots will probably require at least minor modification. Consequently, the following protocol should be considered only as a starting point.

When purifying 124-kd phytochrome by this or any other procedure, the following points must be kept in mind. Proteolytic cleavage of 124-kd to 120-kd phytochrome (1) is relatively specific for Pr (Vierstra and Quail, 1982a; Kerscher and Nowitzki, 1982), (2) is inhibited by PMSF, and (3) occurs rapidly in crude extracts, even near 0°C (Vierstra and Quail, 1982a,b). (4) The presence of substantial quantities of $(NH_4)_2SO_4$ (e.g., 70 mM) and ethylene glycol (e.g., 25% v/v) are reported to minimise proteolysis (Vierstra and Quail, 1983; Litts et al., 1983). It is therefore advisable to emphasise further the need to work rapidly and to maintain phytochrome as close to 0°C as possible. In addition, it is recommended that PMSF, ethylene glycol, and $(NH_4)_2SO_4$ be present in all buffers, and that phytochrome be purified as Pfr, at least until proteolysis is no longer a problem. PMSF should always be added immediately prior to use from a 0.2 M stock solution in isopropanol. The procedure described below assumes extraction from 1 kg of tissue. Other quantities of tissue may be used by appropriate scaling of the procedure.

(1) Extraction and poly(ethyleneimine) precipitation

If freshly harvested tissue is to be extracted, chill seedlings to 4°C and irradiate with saturating red light immediately prior to extraction. If frozen tissue is to be extracted, irradiate the crude extract with saturating red light immediately after homogenization. Refer to Chapter 2 for suggestions concerning appropriate light sources. If desired, the unfiltered output of Sylvania Gro-lux fluorescent lamps is adequate.

Homogenise 1 kg of tissue in a 4-l Waring blender with 750 ml of 0.1 M tris-Cl, 0.14 M $(NH_4)_2SO_4$, 10 mM EDTA, 50% (v/v) ethylene glycol, 20 mM sodium bisulfite (freshly added), pH 8.3 at 4°C. Add PMSF to 4 mM immediately prior to use. Keep the extract as close to 0°C as possible. If frozen tissue was extracted, irradiate the crude homogenate with red light while stirring. Filter the extract through three layers of cheesecloth.

Add per litre 10 ml of a 10% (v/v) poly(ethyleneimine) solution, which is prepared as follows. Mix 100 ml of poly(ethyleneimine) with 700 ml of water. After thorough mixing, cool to 4°C and adjust the pH to 7.8 with 3 N HCl (about 160 ml are required). Allow to stand overnight at 4°C. If necessary, readjust the pH to 7.8 and then bring to final volume. After

---

† I would like to thank Drs. Richard Vierstra and Peter Quail for making the details of their protocol for purification of 124-kd phytochrome available to me in advance of publication.

addition of poly(ethyleneimine) to the crude extract, stir for 15 min, and then centrifuge for 15 min at 25,000 g. Add 10 ml/l of the 0.2 M PMSF stock solution to the supernatant.

(2) Ammonium sulphate fractionation

Add rapidly 250 g finely powdered $(NH_4)_2SO_4$ for each litre of the clarified extract. Keep pH near 7.8 by addition of 2.5 M tris. Stir for 10 min. After ensuring that the $(NH_4)_2SO_4$ is completely dissolved, centrifuge for 15 min at maximum speed (about 30,000 g)in a 6 × 250-ml rotor. Discard the supernatant. Dissolve the pellet in 50 mM tris-Cl, 5 mM EDTA, 14 mM 2-mercaptoethanol, 2 mM PMSF, 25% (v/v) ethylene glycol, pH 7.8 at 4°C. Add sufficient buffer such that the final $(NH_4)_2SO_4$ concentration is 70 mM. To do this, prepare as a standard the buffer with 70 mM $(NH_4)_2SO_4$ and determine its conductivity. Dilute the dissolved pellet with the buffer minus the $(NH_4)_2SO_4$ until the conductivity of the sample is the same as that of the standard. If the pellet is well drained of excess $(NH_4)_2SO_4$, 170 ml of buffer should be sufficient. Alternatively, it should be possible to pass the dissolved sample through a Sephadex G-25 column equilibrated with the buffer containing 70 mM $(NH_4)_2SO_4$, although this approach has not been tested. Clarify the redissolved sample by centrifugation for 10 min at 50,000 g.

(3) Hydroxyapatite chromatography

Make hydroxyapatite as described in Section II. B above. Prepare a 24 × 100-mm column and equilibrate with 50 mM tris-Cl, 5 mM EDTA, 70 mM $(NH_4)_2SO_4$, 14 mM 2-mercaptoethanol, 25% (v/v) ethylene glycol, pH 7.8 at 4°C. The column may be considered equilibrated when the pH of the eluant is stable at 7.8. Add the clarified phytochrome sample to the column, which is run throughout at a flow rate of 1.3 ml/min. After completion of sample addition, wash the hydroxyapatite with 2 column volumes (90 ml) of 50 mM tris-Cl, 5 mM EDTA, 14 mM 2-mercaptoethanol, pH 7.8 at 4°C. Wash further with 5 column volumes of 5 mM potassium phosphate, 5 mM EDTA, 14 mM 2-mercaptoethanol, pH 7.8. Elute phytochrome from the column with 20 mM potassium phosphate, 5 mM EDTA, 14 mM 2-mercaptoethanol, 2 mM PMSF, pH 7.8. Fractions of 5 to 10 ml are convenient. Since phytochrome is in the Pfr form, monitor the eluant at 280 and 730 nm. Although these wash and elution conditions may be considered typical, each batch of hydroxyapatite must be tested individually to ensure that optimal potassium phosphate concentrations are used.

An alternative procedure for eluting phytochrome from the hydroxyapatite column is to use a linear gradient made with 150 ml each of 5 mM and

200 mM potassium phosphate, pH 7.8. The two gradient buffers should also contain 5 mM EDTA and 14 mM 2-mercaptoethanol. By using a gradient for elution, problems arising from variations among batches of hydroxyapatite are effectively eliminated (unpublished data; Cordonnier, personal communication).

The original protocol, as described by Vierstra and Quail (1983), involves overnight elution of the hydroxyapatite column. Modifications to their procedure as described here involve for the most part decreases in incubation times. It is thus practical to complete purification through the hydroxyapatite column on the same day that the tissue is extracted. Time during which proteolysis might occur is correspondingly reduced. This abbreviated procedure has been found to give consistently satisfactory results (Shimazaki and Pratt, unpublished data; Cordonnier, personal communication).

If the purification is to terminate at this step or if the sample is to be purified further by immunoaffinity chromatography, pool fractions containing phytochrome and precipitate by adding 0.65 ml of 3.3 M $(NH_4)_2SO_4$, 50 mM tris, per ml. Stir for 5 min and collect the precipitate by centrifugation for 10 min at 50,000 g. Dissolve the pellet in 0.1 M sodium phosphate, 1 mM EDTA, pH 7.8, to give a phytochrome concentration close to 1 unit/ml. Clarify by centrifugation for 10 min at 50,000 g. Either purify further (Section II. E. 6.) or convert to Pr by irradiation with far-red light and store frozen at or below $-70°C$.

(4) Affi-Gel Blue chromatography
The use of Affi-Gel Blue for phytochrome purification was introduced by Smith and Daniels (1981). The procedure described here is as modified by Vierstra and Quail (1983).

Irradiate the phytochrome-containing pool from the hydroxyapatite column with far-red light. Precipitate with $(NH_4)_2SO_4$ and collect by centrifugation as described above. Dissolve the pellet in 20 ml of 10 mM potassium phosphate, 5 mM EDTA, 14 mM 2-mercaptoethanol, 2 mM PMSF, pH 7.8, and clarify by centrifugation for 10 min at 50,000 g. Add the clarified sample to a 12 × 260-mm Affi-Gel Blue (Bio · Rad, 100–200 mesh, 75–150 μm) column that has been equilibrated previously with this same buffer minus the PMSF. Alternatively, the hydroxyapatite pool may be added to this Affi-Gel Blue column without prior concentration. Operate the column throughout at a flow rate near 0.7 ml/min. Wash the Affi-Gel with 3 to 4 column volumes ($\sim$ 100 ml) of this same equilibration buffer. Wash next with 4 column volumes of 0.1 M potassium phosphate, 5 mM EDTA, 14 mM 2-mercaptoethanol, 1.0 M KCl, pH 7.8, and then with 2 column volumes of 0.1 M potassium phosphate, 5 mM EDTA, 14 mM 2-mercaptoethanol, 0.25 M KCl, pH 7.8. Elute phytochrome with 0.1 M potassium phosphate, 5 mM EDTA, 14 mM 2-mercaptoethanol, 0.25 M KCl, 10 mM flavin mononucleotide, pH 7.8. The eluant should be collected in 5 – 10 ml fractions and monitored at 280 and 667 nm. Suggestions for regenerating the Affi-Gel Blue are provided by the manufacturer.

Pool fractions that contain phytochrome and precipitate by addition of 0.9 ml of 3.3 M (NH$_4$)$_2$SO$_4$, 50 mM tris, per ml. Centrifuge for 10 min at 50,000 g and dissolve the pellet in 5 ml of 0.1 M potassium phosphate, 5 mM EDTA, 14 mM 2-mercaptoethanol, pH 7.8. Clarify by centrifugation for 10 min at 50,000 g.

(5) Gel filtration chromatography

Chromatograph the Affi-Gel Blue pool through a suitable gel filtration column prepared with, for example, Bio-Gel A-1.5M or Sephacryl S-300. A column size of 24 × 500 mm is suitable for a 5-ml phytochrome sample while a 35 × 800 mm column is suitable for a 10-ml sample. The flow rate is adjusted so that, for convenience, the sample may be applied late on one day and recovered the next morning (0.4 ml/min for the smaller column, 1.0 ml/min for the larger). Monitor the elution profile at 280 and 667 nm. Pool the desired phytochrome-containing fractions and concentrate by precipitation with (NH$_4$)$_2$SO$_4$ as described for the Affi-Gel Blue pool. Vierstra and Quail (1983) dissolve the final sample in 0.1 M potassium phosphate, 5 mM EDTA, 14 mM 2-mercaptoethanol, 5% (v/v) glycerol, pH 7.8, and store at −80°C. Phytochrome of this size is also stable in 0.1 M sodium phosphate, 1 mM EDTA, pH 7.8, if stored at or below −70°C.

(6) Immunoaffinity chromatography

The immunoaffinity procedure described above for the purification of 120-kd phytochrome (Section II. D. 6) works with 124-kd phytochrome as well and may be used without modification. In this instance, however, the MgCl$_2$-eluted phytochrome (in contrast to the 120-kd species) is spectrally altered, indicating that it has been denatured in some way (Vierstra and Quail, 1983). Immunoaffinity purified, 124-kd phytochrome is therefore suitable only for applications that permit the use of an altered form of the pigment, such as antibody production (see Chapter 9), amino acid analyses, or denaturing gel electrophoresis (Section III. A).

(7) Comments

As noted already, it is imperative that future characterisation of phytochrome be done with the 124-kd size since this is apparently the native form of the pigment (Vierstra and Quail, 1982a,b). This point is emphasised by the observations that 124-kd and 120-kd phytochrome differ significantly in several parameters already tested. For example, 124-kd phy-

Table III. Flow chart for the purification of 124-kd oat phytochrome from 1 kg of etiolated oat shoots.

Fraction	Phytochrome (units)[a]	SAR[b]	Yield (%)
Crude extract[c]	18.0	0.0046	100
Supernatant after poly(ethlyeneimine) precipitation	15.2	0.0046	85
After $(NH_4)_2SO_4$ fractionation	11.0	0.050	61
Hydroxyapatite pool after $(NH_4)_2SO_4$ concentration	5.0	0.42	28
Affi-Gel Blue pool after $(NH_4)_2SO_4$ concentration	3.4	0.69	19
Bio-Gel A-1.5M pool after $(NH_4)_2SO_4$ concentration	2.3	0.94	13

[a] Phytochrome units were calculated from reported photoreversibility values and the spectral data presented in Fig. 5 in Vierstra and Quail (1983).

[b] SAR values were extrapolated from reported specific activity values, the observation that the final fraction had an SAR of 0.94, and the assumption that SAR exhibits a linear relationship to specific activity.

[c] Phytochrome in the crude extract was measured after clarification by centrifugation of a small aliquot. Data are taken from Vierstra and Quail (1983).

tochrome has a blocked amino terminus while 120-kd phytochrome does not (Litts *et al.,* 1983; Vierstra and Quail, 1983); the larger form of the pigment exhibits spectral properties comparable to phytochrome *in vivo* while the 120-kd size does not (Vierstra and Quail, 1982b); and although 120-kd phytochrome exhibits substantial thermal reversion of Pfr to Pr, 124-kd phytochrome does not (Vierstra and Quail, 1983; Litts *et al.,* 1983).

A summary of what one would hope to obtain when purifying 124-kd phytochrome is given in Table III. Although poly(ethyleneimine) precipitation results in a loss of phytochrome without any purification, Vierstra and Quail (1983) include this step in the procedure for two reasons. Its use (1) facilitates clarification of the crude extract and (2) permits the subsequently obtained $(NH_4)_2SO_4$ precipitate to be dissolved more readily.

III. SIZE DETERMINATION

Since even purified phytochrome samples may contain protease activity, it is essential that the size of the pigment with which one is working be determined upon completion of experiments. In the absence of such a determination, the results must be interpreted with caution. If a sample is considered pure, the simplest and most rigorous approach is sodium dodecylsulphate–polyarcrylamide gel electrophoresis (SDS–PAGE). If a sample is impure, a less direct approach to the problem is required. In this instance, the simplest alternative would be to utilise spectral properties and/ or reversion behaviour of the pigment to indicate its status. This alternative approach, however, would not discriminate between 120-kd and 60-kd phytochrome.

A. Dodecylsulphate–polyacrylamide gel electrophoresis

The large monomer size of phytochrome requires use of polyacrylamide gels of relatively high porosity. Gels prepared from the following recipes, which yield a 3% stacking gel and a 7.5% separating gel, give good results (Laemmli, 1970). The recipes may be scaled up or down as needed. For the separating gel, use: 6 ml of 1.5 M tris-Cl, pH 8.8 at 25°C; 16.4 ml of 11% (w/v) acrylamide, 0.3% (w/v) N,N'-methylenebisacrylamide; 1.1 ml of H_2O; 0.24 ml of 10% (w/v) SDS; 0.24 ml of freshly prepared ammonium persulphate (10% w/v); 24 μl of N,N,N,N'-tetramethylethylenediamine. For the stacking gel, use: 1.4 ml H_2O; 0.75 ml of 0.5 M tris-Cl, pH 6.8 at 25°C; 0.82 ml of 11% (w/v) acrylamide, 0.3% (w/v) N,N'-methylenebisacrylamide; 30 μl of freshly prepared ammonium persulphate (10% w/v); 30 μl of 10% (w/v) SDS; 3 μl of N,N,N,N'-tetramethylenediamine.

Prepare sample buffer as follows: 2.8 ml of H_2O; 0.5 g of sucrose; 0.6 ml of 0.05% bromophenol blue; 0.5 ml of 10% (w/v) SDS; 0.5 ml of 0.5 M tris-Cl, pH 6.8 at 25°C; 0.1 ml of 2-mercaptoethanol. This buffer may be prepared in advance and stored at −20°C. Prepare samples for electrophoresis by mixing them with sample buffer (a ratio of one part sample to four or nine parts sample buffer is satisfactory) in small tubes and incubating for 5 min in a boiling-water bath.

Electrode buffer consists of 57.6 g of glycine, 12 g of tris, and 2 g of SDS made to a volume of 2 l with water. Details of gel preparation, electrophoresis, and gel staining vary with the apparatus used. To discriminate between 60-kd and larger sizes of phytochrome is straightforward because of the large size difference. To discriminate between 124-kd and 120-kd phytochrome is more difficult. Since 120-kd oat phytochrome is in fact a heterogeneous mixture of 118- and 114-kd polypeptides, it will appear as a distinct doublet following electrophoresis by this procedure (Vierstra and Quail, 1982a, 1983; Kerscher and Nowitzki, 1982; Litts *et al.*, 1983; M. -M. Cordonnier and L. H. Pratt, unpublished observations). By contrast, 124-kd phytochrome will migrate as a single molecular-weight species. For critical size analysis, it is best to have the 120-kd and/or 124-kd size of phytochrome available for direct comparison with samples containing phytochrome of unknown size.

B. Spectral properties and reversion behaviour

Since the Pfr form of 124-kd phytochrome has a longer wavelength absorbance maximum than degraded forms ($\sim$730 versus 724 nm) it is possible to make an estimate of the extent of degradation from an absorbance spectrum recorded after saturating red irradiation (Vierstra and Quail, 1982b). However, this simple approach is obviously not very sensitive. A better method is to utilise the more recent observation that 124-kd phytochrome exhibits virtually no thermal reversion of Pfr to Pr, even in the presence of 5 mM dithionite (Vierstra and Quail, 1983; Litts *et al.*, 1983). For example, Vierstra and Quail reported no more than about 5% reversion after 2 h at 3°C for 124-kd phytochrome in the presence of 5 mM dithionite whereas 120-kd phytochrome underwent more than 80% reversion in only 1 h under the same conditions. Since 60-kd phytochrome also exhibits rapid reversion in the presence of dithionite, one cannot determine from this information alone whether the rapidly reverting phytochrome is 120 or 60 kd in size.

IV. DETERMINATION OF PURITY AND HOMOGENEITY

Two expressions are needed to describe phytochrome *in vitro*. Because of its susceptibility to proteases present in plant extracts, it is possible for a sample to have a purity of 100% and yet not be homogeneous owing to proteolysis. Therefore, not only the purity of a sample but also its homogeneity must be determined.

The simplest estimate of purity is the specific absorbance ratio (SAR): A_{667}/A_{280} with phytochrome present as Pr. For phytochrome approaching 100% purity, this ratio has been reported to be 1.3 (see Pratt, 1978, for review) for 60-kd phytochrome, 0.87 for 120-kd oat phytochrome (Hunt and Pratt, 1979b), 0.77 for 120-kd rye phytochrome (Rice *et al.*, 1973), and 0.58 and 0.54 for what was presumably 120-kd zucchini and pea phytochrome, respectively (Cordonnier and Pratt, 1982). Values of 0.97 (Vierstra and Quail, 1983) and 0.88 (Litts *et al.*, 1983) have been reported for 124-kd oat phytochrome. Other values have been reported (see Pratt, 1982), but it is not certain whether they apply to 120- or 124-kd phytochrome. When using this ratio as an estimate of purity, it must be borne in mind not only that the limiting value is different for each size of phytochrome and for each plant species from which it is purified, but also that the ratio can vary considerably for a given purity because the visible extinction of phytochrome is readily decreased by a variety of conditions (for discussions, see Briggs and Rice, 1972; Pratt, 1978, 1983).

If a phytochrome sample is near 100% purity, the simplest estimate of homogeneity is SDS–PAGE. Since a completely pure sample (i.e., all protein present was coded for by mRNA for phytochrome) may be heterogeneous as a consequence of proteolysis, SDS–PAGE alone may underestimate the purity of a sample. Consequently, when a rigorous determination of both purity and homogeneity is required it is best to utilize both SDS–PAGE and an immunochemical method such as immunoelectrophoresis or immunoblotting (see Chapter 9). Assuming that polyclonal antibodies specific to 120- or 124-kd phytochrome are available, the latter method will detect even nonphotoreversible degradation products.

V. RADIOLABELLING

Phytochrome is readily labelled by a variety of methods because it contains a full complement of the 20 amino acids (Rice and Briggs, 1973; Hunt and Pratt, 1980). When it is essential that radiolabelled phytochrome be as "native" as possible, *in vivo* labelling is preferred. When minor alterations in the properties of the molecule are permissible, *in vitro* labelling is more convenient. It must be noted that *in vitro* labelling has so far been done only with 120-kd phytochrome.

A. *In vitro* labelling

(1) ^3H
Phytochrome has been labelled with ^3H by reductive methylation (Hunt and Pratt, 1979a) as described by Rice and Means (1971).

Precipitate about 120 μg (about 100 milliunits) of phytochrome (or of protein in an impure preparation) by addition of 20 g/100 ml $(NH_4)_2SO_4$. Collect the pellet by centrifugation and dissolve it in 100 μl of 0.2 M sodium borate, pH 9.0. Add 40 nmol of formaldehyde (10 μl of 4 mM formaldehyde freshly prepared from paraformaldehyde) and incubate for 5 min on ice.

The following steps must either be performed under an approved fume hood or with proper facilities for scavenging released $^3H_2$. Add 10-μl aliquots of $[^3H]NaBH_4$ (Research Products International, 5–15 Ci/mmol) at 2-min intervals while keeping the sample on ice. Two minutes after the last $[^3H]NaBH_4$ addition, add 800 μl of 0.25 M sodium phosphate, pH 7.0, to neutralise the sample and drive off excess ^3H as $^3H_2$. Incubate on ice for an additional 30 min.

Precipitate labelled protein by addition of an equal volume (1 ml) of saturated $(NH_4)_2SO_4$, pH 7.8. Collect the precipitate by centrifugation and wash it repeatedly by resuspension in 50% saturation $(NH_4)_2SO_4$, pH 7.8, followed by recentrifugation. When the supernatant from an $(NH_4)_2SO_4$ wash contains less than 10% of the precipitable radioactivity, dissolve the pellet in 900 μl of 0.1 M sodium phosphate, pH 7.8. Remove an aliquot for assay by SDS–PAGE and then add 100 μl of sheep serum as a protective agent. Store the labelled phytochrome at or below $-70\,°$C.

[3H]Phytochrome has been used primarily for radioimmunoassay of phytochrome (Hunt and Pratt, 1979a). Phytochrome specific activities of up to 150 cpm/ng (about 40 Ci/mmol phytochrome monomer) have been obtained by reductive methylation when 10 aliquots of [3H]NaBH$_4$ of 1 mCi each were used. [3H]Phytochrome samples have been effectively unchanged even after 1 year of storage at $-70\,°$C. Immunochemical comparison of labelled with unlabelled phytochrome has indicated that the molecule has been little perturbed by reductive methylation (Hunt and Pratt, 1979a).

(2) [125]I

Georgevich *et al.* (1977) have reported the iodination of phytochrome by the lactoperoxidase method to specific activities of 10–300 cpm/ng phytochrome. When higher specific activities were obtained, they found that the [125I]phytochrome was spectrally denatured. Hence 3H labelling would seem to be preferable for routine applications since comparable specific activities are obtained by reductive methylation. Also, phytochrome radiolabelled with 3H is usable over a much longer period of time both because of the longer half-life of 3H and because there is apparently less radiochemical damage to phytochrome from 3H.

To label phytochrome with 125I (Georgevich *et al.*, 1977), prepare the following reaction mixture in a vial kept on ice: 0.3 mg of phytochrome, 0.3 mCi Na125I (17 Ci/mg), and 6 μg of lactoperoxidase in 0.3 ml of 50 mM sodium phosphate, 1 mM EDTA, pH 7.8. Add three 15-μl aliquots of 0.1 mM H_2O_2 at 10-min intervals while keeping the vial on ice. Separate phytochrome from other components of the reaction mixture by filtration through a 10 × 150-mm Bio-Gel A-0.5M column equilibrated with 10 mM sodium phosphate, pH 7.8. The radiolabelled sample may then be assayed by SDS–PAGE followed by autoradiography or scintillation counting.

B. *In vivo* labelling

(1) 3H

To label phytochrome *in vivo* with 3H, add [3H]leucine in 1- to 2-μl drops to coleoptile tips of 3-day-old (at 25 °C) dark-grown oat seedlings (Boeshore and Pratt, 1980). After growth for an additional day, harvest the shoots and extract and purify phytochrome, using a scaled down version of one of the protocols given earlier.

Application of 1 mCi of [³H]leucine (62 Ci/mmole, Schwarz/Mann) to 100 plants yields about 50,000 cpm of ³H incorporated into phytochrome at a specific activity of about 0.4 cpm/ng (about 0.1 Ci/mmole phytochrome monomer).

(2) ³⁵S

To label phytochrome *in vivo* with ³⁵S, transfer 2-day-old dark-grown (at 25°C) oat seedlings, which have roots about 10 mm long, to 100-mm diameter polystyrene Petri dishes (Boeshore and Pratt, 1980; Pratt, 1980). To each dish add 3 ml of H_2O containing carrier-free [³⁵S]H_2SO_4. Grow the oats for an additional two days in darkness at 25°C. Harvest the shoots and extract and purify phytochrome, using a scaled down version of one of the preceding protocols.

Application of 15 mCi of ³⁵S to 50 plants, which gave about 2 g of shoots, yields undegraded [³⁵S]phytochrome with a specific activity of about 2 cpm/ng (about 0.2 Ci/mmole phytochrome monomer) (Pratt, 1980). A total of about 100,000 cpm of ³⁵S are incorporated.

(3) ³²P

Phytochrome has been labelled with ³²P *in vivo* by the following procedure (Quail *et al.,* 1978). Harvest 18 g of 4-day-old dark-grown (at 25°C) oat shoots. Incubate the shoots for 4 h at 25°C with shaking in 20 ml of 2 mM HCl containing 25 mCi [³²P]H_3PO_4. Following the incubation, extract phytochrome and partially purify by one of the methods described earlier.

Because of the massive isotope dilution of ³²P with endogenous P, 25 mCi of ³²P is required to obtain only a few thousand cpm incorporated into phytochrome. Nevertheless, the incorporation of ³²P does imply that phytochrome is a phosphoprotein (Quail *et al.,* 1978), an implication that is supported by direct spectrophotometric assay (Hunt and Pratt, 1980).

VI. STORAGE AND HANDLING

Phytochrome stored at or below −70°C in a simple buffer such as 0.1 M sodium phosphate, pH 7.8, is stable indefinitely. When stored at −20°C, however, phytochrome inevitably denatures irreversibly within a few weeks. It is likely that reports of phytochrome damage attributed to freezing and thawing might sometimes be better attributed to storage at insufficiently low temperatures.

Phytochrome is stable in the pH range of about 7.0 to at least 9.0. Near its pI, however, which for 120-kd oat phytochrome is near 6 (Hunt and Pratt, 1979b) and for zucchini phytochrome is near 6.5 (Cordonnier and Pratt, 1982), 120-kd (and presumably 124-kd also) phytochrome becomes unstable and irreversibly precipitates (Rice *et al.*, 1973).

Handling of phytochrome preparations should always be kept as brief as possible and should be performed as close to 0°C as is practical to minimise proteolysis, which is often a problem even with highly purified samples. The only samples obtained in our laboratory that have been essentially free of protease activity are those purified by the immunoaffinity procedure described above. It is likely, however, that 124-kd phytochrome purified by the Affi-Gel Blue procedure is also free of protease activity.

ACKNOWLEDGMENT

Methods that are derived from the author's research programme and are described herein were developed with support provided by grants from the National Science Foundation.

REFERENCES

Boeshore, M. L. and Pratt, L. H. (1980). *Plant Physiol* **66**, 500–504.
Briggs, W. R. and Rice, H. V. (1972). *Annu. Rev. Plant Physiol.* **23**, 293–334.
Briggs, W. R., Rice, H. V., Gardner, G. and Pike, C. S. (1972). *Recent Adv. Phytochem.* **5**, 35–50.
Butler, W. L., Norris, K. H., Siegelman, H. W. and Hendricks, S. B. (1959). *Proc. Natl. Acad. Sci. U.S.A.* **45**, 1703–1708.
Cordonnier, M. -M. and Pratt, L. H. (1982). *Plant Physiol.* **69**, 360–365.
Correll, D. L. and Edwards, J. L. (1970). *Plant Physiol.* **45**, 81–85.
Correll, D. L., Edwards, J. L., Klein, W. H. and Shropshire, W., Jr. (1968a). *Biochim. Biophys. Acta* **168**, 36–45.
Correll, D. L., Steers, E., Jr., Towe, K. M. and Shropshire, W., Jr. (1968b). *Biochim. Biophys. Acta* **168**, 46–57.
Cundiff, S. C. and Pratt, L. H. (1973). *Plant Physiol.* **51**, 210–213.
Cundiff, S. C. and Pratt, L. H. (1975). *Plant Physiol.* **55**, 212–217.
Gardner, G., Pike, C. S., Rice, H. V. and Briggs, W. R. (1971). *Plant Physiol.* **48**, 686–693.
Georgevich, G., Cedel, T. E. and Roux, S. J. (1977). *Proc Natl. Acad. Sci. U.S.A.* **74**, 4439–4443.
Hopkins, D. W. and Butler, W. L. (1970). *Plant Physiol.* **45**, 567–570.
Hunt, R. E. and Pratt, L. H. (1979a). *Plant Physiol.* **64**, 327–331.
Hunt, R. E. and Pratt, L. H. (1979b). *Plant Physiol.* **64**, 332–336.
Hunt, R. E. and Pratt, L. H. (1980). *Biochemistry* **19**, 390–394.
Jabben, M. and Deitzer, G. F. (1978). *Planta* **143**, 309–313.
Kerscher, L. and Nowitzki, S. (1982). *FEBS Lett.* **146**, 173–176.
Laemmli, U. K. (1970). *Nature (London)* **227**, 680–685.

Litts, J. C., Kelly, J. M. and Lagarias, J. C. (1983). *J. Biol. Chem.* **258**, 11025–11031.

Mumford, F. E. and Jenner, E. L. (1966). *Biochemistry* **5**, 3657–3662.

Pratt, L. H. (1973). *Plant Physiol.* **51**, 203–209.

Pratt, L. H. (1978). *Photochem. Photobiol.* **27**, 81–105.

Pratt, L. H. (1980). *Plant Physiol.* **66**, 903–907.

Pratt, L. H. (1982). *Annu. Rev. Plant Physiol.* **33**, 557–582.

Pratt, L. H. (1983). *Encycl. Plant Physiol., New Ser.* Vol 16, (W. Shropshire, Jr., and H. Mohr, eds.) pp. 152–177. Springer Verlag, Berlin and New York.

Quail, P. H., Briggs, W. R. and Pratt, L. H. (1978). *Year Book—Carnegie Inst. Washington* **77**, 342–344.

Rice, H. V. and Briggs, W. R. (1973). *Plant Physiol.* **51**, 927–938.

Rice, H. V., Briggs, W. R. and Jackson-White, C. J. (1973). *Plant Physiol.* **51**, 917–926.

Rice, R. H. and Means, G. E. (1971). *J. Biol. Chem.* **246**, 831–832.

Roux, S. J., McEntire, K. and Brown, W. E. (1982). *Photochem. Photobiol.* **35**, 537–543.

Siegelman, H. W. and Firer, E. M. (1964). *Biochemistry* **3**, 418–423.

Siegelman, H. W., Wieczorek, G. A. and Turner, B. C. (1965). *Anal. Biochem.* **13**, 402–404.

Smith, W. O., Jr. and Daniels, S. M. (1981). *Plant Physiol.* **68**, 443–446.

Sober, H. A., ed. (1970). "CRC Handbook of Biochemistry: Selected Data for Molecular Biology," 2nd ed. p. C-80. CRC Press, Cleveland, Ohio.

Vierstra, R. D. and Quail, P. H. (1982a). *Proc. Natl. Acad. Sci. U.S.A.* **79**, 5272–5276.

Vierstra, R. D. and Quail, P. H. (1982b). *Planta* **156**, 158–165.

Vierstra, R. D. and Quail, P. H. (1983). *Biochemistry* **22**, 2498–2505.

Vierstra, R. D., Cordonnier, M. -M., Pratt, L. H. and Quail, P. H. (1984). *Planta* **160**, 521–528.

9
*Phytochrome Immunochemistry**
Lee H. Pratt

I. INTRODUCTION

Phytochrome has no known function *in vitro* or when exogenously applied
to a plant that would permit its detection either by bioassay or direct enzy-
matic or other molecular assay. Consequently, assays for phytochrome
initially depended solely upon its unique photoreversible absorbance properties. With the purification of phytochrome to homogeneity and the reali-
sation that this chromoprotein is an excellent antigen (Hopkins and Butler,
1970; Pratt and Coleman, 1971; Rice and Briggs, 1973), application of
immnochemical methods to the study of phytochrome has become possible.

Two approaches for developing antibodies to phytochrome are available.
The simpler approach is to inject highly purified phytochrome as antigen
and then collect and utilize the serum produced. This method produces
what can be considered a polyclonal antibody preparation since it is derived
from the output of a large number of immunologically competent cell lines,
each of which secretes a unique immunoglobulin. Alternatively, one can
obtain monoclonal antibodies, which, as the name implies, are produced
from a clone that is derived from a single antibody-secreting cell (Köhler and
Milstein, 1975). Although a completely pure phytochrome sample is required to obtain monospecific polyclonal antibodies, a monoclonal antibody that is monospecific to phytochrome can be obtained by starting with
only a partially purified phytochrome preparation. Thus, while it is more
work to produce monoclonal antibodies, less effort need be spent on purifi-

---

* Abbreviations: EDTA, ethylenediaminetetraacetic acid; ELISA, enzyme-linked immunosorbent assay; PAGE, polyacrylamide gel electrophoresis; PBS, 10 mM sodium phosphate,
0.14 M NaCl, pH 7.4; SAR, specific absorbance ratio, A_{667}/A_{280} with phytochrome in its red-absorbing form; SDS, sodium dodecyl sulphate; tris, tris(hydroxymethyl)aminomethane.

TECHNIQUES IN PHOTOMORPHOGENESIS
0-12-652990-6

cation of phytochrome. In addition, it will become evident when the application of immunochemical techniques to the study of phytochrome is considered (see Section VI) that each type of antibody preparation (i.e., polyclonal versus monoclonal) has its own set of advantages and disadvantages. No one approach can therefore be considered appropriate for all potential applications.

II. PHYTOCHROME PREPARATION

While it may not be easy to obtain phytochrome of virtually 100% purity, this is an essential prerequisite for producing a monospecific, polyclonal antiserum to phytochrome. As just noted, however, impure preparations can be used to produce monoclonal antibodies. In this case, it is still important to purify phytochrome as completely as is practical since the purer the pigment is when injected, the greater will be the percentage of cell lines obtained that secrete antibodies to phytochrome. Details of phytochrome purification have already been presented in Chapter 8. Only a few general comments will be offered here.

The size of phytochrome used as antigen is not critical for most purposes since an antibody that is made against 60-kd phytochrome (see Chapter 8 for definition and discussion of phytochrome sizes) will almost certainly recognise 120- and 124-kd phytochrome equally well (Pratt, 1973; see Fig. 3). Conversely, antibodies made against 120-kd phytochrome not only bind to 120- and 124-kd phytochrome (Vierstra and Quail, 1982) but also with proteolytic degradation products including the 60-kd photoreversible chromopeptide (Cundiff and Pratt, 1975; see Fig. 3).

Of the large number of protocols that now exist for phytochrome purification, those recommended are described in Chapter 8. Thus the hydroxyapatite, Affi-Gel Blue, gel filtration chromatography sequence developed by Vierstra and Quail (1983) is recommended for preparation of 124-kd phytochrome. If antibodies to phytochrome are already available, however, an immunopurification procedure would be more efficient. Antibodies to phytochrome can also be used to prepare immunoprecipitates from partially purified preparations of the pigment (Cundiff and Pratt, 1973; Cordonnier and Pratt, 1982a). If these immunoprecipitates are injected into the same animal species that was used to produce the antibodies with which the immunoprecipitate was prepared, antibodies specific to phytochrome are readily produced. Since a polyclonal antiserum cross reacts with phytochrome from taxonomically diverse plant species, this approach can be quite useful (Cordonnier and Pratt, 1982a,b). Alternatively, 120-kd phytochrome can be prepared to near 100% purity and then the purification can be completed by polyacrylamide gel electrophoresis (PAGE), as described by Kerscher and Nowitzki (1982). Although not yet reported, it should be possible to complete the purification of 124-kd phytochrome by PAGE if difficulty is experienced in obtaining sufficiently pure phytochrome with the procedure of Vierstra and Quail (1983), as described in Chapter 8. Of course, if monoclonal antibodies are to be produced, then total purification is unnecessary.

The amount of phytochrome needed for immunization is variable. With rabbits, it is desirable (although not necessary, see Kerscher and Nowitzki, 1982) to inject at least 2 mg (Chapter 8, Section I. B) total per animal. Because it is possible that a given rabbit may die prematurely or may not respond properly to the injections, it is also best to begin with at least two animals. For producing monoclonal antibodies with mice, 0.35 mg per animal works well (Cordonnier *et al.*, 1983). For the reasons just given, it is best in this instance to begin with at least three mice. Consequently, for polyclonal and monoclonal antibody production, it is optimal to begin with at least 4 or 1 mg of phytochrome, respectively.

III. POLYCLONAL ANTIBODY PREPARATION

While a variety of animals could be used for raising antiserum, rabbits are most convenient for many reasons, the following being among the most important. (1) Animal care facilities for taking care of rabbits are widely available. (2) A rabbit is relatively small and yet at the same time large enough to bleed easily and to produce a relatively large volume of antiserum. (3) There are a wide variety of commercially available reagents compatible with the use of rabbit immunoglobulins for immunochemical applications. Procedures described here will therefore refer specifically to the use of this animal, although one may choose to work with other animals such as guinea pigs, goats, sheep, or even horses.

A. Phytochrome injections and serum preparation

Emulsify 1 – 1.5 mg of purified phytochrome in a volume of 1 ml of 0.1 M sodium phosphate, pH 7.8, with 1 ml of complete Freund's adjuvant. Emulsification is easily achieved with two 3-ml syringes connected to one another by an 18-gauge needle modified to have a syringe fitting on each end. The phytochrome and adjuvant are then mixed by passing them back and forth between the two syringes until the mixture no longer separates into two phases upon standing for a few moments.

Immunise a rabbit by injection of the emulsion in four equal aliquots, one into each thigh muscle and one subcutaneously behind each shoulder. After 1 month, inject a second 0.5-ml sample of 0.5 – 1.0 mg of phytochrome, emulsified this time with 0.5 ml of incomplete Freund's adjuvant. This second sample is injected in two aliquots, each given subcutaneously behind a shoulder. After an additional 7 – 10 days, take 50 – 60 ml of blood by any method desired (a heart puncture is convenient). One week later, an additional 40 – 50 ml of blood may be taken. Place the blood in 50-ml polycarbonate centrifuge tubes immediately after collection.

After the second bleeding of the rabbit and once it has been given a month or so to recover from the loss of blood, it may be injected with another 0.5 – 1.0 mg of phytochrome as described above for the second injection. Subsequent bleedings are as described above. Additional booster injections and bleedings may follow indefinitely as just described. Alternatively, the rabbit may be bled terminally at the time of the second bleeding, in which case one should obtain a total of about 150 ml of blood from the rabbit.

Allow the blood to clot in the 50-ml centrifuge tubes for 1 h at room temperature. Cut the clot into roughly 1-cm^3 pieces with a knife or spatula. Incubate the blood overnight at 4°C during which time the clot shrinks. Centrifuge in the same 50-ml tubes for 15 min at 4,000 g, preferably in a swinging-bucket rotor. Remove the clarified serum, which should be about one-half the original volume of blood, with a Pasteur pipette. Store either at -20°C or add NaN_3 to 0.02% and store at 4°C. Frozen storage is preferred for long periods of time.

B. Immunoglobulin purification

Fractionation of crude serum is a relatively common (Weir, 1978) but generally unnecessary procedure. Whole serum may be used for all of the applications discussed in this chapter except for the ELISA and for the preparation of antibody columns for phytochrome immunopurification, in which case immunopurified antibodies are needed.

If serum fractionation is desired, the simplest procedure is to precipitate immunoglobulins with $(NH_4)_2SO_4$, leaving the more abundant albumins in solution. To do this, bring the serum to one-third saturation by adding either 0.2 g of $(NH_4)_2SO_4$ or 0.5 ml of saturated $(NH_4)_2SO_4$ solution per milliliter of serum. Collect the precipitate by centrifugation (15,000 g, 15 min), dissolve it in a volume of 10 mM sodium phosphate, 0.14 M NaCl, pH 7.4 (PBS), equivalent to that of the original serum, and clarify by centrifugation (50,000 g, 15 min). Store at -20°C or add NaN_3 to 0.02% and store at 4°C. The fractionation may be carried out either at room temperature or at 4°C. Residual $(NH_4)_2SO_4$ may be removed by dialysis if desired.

If purification of only those immunoglobulins that bind phytochrome is desired, follow the immunopurification procedure described in Chapter 8 (Section II. D. 6).

IV. MONOCLONAL ANTIBODY PREPARATION

It is considerably more difficult to prepare monoclonal antibodies than polyclonal antisera. Nevertheless, this may be desirable for many reasons, the more important of which include the following (Goding, 1980). (1) It is possible to obtain a monospecific monoclonal antibody to phytochrome even with an impure phytochrome preparation as antigen (Cordonnier *et al.,* 1983). (2) Although it is possible to assay for specificity of polyclonal antibodies, it is virtually impossible to have total confidence in specificity. This point is especially important when using sensitive immunochemical techniques such as radioimmunoassay, ELISA, or immunocytochemistry. If a monoclonal antibody binds to phytochrome, however, this provides total confidence in its monospecificity. (3) A polyclonal antibody preparation is heterogeneous. It contains antibodies that bind to different regions of the molecule (Cundiff and Pratt, 1973, 1975) and different antibodies, each of which binds to the same part of the molecule, are to be expected. A monoclonal antibody, by contrast, is chemically homogeneous, recognising only a single region of the molecule and interacting with that determinant in only a single way. Thus with monoclonal antibodies, the most appropriate antibody can be selected for each application, which is something that cannot be done effectively with polyclonal antibodies. For example, an antibody that is best for immunocytochemistry is likely to be different from one that will give best results for phytochrome immunopurification or that will provide the most sensitive ELISA for phytochrome quantitation. (4) Because monoclonal antibodies are chemically homogeneous, they can be used as structural probes of the phytochrome molecule, an application to which polyclonal antibodies are not well suited. (5) Monospecific monoclonal antibodies are much more easily purified than are polyclonal antibodies. (6) Whereas a polyclonal antibody preparation is in limited supply, since the animal producing it has an uncertain and finite lifetime, a monoclonal antibody can be made available (a) in perpetuity by preserving the cell line that produces it in a liquid nitrogen freezer and (b) in quantities limited only by the amount of effort and expense invested in producing it.

Before describing the production of monoclonal antibodies to phytochrome, it will be helpful to consider an overall view of the process as it will be described in more detail below (Fig. 1). The goal is to fuse antibody-secreting cells from the spleen of an immunised mouse with myeloma cells. The product (a hybridoma) possesses the immortality of the myeloma cell as

Myeloma cell (SP2/0) — immortality, no Ig

Spleen cell (BALB/c) — antibody production

GOAL: to obtain a hybrid that exhibits both features simultaneously

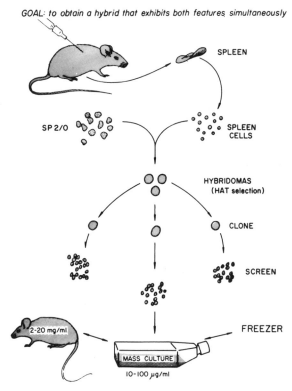

Fig. 1. Outline of the principles of monoclonal antibody production as described in the text. Drawing by Cecile Smith.

well as the antibody secreting ability of the spleen cell. The hybridomas are cultured initially in a selective medium (HAT medium), which contains hypoxanthine, aminopterin, and thymidine. In this medium, neither the parent myeloma nor the spleen cells can grow alone. The two complement each other genetically, however, so the hybrids can replicate (Littlefield, 1964). As described here, the hybridomas are grown as clones to a sufficient cell number so that they can be screened for the ability to secrete antibodies to phytochrome. Those cells that do can then be grown in mass culture. Medium used to maintain the cell line will contain secreted immunoglobulins. As much as 10 – 100 mg of antibody per litre of medium can be obtained. Alternatively, hybridoma cells can be injected into the peritoneal cavity of a mouse to produce an ascites tumor, which will produce ascitic fluid that contains up to 10 or 20 mg of monoclonal antibody per ml. Since 5 – 10 ml of ascitic fluid can be produced from a mouse, this relatively simple procedure is capable of providing large quantities of antibody. If the hybridoma is of interest, it can be preserved by freezing and subsequent storage in liquid nitrogen.

The specific methods described in this section have already proven successful with phytochrome (Cordonnier *et al.*, 1983). For each step in the procedure, however, many variants exist. It is not appropriate here to attempt to justify the specific methods presented nor to discuss these methods relative to the many potential alternatives. For more general discussions of monoclonal antibody technology and of the wide variety of alternative methods available, the reader is referred to one of several edited volumes (Hämmerling *et al.*, 1981; Kennett *et al.*, 1980; Langone and Van Vunakis, 1983; Melchers *et al.*, 1978; Müller, 1979) or reviews (Fazekas de St. Groth and Scheidegger, 1980; Goding, 1980).

A. Laboratory requirements

While it is possible to work in the corner of a laboratory, it is far preferable to have a small, dedicated facility in which to work. Contamination of an established cell line for which stocks exist is not a serious problem. When producing monoclonal antibodies, however, the work involves relatively long periods of time with large numbers of unique cell lines. If a cell line becomes contaminated at any time prior to expansion and preservation by freezing, it will be irretrievably lost. Consequently, it is best to work in a small room that is kept as sterile as possible. Inclusion of UV lights in the room is usual.

In addition to the more routine items found in most laboratories, only four major pieces of equipment are needed. (1) A laminar flow hood is essential to provide a sterile working area for many operations. Since the cell lines are not considered a biohazard, the hood need not be of the containment type. A horizontal flow hood with open front is convenient because it provides easy access to the full working area. (2) An incubator that provides for both temperature and CO_2 control is required for *in vitro* cell culture. (3) An inverted microscope is used both to observe cell growth and to provide a means for selecting hybridoma clones under the laminar flow hood. Phase optics provide the easiest means for distinguishing between living and dead cells. (4) A liquid-nitrogen freezer is used for long term preservation of hybridoma and myeloma cell lines. A freezer that can store at least several hundred 2-ml vials is recommended since the number of cell stocks to be preserved grows rapidly.

B. Phytochrome injections

Although impure phytochrome samples may be used for immunization, it is best to use the purest samples available since the greater the purity when injected, the higher will be the proportion of hybridomas that secrete antibodies to phytochrome. Success has been obtained in making antibodies against pea phytochrome using antigen samples that were only about 20% pure (Cordonnier *et al.,* 1983).

Inject 4-to-6-week-old BALB/c mice with about 0.2 mg each of phytochrome. Total protein injected will vary with the purity of the sample. The sample should be in 0.2 ml of buffer (0.1 M sodium phosphate, 1 mM EDTA, pH 7.8, works well) and is emulsified (see Section III. A) with an equal volume of complete Freund's adjuvant (DIFCO 0638-60 is recommended). Each mouse should receive five equal injections made with a 25-gauge needle: one into the peritoneal cavity and the other four intradermal in the abdominal region. After a minimum of 30 days, inject each mouse again, this time intraperitoneally with 0.1 mg phytochrome in 75 μl buffer emulsified with 75 μl of incomplete Freund's adjuvant (DIFCO 0639-60 is recommended). Ten days later give a third and final injection of 50 μg of phytochrome (without adjuvant) directly into the tail vein. A 30-gauge needle is best for this purpose. Three days later kill the mouse by cervical dislocation or by asphyxiation, remove blood by heart puncture with a 1-ml syringe equipped with a 23- or 25-gauge needle, and remove the spleen aseptically under the laminar flow hood. Isolate cells from the spleen and wash them in PBS in preparation for a fusion.

Best results are obtained by harvesting the spleen when it provides a maximal number of antibody-secreting cells (Stähli *et al.,* 1980). The above immunization series is designed to ensure that the spleen is taken at the optimal time. Timing of the second and third injections relative to the time at which the spleen is harvested is critical. A well-immunised mouse goes into mild anaphylactic shock for 60 min or so after the third injection. If too much phytochrome is injected at this time, the mouse will probably die within that period.

C. Cell culture and freezing

Two different culture media are used: RPMI-1640 (e.g., GIBCO 430-1800) or IMDM (e.g., GIBCO 430-2200). Both are prepared according to the manufacturer's instructions and both are supplemented with penicillin (100 units/ml) and streptomycin (100 μg/ml). The terms RPMI and IMDM will refer to these media as supplemented with the antibiotics. Foetal calf serum is added as indicated.

The SP2/0-Ag14 myeloma cell line is used for fusions since it does not produce any immunoglobulin gene products (Shulman *et al.,* 1978). Immunoglobulins secreted by a hybridoma or ascites tumor can then be purified with confidence that they are not contaminated by inactive protein.

Both myeloma and hybridoma cell lines are cultured routinely in RPMI with 5% foetal calf serum at 37°C in air supplemented to 5% CO_2. Cells are maintained routinely in 75-cm² culture flasks and are kept in mid-log growth phase by feeding on a daily basis.

To freeze cells, harvest the contents of one 75-cm² culture flask (about 50 ml) during mid-log growth phase at which time they should be at a concentration of ∼5 × 10⁵/ml. Collect by centrifugation for 5 min at 200 *g*. Resuspend the pelleted cells in 1 ml of IMDM with 20% foetal calf serum. Transfer to a sterile 2-ml cryotube, add 0.15 ml of dimethyl sulphoxide (by virtue of its toxicity, it may be considered sterile as it comes from a fresh reagent bottle that is handled aseptically), mix rapidly, and freeze by burying the cryotube in powdered dry ice. When frozen, transfer the tube to a −70°C freezer for 1 or 2 days and then transfer further to a liquid-nitrogen freezer. Ensure that the cells do not become warmer than −70°C during transfer.

To reinitiate growth of frozen cells, thaw the contents of a cryotube rapidly by swirling the tube gently but continuously in a 37°C water bath. As soon as the cells are thawed, transfer the contents of the tube to a 50-ml sterile centrifuge tube that already contains 19 ml of IMDM or RPMI without serum. Mix well by swirling. Determine the concentration of viable cells by counting under a microscope with a haemacytometer. By addition of 1 μl of Evan's Blue solution (0.1 g in 1 ml of H_2O) to a 100-μl aliquot of the cells, one can readily discriminate between living and dead cells since living cells exclude the dye. While counting viable cells in this small aliquot, collect the bulk of the cells in the 50-ml tube by centrifugation for 5 min at 200 *g*. Suspend the collected cells in enough IMDM or RPMI with 20% foetal calf serum to give a concentration of viable cells of 3 × 10⁵/ml. Transfer in 1-ml aliquots to one or more 24-well culture plates. Feed the cells with IMDM or RPMI and 5% foetal calf serum as needed and transfer to 75-cm² flasks when they are growing well.

D. Fusion and cloning

Fusions are performed by minor modification of the procedure described by Staehelin *et al.* (1981). Wash cells (about 5 × 10⁷) from one spleen in RPMI and then mix with 10⁷ SP2/0-Ag14 cells that were harvested during mid-log growth. Co-sediment cells in a 50-ml conical tip centrifuge tube for 5 min at 200 *g*. Pour off the supernatant completely. Place the centrifuge tube in a

37°C water bath and add with a 1-ml syringe 0.7 ml of poly(ethylene glycol) solution that had been prewarmed to 37°C. [This solution is made by autoclaving for 30 min 10 g of poly(ethylene glycol) (e.g., Sigma P-3640). After cooling to 80°C, add 10 ml of serum-free RPMI and mix well. Store in sterile 1-ml aliquots at −20°C. Do not adjust pH.] Addition of the poly(ethylene glycol) solution should take about 60 sec. Let stand for 90 sec in the 37°C water bath. Dilute with 10 ml of RPMI (serum-free) by adding in drops over a 3-min period, beginning slowly and adding at an increasing rate while swirling constantly. Collect cells by centrifugation for 5 min at 200 g.

Hybridomas are grown and cloned simultaneously by the procedure developed by Davis et $al.$ (1982). Suspend the pelleted fusion products in 15 ml of the following solution: 26.7 ml of foetal calf serum + 20.6 ml of IMDM + 1.33 ml of HT (136 mg of hypoxanthine + 37.8 mg of thymidine in 100 ml of H_2O; warm to ∼50°C to dissolve) + 1.33 ml of aminopterin (add 1.76 mg to 99 ml of H_2O, add 0.5 ml of 1 N NaOH to dissolve and then 0.5 ml of 1 N HCl to neutralize) + 1.33 ml of lipopolysaccharide (DIFCO 3122-25) from a 5 mg/ml stock solution. Add 25 ml of 2% methylcellulose with a syringe (see below for recipe). Cap tube tightly and mix contents thoroughly by inverting several times. Use a 12-ml syringe with an 18-gauge needle to distribute the viscous mixture in 1-ml aliquots into 40 35-mm-diameter Petri dishes, which are then covered. Place two such Petri dishes into a covered 100-mm-diameter Petri dish. Add a third 35-mm-diameter Petri dish that contains 1 ml of sterile water. Leave the lid off this third dish, which is intended to ensure saturating humidity within the 100-mm-diameter Petri dish. Place the Petri dishes in an incubator and leave them undisturbed for 7–10 days, during which time hybridoma colonies develop.

When the colonies are visible to the eye (about 1 mm in size) they may be picked for transfer to IMDM with 20% foetal calf serum in 24-well culture plates. Heat-sterilized 20-μl glass micropipettes equipped with a rubber bulb are convenient for picking and transferring cells. Examine the Petri dishes with an inverted microscope equipped with a four-power objective to ensure that each colony picked is monoclonal (Davis et $al.$, 1982). Feed the cells as necessary. They are ready for screening 2 days after selection.

Preparation of the 2% methylcellulose medium is simple only if the recipe given here is followed carefully. Separately sterilize by autoclaving a 150-ml Fleaker (Corning Glass Works) and cap. The Fleaker should contain a magnetic stir bar. Work as much as possible in a laminar flow hood. Use a top-loading balance that will permit addition of reagents by weight. Determine the tare weight of the Fleaker and stir bar. Add 30.2 g of H_2O to the Fleaker, heat to boiling over a Bunsen burner and boil for 5 min with an aluminium foil cap over the Fleaker. Add 1.2 g of methylcellulose to the hot water. Ensure that all of the methylcellulose falls into the water, with none being left behind on the walls of the container. If you do not succeed at this step, it is best to begin again with a clean, sterile Fleaker. Allow the mixture to boil gently for 5 min. Be careful because it boils over easily. Place the Fleaker in an ice bath and cool the contents rapidly, swirling the Fleaker continuously. The solution clears and becomes quite viscous at this stage. Add 30.3 g of double strength IMDM to a Nalgene filter unit (120-0020) and sterilize by vacuum filtration. Add the sterilized medium to the Fleaker while mixing. The amount of medium filtered is calculated to include that which is left behind in the filter unit. Add sterile water to the Fleaker to attain a final weight of 61.4 g for the contents. Close the Fleaker with the presterilized Fleaker cap and slowly stir the contents with the magnetic stirrer overnight at 4°C. The preparation is now ready for use. Different batches of methylcellulose (e.g., Sigma M-0512; 4000 centipoise for a 2% solution at 25°C) should be tested. It is inexpensive, so it is best to buy relatively large batches (e.g., 250 g) so that when a suitable lot is found it will be available in large quantity. The methylcellulose should be tested for its effect, if any, on cell growth and also for its appearance. Some lots contain more insoluble debris than others and how much one is willing to tolerate is a personal decision. The methylcellulose solution may be stored for several months at 4°C or indefinitely at −20°C.

E. Screening

A successful fusion will yield several hundred hybridomas, all of which should be screened within a few days. An efficient procedure is therefore essential. An ELISA (see Voller *et al.*, 1980, for review) is convenient since phytochrome binds readily to vinyl assay plates (e.g., Costar 2596). The assay as used for screening (Cordonnier *et al.*, 1983) is illustrated in Fig. 2. As described elsewhere (Cordonnier *et al.*, 1984), this ELISA can also be used to characterize monoclonal antibodies to phytochrome.

Prepare 96-well vinyl plates by filling completely with 95–100% ethanol and allowing to stand for 2 h at room temperature. Empty the plates and rinse well with 95–100% ethanol from a squeeze bottle. This procedure enhances uniform binding of phytochrome to the wells. Allow the plates to air dry. The evening before a screening, add to each well in alternate rows (uncoated rows will serve as controls) 50 μl of phytochrome in 0.2 M sodium borate, 75 mM NaCl, pH 8.5. A phytochrome concentration of 5 μg/ml (Chapter 8. I. B) works well. While phytochrome that is as little as 20% pure may be used, the purer the sample is, the less likely will be the detection of cell lines secreting antibodies to potential contaminating antigens. Incubate the plate overnight at 4°C.

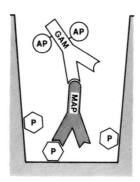

Fig. 2. A scheme for the detection of monoclonal antibodies to phytochrome by ELISA. MAP = monoclonal antibody to phytochrome; GAM = goat antibody to mouse IgG; P = phytochrome; AP = alkaline phosphatase, which is covalently linked to GAM. Details of the assay are described in the text.

The following morning, invert the plate over a sink and empty it by flicking forcefully. Wash once by filling with 10 mM tris-Cl, pH 8.0 at 25°C, 0.05% Tween 20, 0.02% NaN$_3$ from a squeeze bottle. Empty as before. To block further nonspecific binding, fill the wells with 1% bovine serum albumin in PBS and incubate for 30 min at room temperature. Empty and wash once as before. Add one drop of hybridoma medium or 50 μl of other test reagent (e.g., diluted serum from the mouse used for the fusion) to a well that had been coated with phytochrome and to an adjacent well as a control for nonspecific binding. Incubate for 2 h at 37°C. Empty and wash as before, this time washing three times. Add 50 μl of alkaline phosphatase-coupled goat antibody to mouse IgG (e.g., TAGO 6550 or Sigma A-5153), which is diluted 500-fold with PBS containing 0.05% Tween 20, 0.02% NaN$_3$, and 1% bovine serum albumin. Incubate 2 h at 37°C. Empty and then wash three times as before. Add 50 μl of substrate solution (15 mg of p-nitrophenyl phosphate in 25 ml of the following buffer: 48 ml of diethanolamine, 50 mg of MgCl$_2$ in 400 ml of H$_2$O, adjust pH to 9.8 with 1 N HCl, bring to final volume of 500 ml with H$_2$O) and incubate for 30 min at room temperature. Stop the reactions by addition of 50 μl of 3 N NaOH. The assay easily detects monoclonal antibodies at concentrations as low as 10–100 ng/ml. Results can be evaluated visually or quantitated by spectrophotometric measurement at 400 nm.

Negative cell lines should be discarded immediately. Positive cell lines should be expanded into additional 24-well culture plates for protection against possible contamination. As soon as the cells are growing well they should be transferred to 75-cm^2 flasks for mass culture. It is advisable to freeze cell stocks as soon as practical since there is a chance that the cell line will, with time, lose the ability to secrete immunoglobulins. It is also best to produce about 1 litre of medium for each positive cell line at the outset so that sufficient antibody will be immediately available for evaluation.

The above procedure for initial growth of hybridomas in a semisolid medium has many advantages. No feeding or other attention is required before picking clones. Nor is there any requirement for commonly used "feeder" cells (Kennett *et al.*, 1980). In addition, the need for routine cloning after screening is eliminated, thereby saving considerable labour. Finally, screening is done at a time when each positive cell line is already monoclonal. Therefore, when searching for an antibody with a specific property, this search may be made even during the initial screening.

F. Production of antibodies *in vitro* and *in vivo*

To produce antibodies in quantity by *in vitro* culture of hybridomas, the cells need only be grown in 75-cm^2 flasks with RPMI containing 1 – 5% foetal calf serum (it is most economical to use the lowest percentage of serum that will maintain suitable growth; this value varies with each cell line). The medium is harvested on a regular basis and clarified by centrifugation for 5 min at 200 g. It is most convenient to concentrate the used medium by precipitation with 313 g/litre $(NH_4)_2SO_4$ and centrifugation at 15,000 g for 15 min. The resulting pellet may be dissolved in 5% of the original volume of 0.1 M sodium phosphate, 1 mM EDTA, pH 7.8, and stored at $-20°C$.

To produce antibodies *in vivo* via ascites tumors, inject $10^6 - 10^7$ cells harvested in mid-log growth into the peritoneal cavity of a young (e.g., 4-week-old) BALB/c mouse. If an ascites tumor develops, the abdomen of the mouse becomes distended within 10 days to 2 weeks. Fluid may be taken repeatedly or only once if desired. The mouse dies within several days after the tumor becomes apparent and fluid taken even a few hours after death is not usable. If wished, ascites cells may be harvested from the mouse and injected into another mouse to continue the tumor and produce more fluid. Harvested ascitic fluid is clarified by centrifugation for 15 min at 1000 g and then stored at $-20°C$.

G. Immunoglobulin purification

Purification of monoclonal antibodies either from concentrated medium or from ascitic fluid is a relatively simple procedure. By the method described here (Cordonnier *et al.*, 1983), however, substantial quantities of immuno-purified rabbit antibodies to mouse IgG must first be prepared. To do this, immunise rabbits with commercially obtained mouse IgG (e.g., Sigma I-5381) and prepare antiserum as already described for phytochrome (Section III). Immobilize 10 mg of mouse IgG on CNBr-activated agarose as described for phytochrome (Chapter 8, Section II. D. 6). Use this mouse-IgG column to immunopurify rabbit antibodies to mouse IgG from the crude serum as described for the purification of rabbit antibodies to phytochrome (Chapter 8, Section II. D. 6). The immunopurified antibodies to mouse IgG are then coupled to CNBr-activated agarose as before (Chapter 8, Section II. D. 6). Either hybridoma medium or ascitic fluid can be passed through the affinity column in order to adsorb monoclonal antibodies, regardless of the cell line that produced them. Save the adsorbed medium or fluid. Wash the column with PBS until A_{280} decreases below 0.03. Layer over the column one column volume of 0.1 M glycine-Cl, pH 2.5. Follow immediately with PBS, since prolonged exposure to low pH denatures the antibodies that are eluted. Collect 1-ml fractions in test tubes that already contain 0.2 ml of 0.5 M tris-Cl, pH 8.3 at 25°C, which will neutralize the acid. Pool all fractions that contain significant A_{280}. Precipitate the antibodies by addition of an equal volume of saturated $(NH_4)_2SO_4$ solution and collect by centrifugation at 30,000 g for 15 min. Dissolve in enough 0.1 M sodium phosphate, 1 mM EDTA, pH 7.8, to give a monoclonal antibody concentration of 1–5 mg/ml. Store at −20°C or add NaN$_3$ to 0.02% and store at 4°C. A 1-mg/ml solution of mouse IgG has an A_{280} of about 1.4 for a 1-cm light path. The adsorbed medium or fluid may be passed through the mouse-IgG column repeatedly until all antibody has been adsorbed.

V. ASSESSMENT OF ANTIBODY SPECIFICITY

The specificity of an antibody preparation must be established empirically. To do this for a monoclonal antibody it is only necessary to demonstrate that the antigen to which it binds is phytochrome, since the antibody is by definition monospecific. [Nevertheless, caution is still required since even a monospecific antibody can cross react with another protein if the latter possesses an epitope related to that on phytochrome.] For a polyclonal antibody preparation, however, specificity must be demonstrated by a method that is appropriate to its use. It is possible here only to indicate methods for general assessment of specificity that may not be adequate for all purposes.

A. Polyclonal antibodies

A specific antiserum to phytochrome should produce a single precipitation band both by double diffusion (e.g., Fig. 3, see Section VI. A for method) and immunoelectrophoresis (e.g., Fig. 4, see Section VI. B for method) when

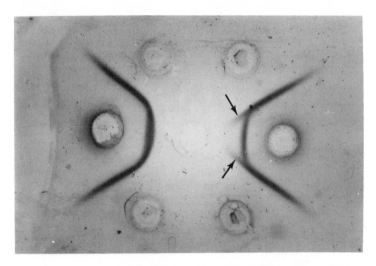

Fig. 3. Ouchterlony double immunodiffusion plate which indicates that antiserum against 120-kd oat phytochrome (right well) recognizes 60-kd (center well) and 120-kd (top and bottom wells) oat phytochrome differently (see spurs marked by arrows). In contrast, antiserum against 60-kd oat phytochrome (left well) recognizes the two sizes of phytochrome identically. (Plate courtesy of R. Hunt.)

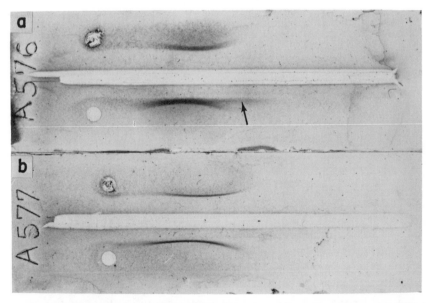

Fig. 4. Immunoelectrophoresis plates which indicate that while crude antiserum to zucchini phytochrome (trough in a) exhibits a possible nonspecificity (arrow) when tested against brushite-purified zucchini phytochrome (lower well in a), immunoglobulins to zucchini phytochrome (trough in b) purified from this serum are specific for phytochrome by this assay when tested against the same brushite-purified zucchini phytochrome (lower well in b). The top well in each plate was filled with immunoaffinity-purified zucchini phytochrome. (From Cordonnier and Pratt, 1982a.)

tested against partially purified phytochrome. Unfortunately, while it is possible to infer that a serum is specific when only single bands are observed, proteolysis will lead to heterogeneity of phytochrome *in vitro* (see Chapter 8) and may give rise to multiple precipitation reactions, especially as observed by immunoelectrophoresis (Cundiff and Pratt, 1975). The presence of multiple bands indicates, therefore, only possible nonspecificity.

Even when an antiserum is apparently nonspecific (e.g., Fig. 4a), it is possible to immunopurify monospecific immunoglobulins to phytochrome from that serum (Fig. 4b; see Chapter 8, Section II. D. 6) if the immobilised phytochrome that is used does not contain the contaminating antigen.

B. Monoclonal antibodies

With a monoclonal antibody, it is only necessary to demonstrate that it is to phytochrome. Two direct approaches are available (Cordonnier *et al.,* 1983). (1) Incubate the antibody with phytochrome, precipitate the antibody, and determine spectrophotometrically whether phytochrome has also been precipitated. (2) Incubate the antibody with phytochrome, precipitate the antibody, and determine by SDS–PAGE whether the precipitate includes phytochrome. Immunoblotting (see Section VI, F) could also be used as a test of specificity. It is, however, not a generally applicable procedure since as anticipated not all antibodies to phytochrome will bind to it when it is immobilized on nitrocellulose following SDS-PAGE (Daniels and Quail, 1984; M.-M. Cordonnier, personal communication).

For the first approach, prepare a solution that contains excess antibody (e.g., mix 50 μg of phytochrome with 100 μg of purified antibody or with 500 μl of 20-fold concentrated hybridoma medium). As a control, prepare a solution that contains either a monoclonal antibody that is not to phytochrome, commercially obtained mouse IgG, or 20-fold concentrated nonimmune hybridoma medium. Incubate at 4°C for 60 min. Wash *Staphylococcus aureus* cells (e.g., Calbiochem-Behring Corp. 507858) three times by centrifugation and resuspension in PBS. Add 300 μl of a 10% suspension of the washed cells and incubate for a further 15 min. The pH during this latter incubation must be 8.0 or above (Ey *et al.,* 1978). Collect the cells and associated mouse antibody by centrifugation. Assay the supernatants for phytochrome by spectrophotometry (see Chapter 6). If a monoclonal antibody is to phytochrome, it will usually, but not always (Cordonnier *et al.,* 1983; Shimazaki *et al.,* 1983), precipitate it under these conditions.

For the second approach, prepare a similar solution but this time with equal quantities by weight of phytochrome and antibody. Incubate as before for 60 min at 4°C. Pass the mixture through a small column (e.g., 0.2 ml) of agarose to which rabbit antibodies to mouse IgG have been coupled. Wash the column with excess PBS. Elute bound antibodies with antigen attached, using one column volume of 0.1 M glycine-Cl, pH 2.5, followed by 2 column volumes of PBS. Precipitate eluted protein by adding an equal volume of saturated $(NH_4)_2SO_4$ solution and centrifuging. Assay the precipitate by SDS-PAGE as described in Chapter 8 (Section III. A) to see if it contains phytochrome, in which case one can be certain that the antibody is to phytochrome (assuming that a nonimmune mouse IgG control gives a negative result as anticipated).

VI. IMMUNOCHEMICAL APPLICATIONS

Only those immunochemical methods that have been applied to the study of phytochrome and that are likely to find continued application are described here. The reader should be aware that many additional methods, as well as variations of each technique described here, are available (e.g., Weir, 1978; Van Vunakis and Langone, 1980; Langone and Van Vunakis, 1981a,b, 1982, 1983a,b).

A. Ouchterlony double immunodiffusion

The following procedure is tailored to the use of a Gelman apparatus. Label micro slides at one end with a diamond or tungsten carbide marking pencil. Dip the slides into hot 1% agarose that contains 1 drop of glycerol per 25 ml. Drain well and allow to air-dry thoroughly. Layer enough agarose medium (1 g of agarose in 100 ml of 10 mM potassium phosphate, pH 7.8, 0.15 M NaCl, 0.02% NaN_3) over the micro slides, which must be absolutely level, to form a layer 1 mm thick. Allow to gel at room temperature and then transfer to 4°C for storage in a humidified container. Punch holes into the agarose plate to form the desired pattern and remove agar from the holes to form sample wells. A convenient pattern is one of 6 wells equally spaced around a central well (Fig. 3). Each well is filled either with antibody (serum or purified immunoglobulins) or antigen (phytochrome), and diffusion is allowed to occur to completion at 4°C in a humid chamber. With wells of 2 mm diameter (about 5 μl capacity) spaced 7 mm apart (Fig. 3), 24 h is sufficient. Antiserum is normally used undiluted, while phytochrome is used at a concentration of 0.2 – 0.3 mg/ml and purified immunoglobulins at a concentration of about 1 mg/ml.

Precipitin bands may be viewed by scattered light against a dark background or, for maximum sensitivity, after staining with a suitable dye such as Coomassie brilliant blue R. To stain precipitin bands, first wash the plates for 24 h by diffusion against 2 changes of 0.15 M NaCl, 0.02% NaN_3, to remove extraneous protein. Wash for an additional 1 h by diffusion against H_2O to remove NaCl. Air-dry overnight, which results in a thin film that adheres to the glass slide. Place the dried plate in stain (1 g of Coomassie brilliant blue R, 180 ml of methanol, 180 ml of H_2O, and 37 ml of glacial acetic acid) for 15 min and then destain for 10–20 min in 45% (v/v) methanol, 10% (v/v) glacial acetic acid in H_2O. Dry the plate.

B. Immunoelectrophoresis

Agarose plates are prepared as for double diffusion except that the agarose is dissolved in 75 ml of H_2O + 25 ml of barbital buffer (15.5 g of sodium diethylbarbiturate + 1.38 g of diethylbarbituric acid + 10.2 g of sodium acetate + 1485 ml of H_2O). After dissolving the agarose, 1 ml of 2% (w/v) NaN_3 is added. It is important to use agarose rather than agar because the latter, which contains charged groups, decreases the electrophoretic mobility of phytochrome.

Prepare wells as for double diffusion and cut troughs, but do not remove the agar from the troughs (Fig. 4). Add phytochrome to the wells at a concentration of between 0.1 and 1 unit/ml; the higher concentration gives better results. Fill the electrode chambers with barbital buffer and electrophorese at 4°C in the direction of the positive electrode for 2 h at 15 V/cm. Following completion of electrophoresis, remove the agar from the troughs and fill with undiluted antiserum or with purified immunoglobulins at a concentration of about 1 mg/ml. Permit diffusion to occur for 24 h at 4°C in a humid chamber. Observe precipitin bands by scattered light or after staining them as described for Ouchterlony double-diffusion plates.

C. Immunoaffinity purification of phytochrome

The use of antibodies for the purification of phytochrome is described in Chapter 8 (Section II. D. 6).

D. Phytochrome quantitation by enzyme-linked immunosorbent assay (ELISA)

Immunochemical quantitation of phytochrome has been done by both radial immunodiffusion (Pratt *et al.,* 1974) and radioimmunoassay (Hunt and Pratt, 1979). Since it is far simpler and more sensitive to quantitate phytochrome by ELISA, only this more recently used procedure will be described here (Shimazaki *et al.,* 1983). The ELISA variant developed for phytochrome quantitation requires antibodies from two different animals (Fig. 5). By adsorbing excess antibody from one animal (RAP) to the ELISA

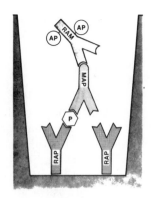

Fig. 5. A scheme for the quantitation of phytochrome by an indirect, sandwich-type ELISA. MAP = monoclonal antibody to phytochrome; RAM = rabbit antibody to mouse IgG; RAP = rabbit antibody to phytochrome; P = phytochrome; AP = alkaline phosphatase, which is covalently linked to RAM.   Details of the assay are described in the text.

plate in excess, it is possible subsequently to adsorb phytochrome quantitatively from a crude plant extract.   Addition of an antibody to phytochrome from the second animal (MAP) will then produce a complex that can be quantitated by addition of labelled antibody to that from the second animal (AP-RAM; Fig. 5).   Since the labelled antibody is specific, it exhibits negligible binding to the antibody that was used to coat the plate.   This variant has the advantages of being very sensitive, of giving very low levels of reaction product in the absence of added antigen, and of being a direct assay.   Thus phytochrome is detected by an increase in reaction product over a very low background rather than by a decrease in reaction product over a high background, as is the case with more conventional competitive immunoassays (Hunt and Pratt, 1979; Voller *et al.,* 1980).

   Prepare crude extracts for assay by grinding frozen plant tissue in liquid nitrogen with a mortar and pestle.   Add 0.63 ml of H_2O, 70 μl of 0.5 M tris-Cl, pH 8.5 at room temperature, and 10 μl of 2-mercaptoethanol per g of tissue.   Clarify the crude extract by centrifugation for 10 min at 50,000 g. The clarified extract is used without further preparation.

   Wash ELISA plates with ethanol as already described (Section IV. E). Add 50 μl of 5 μg/ml immunopurified rabbit antibodies to either oat or pea phytochrome (see Chapter 8, Section II. D. 6).   The antibodies are diluted in 0.2 M sodium borate, 75 mM NaCl, pH 8.5.   Incubate overnight at 4°C. Empty and wash the plate once as already described (Section IV. E).   To block remaining nonspecific binding sites, fill the wells with 1% bovine serum albumin in PBS.   After a 30-min incubation at room temperature, empty and wash once as before.   Add 50 μl of a known quantity of purified phytochrome for a standard curve or 50 μl of crude extract, which should be freshly prepared as described in the preceding paragraph. Normally the crude extract will have to be diluted substantially (as much as 1000-fold if from etiolated tissue) prior to assay, in which case it should be diluted with Diluent (PBS containing 0.05% Tween 20, 0.02% NaN₃, and 1% bovine serum albumin). Incubate in white or green light as desired for 4 h at 4°C. Wash three times as before. Add 50 μl of monoclonal antibody (or immunopurified antibody from any second animal such as a guinea pig). A mixture of monoclonal antibodies will provide a more sensitive assay than just one used alone provided that they bind to different epitopes on phytochrome. The antibodies should be at 10 μg/ml each in Diluent. Incubate for 45 min at 37°C. Wash three times as before and add 50 μl of alkaline phosphatase-conjugated antibody to mouse IgG. (If polyclonal antibodies from another animal were used, then the conjugated antibody must be to the IgG of that other animal.) A 500-fold dilution of commercially obtained conjugated antibody is satisfactory (see Section IV. E). Incubate for 2 h at 37°C. Wash three times. Add substrate solution, incubate, and stop the reaction as described previously (Section IV. E). With antibodies of high affinity, less than 250 pg (1 femtomol of dimer) of phytochrome can be detected.

Crude plant extracts can exhibit nonspecific inhibitory activity in this ELISA. This inhibition can be seen most easily by quantitating pea phytochrome after dilution of a pea phytochrome standard with a crude extract from oat shoots (Shimazaki *et al.*, 1983). Since the monoclonal antibodies to pea phytochrome that are used in the assay do not cross react significantly with oat phytochrome, the observed inhibition in activity may be considered nonspecific. Nonspecific inhibition can be compensated for by assaying the effect on a pea phytochrome assay of an oat extract at the dilution used for oat phytochrome quantitation (Shimazaki *et al.*, 1983). Since virtually all steps in the ELISA can be automated, it is easy to perform this correction.

E. Immunocytochemistry

The double-indirect, peroxidase–antiperoxidase method as applied to phytochrome was used originally for its microscopic visualisation *in situ* at both the light and electron microscope levels (see Pratt *et al.*, 1976, for review). Newer approaches, which yield better structural preservation of the tissue to be examined as well as more sensitive immunocytochemical staining, have since become available. Only these newer approaches will be described here (Fig. 6; Epel *et al.*, 1980; Verbelen *et al.*, 1982; Saunders *et al.*, 1983).

Fix small tissue pieces in 3% formaldehyde, 0.5% glutaraldehyde, 0.1 *M* sodium phosphate, pH 7.4, for 2–4 h at 25°C in darkness. Wash overnight

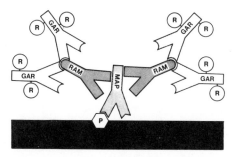

Fig. 6. A scheme illustrating a method for the indirect immunofluorescence visualisation of phytochrome with monoclonal antibodies. P = phytochrome in a section of fixed tissue; MAP = monoclonal antibody to phytochrome; RAM = rabbit antibody to mouse IgG; GAR = goat antibody to rabbit IgG; R = rhodamine, which is covalently linked to GAR.

at 4°C in 0.3% formaldehyde, 0.1 M sodium phosphate, pH 7.4. The procedure may then be continued under white light. Cut the fixed tissue into 0.5-mm cubes. Add a few grains of $NaBH_4$ in 0.1 M sodium phosphate, pH 7.4, and incubate at room temperature for 15 – 30 min to eliminate possible glutaraldehyde-induced fluorescence. Infiltrate the tissue with 1.3 M sucrose, 0.1 M sodium phosphate, pH 7.4, by incubation overnight at 4°C and then mount the pieces on a copper block for sectioning in a cryoultramicrotome at about − 50°C. The precise temperature needed to get satisfactory results must be determined by trial and error. Pick up the sections (1 μm thick is satisfactory) with a small wire loop containing a drop of 2.3 M sucrose, bring to room temperature, and transfer to a glass micro slide that has been treated previously with Slipicone (Dow Corning, Cat. No. 316) to make the glass surface hydrophobic and thereby facilitate section adhesion. All further steps are carried out at room temperature.

Treat the sections with 2% gelatin in PBS for 10 min to make them hydrophilic. Wash for 10 min with PBS and then for another 10 min with 20 mM glycine in PBS. Cover the sections with sheep serum and incubate for 15 min. Wash twice with PBS containing 10% (v/v) sheep serum. Add either 10 μg/ml of a suitable monoclonal antibody to phytochrome (not all work well for this application) (MAP in Fig. 6) or 30 μg/ml of immunopurified polyclonal antibody to phytochrome. In either case, the antibodies are diluted with PBS-10% sheep serum. Incubate for 2 h. Rinse with PBS-10% sheep serum and then cover with sheep serum for 15 min. Wash twice with PBS-10% sheep serum. If monoclonal antibodies are used, next add immunopurified rabbit antibodies to mouse IgG (see Section IV. G) at 10 μg/ml in PBS-10% sheep serum (RAM in Fig. 6). Incubate for 45 min, rinse, add sheep serum, incubate for 15 min, and then wash twice with PBS-10% sheep serum. Add a suitable dilution of rhodamine-labelled antibodies to rabbit IgG (e.g., 400-fold dilution of Miles-Yeda, Ltd., Cat. No. 61-266) prepared in PBS-10% sheep serum (R-GAR in Fig. 6). Incubate for 45 min and then wash twice with PBS. Add 10% PBS, 90% glycerol as a mounting medium and cover with a cover slip. Phytochrome-associated fluorescence may then be observed with a fluorescence microscope equipped for rhodamine visualisation.

Controls for nonspecific and endogenous activity should be performed on parallel sections by substituting nonimmune rabbit or mouse IgG for polyclonal rabbit or monoclonal mouse antibody, respectively. Satisfactory controls exhibit negligible fluorescence.

A rhodamine-labelled antibody to mouse IgG is not used in conjunction with monoclonal-antibody labelling of phytochrome because the inclusion of the rabbit antibody to mouse IgG (RAM in Fig. 6) as a bridge enhances sensitivity. This enhancement is derived presumably from a subsequent attachment of a greater number of rhodamine molecules per antigenic site.

A modification of this procedure has been reported for the visualization of phytochrome by transmission electron microscopy. Verbelen et al. (1982), who worked with polyclonal rabbit antibodies, substituted ferritin for rhodamine as the label, thus making the stain electron dense. A potentially better label, however, should be colloidal gold because of its greater contrast.

The procedures for immunocytochemistry described here were adapted from methods described by Bourguignon et al. (1978) and Tokuyasu and Singer (1976). Further details about the above and other methods may be obtained from a review by Tokuyasu (1980) and a monograph by Sternberger (1979).

F. Immunoblotting

Immunoblotting is effectively a modification of immunocytochemical methods, the modification being that the antigen is visualized on nitrocellulose or some other equivalent medium rather than *in situ*. Immunostaining of nitrocellulose, following transfer of polypeptides to it from SDS polyacrylamide slab gels (Towbin et al., 1979), has recently been applied to the detection of phytochrome (Kerscher and Nowitzki, 1982; Daniels and Quail, 1983; Vierstra et al., 1984). If one can have confidence in the monospecificity of antibodies used for immunostaining, then it is practical to use this technique to characterize phytochrome as to monomer size (Kerscher and Nowitzki, 1982; Vierstra et al., 1984) and heterogeneity (see Chapter 8, Section IV.), and to identify phytochrome peptides resulting from proteolytic degradation of the native monomer (Daniels and Quail, 1984; M.-M. Cordonnier, personal communication).

As with other immunochemical methods, many variants of the basic procedure exist. The specific protocol presented here is provided by M.-M. Cordonnier (personal communication) and is the same as that used in the author's own laboratory. The reader is referred to a review by Gershoni and Palade (1983) for a more general discussion of this technique.

The first step is to separate polypeptides by SDS–PAGE. The methods described in Chapter 8 (Section III. A.) are satisfactory for this purpose, although use of different acrylamide concentrations and/or use of an acrylamide gradient might be more appropriate depending upon the sizes of the polypeptides to be examined.

Transfer polypeptides from the slab gel to nitrocellulose (e.g., Bio-Rad Nitrocellulose Membrane, 0.45 μm). Transfer is accomplished most easily with a commercial apparatus, following the detailed instructions supplied with it (e.g., Bio-Rad Trans-Blot apparatus). To effect transfer, construct a sandwich consisting of the following components, given in order: Scotch-Brite or a thick sponge pad, blotting paper (e.g., Whatman #1), nitrocellulose (include this sheet only if transfer is to be by diffusion — see below), polyacrylamide gel, nitrocellulose, blotting paper, and Scotch-Brite or another thick sponge pad. Hold the sandwich together tightly in an appropriate press. Ensure that there are no air bubbles between the nitrocellulose and gel because they will interfere with transfer of polypeptides. For diffusion transfer, place the sandwich in a large volume (e.g., 2 litres) of transfer buffer (25 mM tris, 192 mM glycine, 20% methanol, pH adjusted to 8.3 with HCl) for at least 48 h. For electrotransfer, place the sandwich in a suitable electrophoretic transfer chamber filled with transfer buffer. Transfer is virtually complete within 3 h at 40 V using the Bio-Rad apparatus. For electrotransfer ensure that the nitrocellulose is on the anode side of the gel.

Following transfer of polypeptides to the nitrocellulose, remove the sandwich from the press. At this stage, mark the nitrocellulose so that both its orientation with respect to the gel, and the surface that was appressed to the gel, can be determined later. Use a pen with water-insoluble ink for this purpose. If desired, transiently stain the nitrocellulose by immersion in Ponceau S (0.2% in 3% trichloroacetic acid) followed by a brief rinse with water. Since the Ponceau S elutes from the protein during preparation for immunochemical staining, it provides a convenient method for visualizing total protein on the nitrocellulose prior to immunostaining. Place the nitrocellulose in PBS with 1% bovine serum albumin, 1% sheep serum, and 0.02% NaN_3, which minimizes further nonspecific binding of reactive protein to the nitrocellulose. The nitrocellulose may be used anywhere from 30 min to several months later. Long term storage should be at 4°C.

Immunostaining is conveniently done at room temperature on a completely level surface covered with Parafilm. Place the nitrocellulose, which may be cut into strips of any size desired, on the Parafilm such that the side with adsorbed protein (the side that was originally adjacent to the gel) is up. Add primary antibody at 1 μg/ml if purified, or at an approximately equivalent concentration if present in serum, ascitic fluid or hybridoma medium. Dilute the primary antibody, as well as all subsequent antibodies, in the same diluent as that used for ELISA (Section VI. D.), except that sheep serum has been added to 10%. Add primary antibody and all subsequent reagents in sufficient quantity to form a layer about 1 mm thick on the nitrocellulose. The Parafilm is sufficiently hydrophobic that liquid will not flow away from the nitrocellulose. After a 1-h incubation, transfer the strips to appropriately sized containers and wash them for 2 15-min periods in ELISA wash buffer (Section VI. D.) supplemented with 1% sheep serum and 0.1% bovine serum albumin. Agitate the strips gently during washing. Transfer the washed strips back to clean Parafilm.

If the primary antibody is monoclonal, use as second antibody 1 μg/ml of rabbit antibody to mouse IgG, which enhances sensitivity about three-fold. Incubate for 1 h. Wash the strips as described above.

The final antibody is alkaline phosphatase conjugated antibody to rabbit IgG. If the primary antibody was a polyclonal antibody from an animal other than a rabbit, then the enzyme conjugate should of course be directed to IgG from that other animal. Optimal results are usually obtained with commercial conjugates (e.g., Sigma A-8025) diluted 1000-fold. Incubate 1 h and wash as before. Prior to addition of substrate, equilibrate the nitrocellulose for 15 min in substrate buffer (0.1 M tris-Cl, 0.5 mM MgCl$_2$, pH 9.0). Transfer the strips to Parafilm in boxes that can be tightly sealed (e.g., Nalgene 5700-0500). Add substrate solution (0.5 mg/ml bromochloroindolyl phosphate in substrate buffer; see Knecht and Dimond, 1984) and incubate as long as desired. Overnight at room temperature is convenient. Rinse the nitrocellulose briefly with water, blot it dry with paper towels, and air dry at room temperature by placing it on a clean, nonadherent surface. The blue reaction product will fade in sunlight, so it is recommended that the nitrocellulose be stored out of the light.

G. Other methods

As noted already, a large variety of other immunochemical methods are available, some of which have been applied to phytochrome but not described here. In addition to those already mentioned above, others include micro complement fixation (Hopkins and Butler, 1970; Pratt, 1973; Pratt et al., 1974; Cundiff and Pratt, 1975). Micro complement fixation has been used to compare Pr and Pfr immunologically, to clarify the relationship between large and small sizes of phytochrome, and to determine immunological similarities among different phytochrome preparations obtained from different plant species. Since this method does not require any special care when applied to phytochrome, it is appropriate to conserve space here by referring the interested reader to these original publications for further information.

ACKNOWLEDGMENT

Methods that are derived from the author's research programme and are described herein were developed with support provided by the National Science Foundation and the Department of Energy.

REFERENCES

Bourguignon, L. Y. W., Tokuyasu, K. T. and Singer, S. J. (1978). *J. Cell. Physiol.* **95**, 239–258.
Cordonnier, M.-M. and Pratt, L. H. (1982a). *Plant Physiol.* **69**, 360–365.
Cordonnier, M.-M. and Pratt, L. H. (1982b). *Plant Physiol.* **70**, 912–916.
Cordonnier, M.-M., Smith, C., Greppin, H. and Pratt, L. H. (1983). *Planta* **158**, 369–376..
Cordonnier, M.-M., Greppin, H., and Pratt, L. H. (1984). Characterization by enzyme-linked immunosorbent assay of monoclonal antibodies to *Pisum* and *Avena* phytochrome. *Plant Physiol.* **74**, 123–127.
Cundiff, S. C. and Pratt, L. H. (1973). *Plant Physiol.* **51**, 210–213.
Cundiff, S. C. and Pratt, L. H. (1975). *Plant Physiol.* **55**, 207–211.
Daniels, S. M., and Quail, P. H. (1984). Use of monoclonal antibodies to identify and characterize distinct structural regions of 124 kd phytochrome from *Avena*. *In* "Hybridoma Technology in Agricultural and Veterinary Research" (N. J. Stern, ed.), Rowman and Allanheld (in press).
Davis, J. M., Pennington, J. E., Kubler, A.-M. and Conscience, J.-F. (1982). *J. Immunol. Methods* **50**, 161–171.
Epel, B. L., Butler, W. L., Pratt, L. H. and Tokuyasu, K. T. (1980). *In* "Photoreceptors and Plant Development" (J. De Greef, ed.), pp. 121–133. Antwerpen Univ. Press, Antwerp.
Ey, P. L., Prowse, S. T. and Jenkin, C. R. (1978). *Immunochemistry* **15**, 429–436.
Fazekas de St. Groth, S. and Scheidegger, D. (1980). *J. Immunol. Methods* **35**, 1–21.
Gershoni, J. M., and Palade, G. E. (1983). Protein blotting: Principles and applications. *Anal. Biochem.* **131**, 1–15.
Goding, J. W. (1980). *J. Immunol. Methods* **39**, 285–308.
Hämmerling, G. J., Hämmerling, U., and Kearney, J. F. (eds.) (1981). "Monoclonal Antibodies and T-Cell Hybridomas," Elsevier, Amsterdam.
Hopkins, D. W. and Butler, W. L. (1970). *Plant Physiol.* **45**, 567–570.
Hunt, R. E. and Pratt, L. H. (1979). *Plant Physiol.* **64**, 327–331.
Kennett, R. H., McKearn, T. J. and Bechtol, K. B., eds. (1980). "Monoclonal Antibodies." Plenum, New York.
Kerscher, L. and Nowitzki, S. (1982). *FEBS Lett.* **146**, 173–176.
Knecht, D. A., and Dimond, R. L. (1984). Visualization of antigenic proteins on Western blots. *Anal. Biochem.* **136**, 180–184.
Köhler, G. and Milstein, C. (1975). *Nature (London)* **256**, 495–497.

Langone, J., and Van Vunakis, H., eds. (1981a). "Methods in Enzymology," Vol. 73, Part B. Academic Press, New York.

Langone, J. and Van Vunakis, H., eds. (1981b). "Methods in Enzymology," Vol. 74, Part C. Academic Press, New York.

Langone, J. and Van Vunakis, H., eds. (1982). "Methods in Enzymology," Vol. 84, Part D. Academic Press, New York.

Langone, J., and Van Vunakis, H. (eds.) (1983a). Immunochemical techniques. Part E. *Methods Enzymol.* **92,** 624 pp.

Langone, J., and Van Vunakis, H. (eds.) (1983b). Immunochemical techniques. Part F. *Methods Enzymol.* **93,** 448 pp.

Littlefield, J. W. (1964). *Science* **145,** 709–710.

Melchers, F., Potter, M. and Warner, N., eds. (1978). "Current Topics in Microbiology and Immunology," Vol. 81. Springer-Verlag, Berlin and New York.

Müller, G., ed. (1979). "Immunological Reviews," Vol. 47. Munksgaard, Copenhagen.

Pratt, L. H. (1973). *Plant Physiol.* **51,** 203–209.

Pratt, L. H. and Coleman, R. A. (1971). *Proc. Natl. Acad. Sci. U.S.A.* **68,** 2431–2435.

Pratt, L. H., Kidd, G. H. and Coleman, R. A. (1974). *Biochim. Biophys. Acta* **365,** 93–107.

Pratt, L. H., Coleman, R. A. and Mackenzie, J. M., Jr. (1976). *In* "Light and Plant Development" (H. Smith, ed.), pp. 75–94. Butterworth, London.

Rice, H. V. and Briggs, W. R. (1973). *Plant Physiol.* **51,** 939–945.

Saunders, M. J., Cordonnier, M.-M., Palevitz, B. A., and Pratt, L. H. (1983). *Planta* **159,** 545–553.

Shimazaki, Y., Cordonnier, M.-M. and Pratt, L. H. (1983). *Planta* **159,** 534–544.

Shulman, M., Wilde, C. D. and Köhler, G. (1978). *Nature (London)* **276,** 269–270.

Staehelin, T., Durrer, B., Schmidt, J., Takacs, B., Stocker, J., Miggiano, V., Stähli, C., Rubinstein, M., Levy, W. P., Hershberg, R. and Pestka, S. (1981). *Proc. Natl. Acad. Sci. U.S.A.* **78,** 1848–1852.

Stähli, C., Staehelin, T., Miggiano, V., Schmidt, J. and Häring, P. (1980). *J. Immunol. Methods* **32,** 297–304.

Sternberger, L. (1979). "Immunocytochemistry," 2nd ed. Wiley, New York.

Tokuyasu, K. T. (1980). *Histochemistry* **12,** 181–203.

Tokuyasu, K. T. and Singer, S. J. (1976). *J. Cell Biol.* **71,** 894–906.

Towbin, H., Staehelin, T., and Gordon, J. (1979). Electrophoretic transfer of proteins from polyacrylamide gels to nitrocellulose sheets: procedure and some applications. *Proc. Natl. Acad. Sci. USA* **76,** 4350–4354.

Van Vunakis, H. and Langone, J., eds. (1980). "Methods in Enzymology," Vol. 70, Part A. Academic Press, New York.

Verbelen, J.-P., Pratt, L. H., Butler, W. L. and Tokuyasu, K. (1982). *Plant Physiol.* **70,** 867–871.

Vierstra, R. D. and Quail, P. H. (1982). *Proc. Natl. Acad. Sci. U.S.A.* **79,** 5272–5276.

Vierstra, R. D. and Quail, P. H. (1983). *Biochemistry* **22,** 2498–2505.

Vierstra, R. D., Cordonnier, M.-M., Pratt, L. H., and Quail, P. H. (1984). Native phytochrome: immunoblot analysis of relative molecular mass and *in vitro* proteolytic degradation for several plant species. *Planta* **160,** 521–528.

Voller, A., Bidwell, D. and Bartlett, A. (1980). *In* "Manual of Clinical Immunology" (N. R. Rose and H. Friedman, eds.), 2nd ed., pp. 359–371. Am. Soc. Microbiol., Washington D.C.

Weir, D., ed. (1978). "Handbook of Experimental Immunology," 3rd ed. Blackwell, Oxford.

10

Model Compounds for the Phytochrome Chromophore

Hugo Scheer

I. INTRODUCTION

A. Usefulness of model studies

The use of model compounds is a general technique in the elucidation of the structures of natural products. It is particularly helpful when dealing with products which occur only in small amounts and are covalently bound to a protein. Phytochrome belongs to this group. Model studies have contributed substantially to the elucidation of the molecular structure of the phytochrome chromophores in both the Pr (Rüdiger *et al.*, 1980; Lagarias and Rapoport, 1980) and the Pfr form (Thümmler *et al.*, 1983), their mode of linkage to the apoprotein and the analysis of the noncovalent interactions with the latter (for references to the major literature, see Rüdiger and Scheer, 1983; Scheer, 1981).

Model compounds may be expected to find new applications in the investigations of phytochrome intermediates, but they are equally useful in areas beyond structural work. The most important application is to test the reactivity of the chromophore, which would otherwise require impractically large amounts of phytochrome. Selective modifications of either the protein or the chromophore can be developed (see, e.g., Kufer and Scheer, 1979), and the optimum conditions for spectroscopic studies have often been selected by this means (Köst *et al.*, 1975; Scheer, 1976; Thümmler *et al.*, 1983).

This chapter outlines the preparation of model compounds and some related techniques which have been important in previous work on phytochrome. Each of the models represents only certain limited aspects of the

TECHNIQUES IN PHOTOMORPHOGENESIS
0-12-652990-6

phytochrome proper, and they differ also considerably in their accessibility. The choice of any particular model will depend on the actual problem under investigation, and also on the particular experience and equipment of the laboratory. Some (biased) comments on these problems are included, along with some unpublished, but useful, procedural modifications.

B. General considerations for the handling of models

Most of the models described here are bile pigments characterised by beautiful colours and a high reactivity. The former is often deceiving, the latter at best annoying, and both require some precautions for the handling of bile pigments.

It is important to prepare each pigment freshly, or at least to repurify it prior to use, because many of the pigments are readily oxidised within hours. Alkaline conditions and/or the presence of heavy metals and/or strong light, which all tend to accelerate oxidation, should be avoided. As an example, the dihydrobilindione 5 and the chromopeptides of phytochrome and phycocyanin are oxidised completely within minutes in the presence of base and zinc (Section VI. C). It is important to store the pigments in the cold, in the dark, and under nitrogen, preferably in sealed ampoules. Finally, it is extremely important never to trust one's eyes. Although visual inspection can be very sensitive to small variations in colour, it can also completely fail. As an example, the free base dihydrobilindione 5 (as well as the chromopeptides of phycocyanin and phytochrome) look identically bright blue as does the anion of one of the common oxidation products as well as the cation of another one. Spectral and chromatographic tests are, therefore, always recommended before and after any experiment. Because of the ease of protonation, deprotonation, and complexation with metals, all spectroscopic characterisations should be done under defined conditions. Acidic conditions (pH ≤ 1.5 in aqueous solutions, or 5% methanolic sulphuric acid) are useful, because oxidation and complexation with traces of metals is impeded, most bile pigments are fully protonated, and the UV-visible absorption band half-widths and maxima are narrowed and increased.

C. Nomenclature of bile pigments

The nomenclature is currently not very well defined. The recent IUPAC recommendation (Bonnett, 1978; IUPAC-IUB Joint Commission on Biochemical Nomenclature, 1979) gives the numbering scheme which is shown in Fig. 1B, in comparison with the old and still widely used Fischer system (Fig. 1A). This recommendation also lists a number of trivial names along

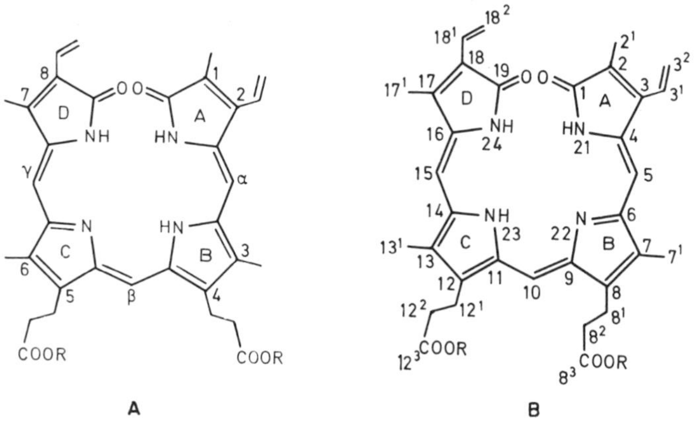

Fig. 1. IUPAC recommendation for numbering of compounds.

with a rational nomenclature, which is again yet only partly accepted. To avoid confusion, formulae to all compounds have been included, and the numbering follows the IUPAC recommendation of Fig. 1B.

II. SYNTHETIC BILE PIGMENTS

Only two bile pigments, bilirubin (8) and biliverdin (7) are commercially available in moderate amounts, and only a few companies (e.g., Porphyrin Products) supply some others in analytical quantities. Since 7 and 8 are only of limited value as phytochrome models, the useful pigments must be obtained from other sources. The total synthesis of bile pigments by linear condensation of pyrroles has been greatly advanced (see, for example, Gossauer et al., 1981a,b), but is, for unsymmetrically substituted target molecules, still very demanding. A comparably quick synthesis of the symmetrically substituted biliverdin-IVγ has been published by Falk and Grubmayr (1977); this has been used, for example, as the starting material for the bridged bilindione 6 (Section II. D).

Another source of bile pigments is the oxidative cleavage of cyclic tetrapyrroles (porphyrins), which is also the biosynthetic route to the naturally occurring bile pigments (O'Carra, 1975; MacDonagh, 1979; Brown et al., 1981). The accessibility of bile pigments by the degradative route is limited by that of the parent porphyrins. Of the two common synthetic porphyrins, [e.g., 2,3,7,8,12,13,17,18-octaethylporphyrin and 5,10,15,20(=meso)-tetraphenylporphyrin], only the latter (2) has been used for model studies, owing to its somewhat similar substitution pattern with the naturally occurring bile pigments.   The octaethylbilindiones **1** and **4** (Sections II. A–C) are derived from this source.   Oxidative cleavage applied to the natural porphyrins is also possible, and examples are given in Section III.   Finally, very useful models have been obtained from the readily accessible phycobiliproteins, which also serve by themselves as excellent models for native phytochrome.   Examples of these are given in Section IV.

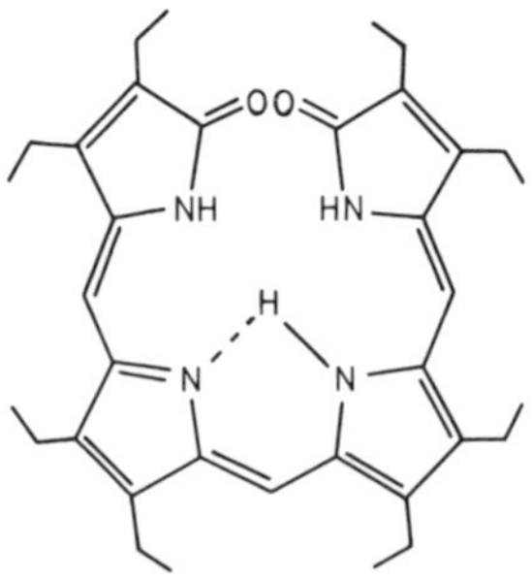

1a

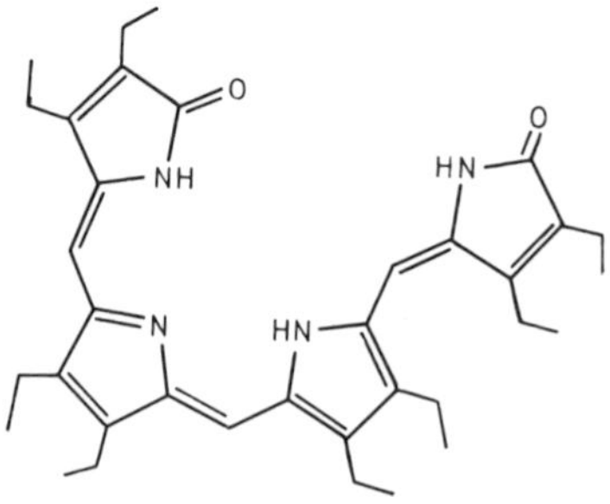

1b

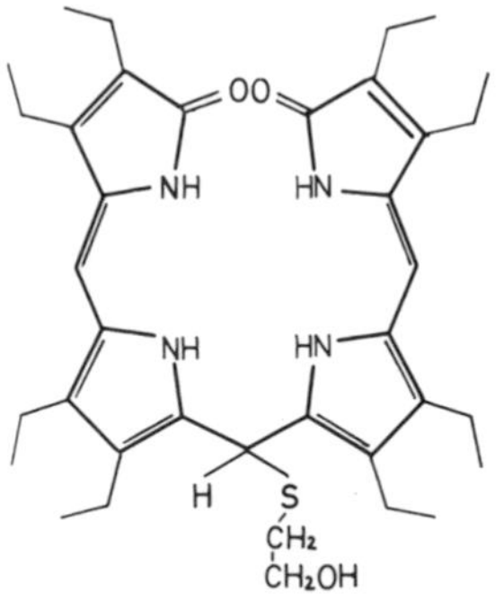

1c

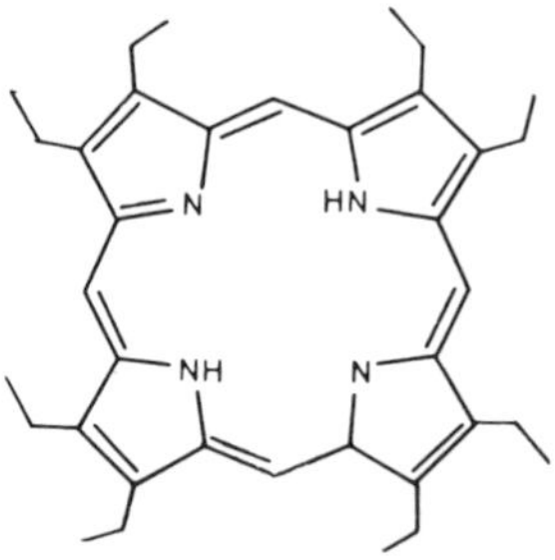

2

A.   (*Z,Z,Z*)-2,3,7,8,12,13,17,18-Octaethylbilindione

The following procedure is modified from Cavaleiro and Smith (1973).   The starting material, octaethylporphyrin (**2**), is commercially available from, for example, Porphyrin Products and Strem.   Compound **1a** is also obtained as a by-product in varying yield during the preparation of the dihydrobilindione **4a**.

Octaethylporphyrin (**2**) is first converted to octaethylhemin by reaction with anhydrous ferric chloride in glacial acetic acid (95–100% yield). The hemin is cleaved to the bilin **1a** by coupled oxidation with ascorbate and oxygen in aqueous pyridine. Compound **2** (200 mg) is dissolved in pyridine (60 ml, analytical grade) and heated to $50 \pm 1°C$. Under vigorous mechanical stirring, a solution of ascorbic acid (7.2 g) in water (160 ml) is added from a dropping funnel. While the temperature is kept at 50°C and the pH of the solution is adjusted to about 7.5, the vigorously stirred solution is flushed with oxygen for 3 h. After workup (extraction with methylene chloride, washing with water, drying over sodium chloride, and evaporation to dryness), the green residue is dissolved in 5% methanolic sulphuric acid (25 ml) and kept for 12 hr in the dark at ambient temperature. The product mixture is worked up and chromatographed on silica with carbon tetrachloride–acetone (93:7, v/v) to yield the green octaethylbilindione **1a** in 40–50% yield. The violet by-product is 4,5-dihydro-4,5-dimethoxyoctaethylbilindione, the dimethylether of (**3**) (10–15% yield).

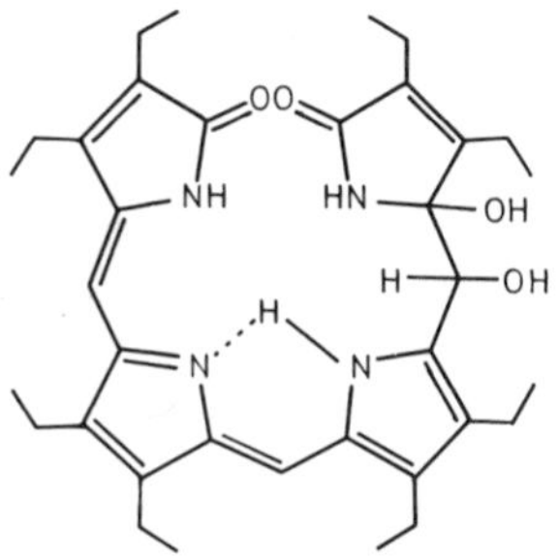

3

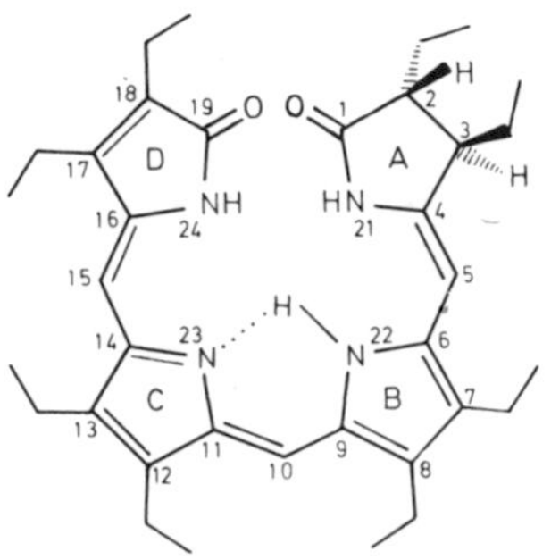

4a

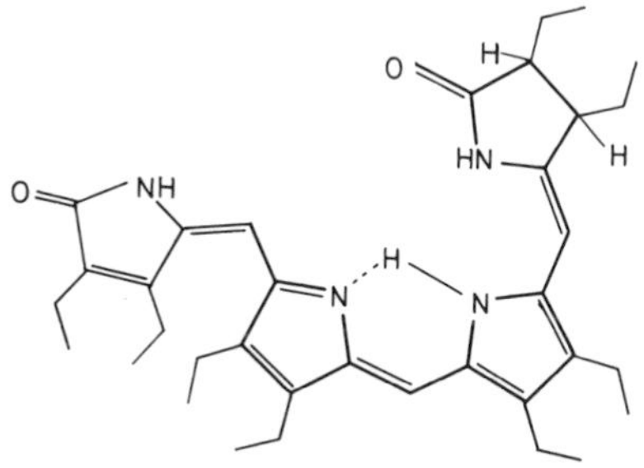

4b

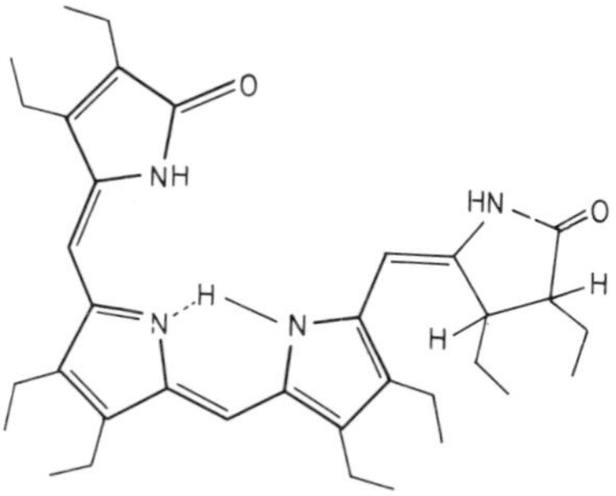

4c

5

B. (*Z,Z,Z*)-*trans*-2,3,-Dihydro-2,3,7,8,12,13,17,18-octaethylbilindione

The dihydrobilindione **4a** is prepared from commercial octaethylporphyrin **2** by the method of Cavaleiro and Smith (1973). It involves the conversion to the hemin and reduction to octaethylchlorin **5** with sodium in isoamyl alcohol according to Whitlock *et al.* (1969). The chlorin **5** itself, or its zinc complex (prepared with zinc acetate in methanol–chloroform (1:1) at 60°C) is then treated with thallium trifluoroacetate in a dry mixture of methylene chloride and tetrahydrofurane. After workup of the reaction mixture, it is chromatographed with benzene over partially deactivated alumina. The first eluting band contains the green *meso*-trifluoroacetoxychlorin (or its zinc complex, respectively), which is converted on standing to the blue dihydrobilindione **4a**. The latter is purified by chromatography over partially deactivated alumina. The yield is rather variable (5–20%), the main factor being the purity of the thallium trifluoroacetate. Following a suggestion of A. McKillop (private communication), we have obtained the best yields with homemade rather than the often very impure commercial thallium oxide. This is prepared by dissolving thallium sulphate (10 g) in aqueous sodium hydroxide (2 N, 50 ml) followed by dropwise addition of hydrogen peroxide (30%, 4.7 ml). The dark precipitate is collected by centrifugation and repeatedly washed with aqueous sodium hydroxide (0.5 N) followed by water, until neutral. The product is dried over sodium hydroxide and converted to the thallium trifluoroacetate by the procedure of McKillop *et al.* (1971).

C. *Z* to *E* Isomerisation of bilindiones — example:
(*E,Z,Z*)-2,3,7,8,12,13,17,18-octaethylbilindione

The "outer" methine double bonds Δ4,5 and Δ15,16 in bilindiones are localised and capable of the formation of stable *Z,E* isomers (4Z,9Z,15Z;

$4E,9Z,15Z$; $4Z,9Z,15E$ and $4E,9Z,15E$).   No stable isomers have yet been observed for the "inner" methine double bond(s) which is probably due to a delocalisation over the C-9,10 and C-10,11 bonds.   All E isomers are thermodynamically less stable than the Z,Z,Z isomers (Falk and Müller, 1981), to which they revert photochemically and thermally in a reaction catalysed by, e.g., acid and redox reagents.   The photoisomerisation works in most free bilindiones (biliverdins) only in one direction ($E \rightarrow Z$), and these pigments are then generally only isolated in the Z,Z,Z form (Falk *et al.*, 1979). The 10,23-dihydrobilindiones (bilirubins) are capable, by contrast, of photoisomerisations in both directions and thus form, upon irradiation with visible light, photostationary mixtures containing moderate amounts of the E,Z,Z and Z,Z,E isomers and a minor amount of the E,Z,E isomer (MacDonagh *et al.*, 1982).   Falk *et al.* (1980) have devised a general method which is based on this fact.   The biliverdin is first converted to a rubinoid pigment by addition of a nucleophile to C-10, which is then subjected to photoisomerisation.   Workup under conditions which remove the nucleophile lead then to a mixture of the corresponding Z,E-isomeric biliverdins. The preparation of the E,Z,Z-bilindione **1b** given below (Kufer *et al.*, 1982a) follows essentially the original procedure of Falk *et al.* (1979).   Owing to the symmetry of the Z,Z,Z-adduct **1a**, only one E,Z,Z isomer (**1b**) can be formed.

Although the method is principally applicable to all biliverdins, practical problems may arise from unsymmetric products (separation, regioselectivity), from labile substituents (e.g., irreversible addition of the nucleophile) and/or from biliverdins which do not readily add nucleophiles at C-10. These problems are of considerable interest with regard to the photoisomerisation of the native phytochrome chromophore.   The general procedure may then have to be modified.   For example, the unsymmetric and more unstable dihydrobilindione **4a** requires a larger amount of the thiol to form the rubinoid addition product, and forms only one ($15E$) (**4b**) of the two possible E isomers (**4b,c**), which is also more labile under the workup conditions (Kufer *et al.*, 1982a).

The Z,Z,Z-bilindione **1a** is treated with 2-mercaptoethanol in dimethyl sulphoxide (7.5% v/v) to yield a yellow solution of the addition product.   It is irradiated for 10 min with blue light under nitrogen.   All the following steps must be carried out under a dim green safelight or, where possible, in darkness.   The solution is first poured into chloroform, and the 2-mercaptoethanol is extracted with dilute aqueous potassium hydroxide.   The resulting green chloroform solution is worked up, and the residue is chromatographed on silica H plates with chloroform – methanol (20:1).   The yield of the E,Z,Z isomer **1b** is 10%.

D. 21*N*,24*N*-Methyleneetiobiliverdin-IVγ

One of the problems in bile pigment chemistry is their large conformational freedom.   21*N*,24*N*-Methyleneetiobilindione-IVγ (**6**) (Falk and Thirring, 1981) is restricted to cyclic conformations.   Based on substantiated molecular orbital calculations and spectroscopic data, a predominantly cyclic–helical conformation has been assigned to biliverdin **9** and many other free bilindiones (see Scheer, 1981, for leading references).   Another conformationally restricted bilindione, the "extended" isophorcabilin **19** is described in Section III. I.   Compound **6** has been prepared by Falk and Thirring (1981) from its open-chain parent, etiobiliverdin-IVγ, by insertion of a methylene group.   The procedure should be applicable to all octaalkylbilindiones.   The reaction is (probably for steric reasons) regioselective, and only small amounts of the isomers bridged between the neighbouring N-21 and N-22 (=N-24 and N-23) or N-22 and N-23 atoms are formed.

The anion of etiobiliverdin-IVγ is treated in dimethyl sulphoxide–potassium hydroxide with diiodomethane under argon at 100°C.   Compound **6** is purified and separated from its isomers by chromatography on silica G with chloroform–methanol (50:1).   The yield is 20%, the two isomers are isolated in 1 and 2% yield, respectively.

III. SEMISYNTHETIC BILE PIGMENTS

A. Biliverdins from bilirubins — example: biliverdin-IXα

Biliverdin-IXα (**7**) is prepared from commercial bilirubin (=bilirubin-IXα) (**8**) by oxidation with high potential quinones.   Several other oxidants have been described in the literature (see MacDonagh, 1979), but quinones are now most widely used (Stoll and Gray, 1977; MacDonagh and Palma, 1980; Manitto and Monti, 1979a) and worked best in our hands.   The following procedure is similar to the one published by Stoll and Gray (1977) and Manitto and Monti (1979a).   It can be applied to the oxidation of most 10,21-dihydrobilindiones (bilirubins) to the corresponding bilindiones (biliverdins), provided none of the substituents is attacked by the oxidant more readily than the tetrapyrrole itself.   The reaction can be checked spectrophotometrically (increase around 650 nm, decrease around 450 nm).   The workup described here is tailored to biliverdin-IXα and other biliverdins containing free carboxyl groups.   With other bilindiones the chromatography must be changed according to the specific needs.   A solution of bilirubin (**8**) (50 mg) in dimethyl sulphoxide (50 ml) is flushed with nitrogen for 15 min.   Over a period of 15 min, 2,3-dichloro-5,6-dicyano-*p*-benzoquinone (sublimed, 50 mg) in dimethyl sulphoxide (10 ml) is added under nitrogen. After a further 15 min, the continuously stirred mixture is partitioned between chloroform (500 ml) and water (350 ml). A little precipitate, which often forms, is discarded. The organic phase is washed three times with water, dried over sodium chloride, and evaporated to dryness (temperature below 40°C). The yield is 32–38%.

The crude product is chromatographed on silica plates with the upper phase of toluene–acetic acid–water (5:5:1). The green zone ($R_f \approx 0.1$) is eluted from the scraped-off material with acetic acid and then separated from residual silica by centrifugation. The solution is then partitioned between chloroform and water as described above. The pure 7 is crystallised by dissolution in chloroform (3 ml) containing a few drops of methanol, addition of n-hexane and standing at −20°C. The yield from bilirubin is about 10% but is much higher when working with esters.

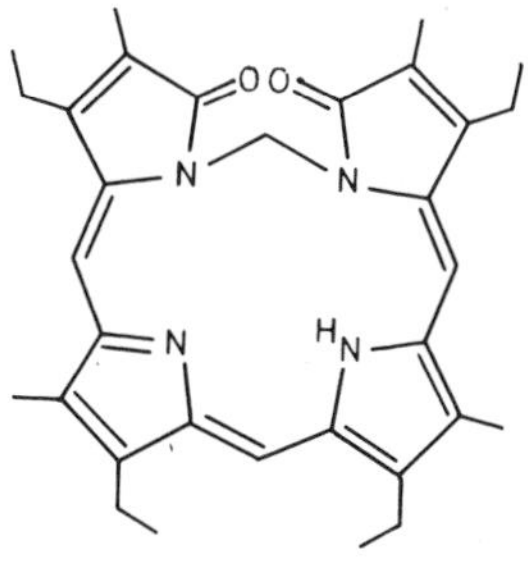

6

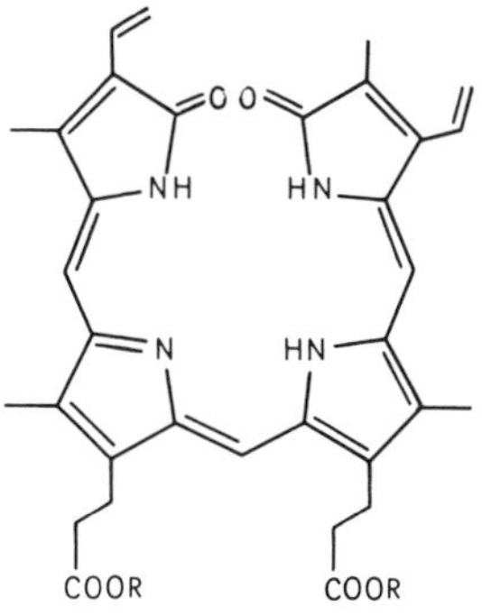

7: R = H
9: R = CH$_3$

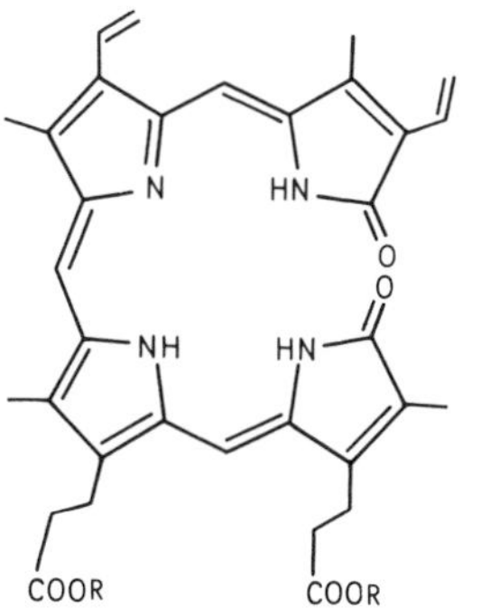

11: R = H
14: R = CH$_3$

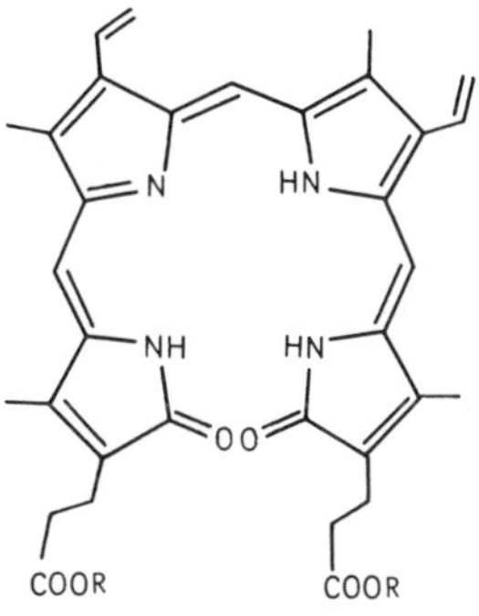

12: R = H
15: R = CH$_3$

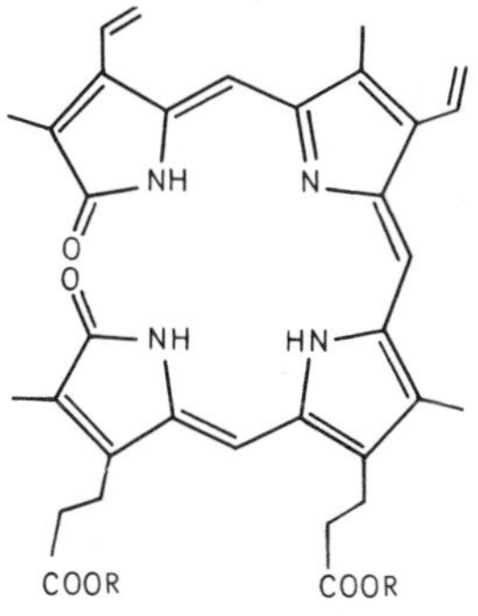

13: R = H
16: R = CH$_3$

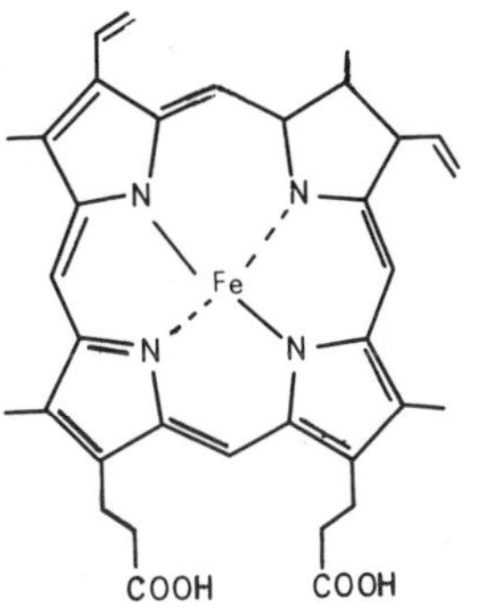

17

B. Esterification of biliverdins with free carboxyl groups — example: biliverdin-IXα dimethyl ester

Esterification of biliverdin-IXα to its dimethyl ester (9) is carried out with ethanol under acid catalysis. It can be applied to all bile pigments which are stable to acid, e.g., to biliverdins, but not to unsymmetric bilirubins (scrambling). For the latter, diazomethane is the reagent of choice (see Section III. C). The often cited methanolic hydrochloric acid (requiring gaseous hydrogen chloride for its preparation) can, according to our experience, always be replaced by methanolic sulphuric acid. An alternative is the use of boron trifluoride in methanol, introduced by Cole *et al.* (1967). In most cases, the yields are comparable, but it may be useful to run a test first. It may also be useful to change the temperature. The procedure described below requires 10 min refluxing at 64°C, but similar yields are obtained (with 7) if the reaction is carried out in the refrigerator overnight; this is a superior method for heat-sensitive pigments. Biliverdin-IXα (7) (10 mg) is refluxed in methanol (150 ml) under nitrogen. After addition of boron trifluoride in methanol (20%, 60 ml), the solution is refluxed for an additional 10 min under nitrogen. The reaction mixture is partitioned between chloroform and water, and the organic phase is washed until neutral and then evaporated to dryness. Chromatography on silica plates with chloroform – acetone (95:5) yields 85 – 90% 9.

C. Esterification of bilirubins with free carboxyl groups — example: bilirubin-IXα dimethyl ester

Bilirubins "scramble" under various conditions, especially in acid (see, e.g., MacDonagh, 1979). This means that the two halves of the molecule are interchangeable and eventually form a statistical mixture of the possible isomeric bilirubins. Asymmetric bilirubins cannot therefore, be esterified by acid catalysis. In these cases diazomethane is the best reagent, but it should be kept in mind that it can react with other functional groups in addition.

A suspension of bilirubin-IXα (8) (30 mg) in chloroform (10 ml) is treated with an ethereal solution of diazomethane (0.5 ml) and kept for 12 h in the dark at 4°C. The solution is washed with aqueous sodium carbonate (10%), dried, and evaporated at ≤ 30°C. The crude product is chromatographed on neutral alumina (activity "super 1"). After elution of two minor yellow bands with chloroform, bilirubin-IXα dimethyl ester (10) is eluted with chloroform – methanol (9:1) in 65% yield.

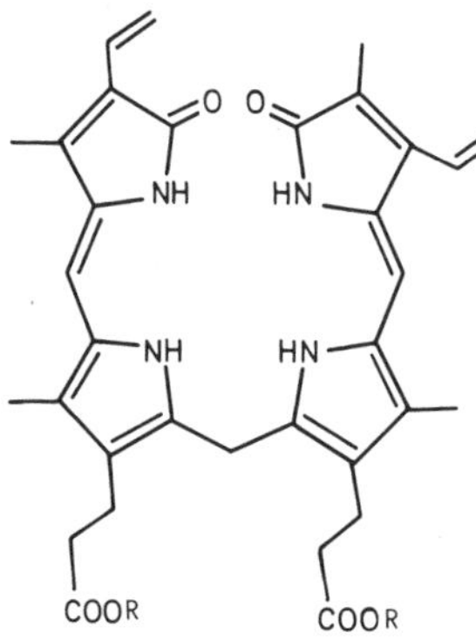

8: R = H
10: R = CH$_3$

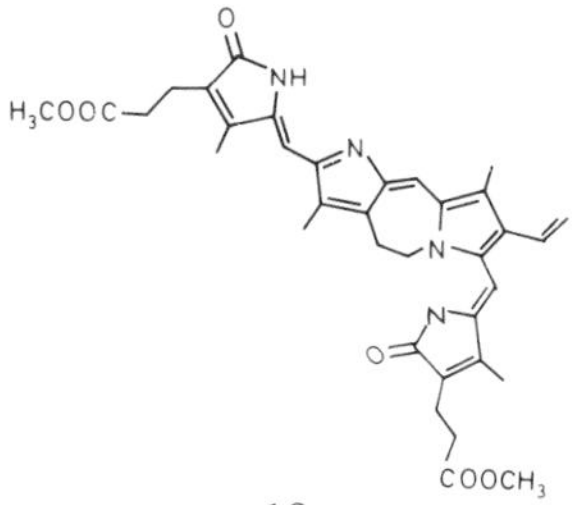

18

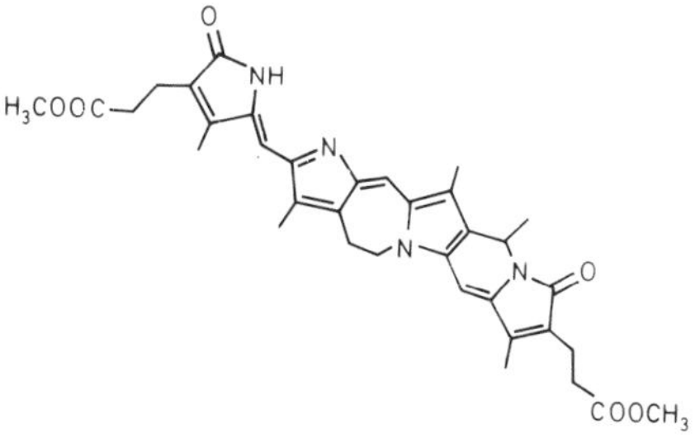

19

D. Isomeric biliverdins-IXα,β,γ, and δ and their dimethyl esters by coupled oxidation of protohemin

The isomeric pigments are obtained in a roughly 1:1:1:1 mixture by coupled oxidation of protohemin (17) in pyridine. Bonnett and MacDonagh (1973) give two procedures which use either ascorbate or hydrazine with oxygen. Both reactions work best with small amounts of hemin (≤ 100 mg batches). The reaction with hydrazine can be scaled up, but we obtained the best results by a series of smaller batches rather than one large one. The ring opening of the resulting oxophlorins to the four isomeric biliverdins-IXα,β,γ, and δ (7, 11, 12, and 13, respectively) and the esterification of the latter are achieved by treatment with methanolic sulphuric (or hydrochloric, see Section III. B) acid, or with methanolic boron trifluoride.

The corresponding four isomeric esters (9, 14, 15, and 16) then must be separated, which is the most time-consuming step in the preparation. The strategy of the separation depends on the desired isomer. The isolation of the IXγ isomer (15), which is the starting material for the phorca- (18) and isophorcabilins (19) (Sections III. H and I), works well on a preparative scale with a two-step chromatography on silica plates. The first, which can also be done on a column, involves chloroform–methanol (97:3) to yield three zones containing, in increasing order of mobility, the δ isomer (16), a mixture of the α and γ isomers (9 and 15), and the β isomer (14). The α,γ mixture is then separated with toluene–2-butanone–acetic acid (10:5:0.5) (O'Carra and Colleran, 1970). The upper zone contains the γ isomer (15) and the yield is about 7 mg from 100 mg of 17, which corresponds to 33% if one assumes a random opening at all four methine bridges. The separation of the free acids (7, 11, 12, and 13) is possible, but impractical on a preparative scale.

Smith *et al.* (1980) have recently achieved a better preparation of various isomeric bilindione dimethyl esters, which is also applicable to the free acids. The isomers are separated prior to the ring opening proper at the stage of the oxophlorins, which are also prepared by a different technique. It starts with the zinc complex of the porphyrin, which is accessible via demetalation of the hemin and remetalation with zinc in refluxing methanol–chloroform. The zinc porphyrin is first treated with thallium trifluoroacetate (see Section II. B) to yield, after acidic workup, the mixture of the four isomeric oxophlorins. They are separated on silica plates with methylene chloride–methanol (97:3) (two developments) to yield, in increasing order of mobility, the β, α, δ, and γ isomers. Iron reinsertion and ring opening of the separate isomers with pyridine–oxygen yields the biliverdins. No yield is stated. The reaction has been applied to deuteroporphyrin and several 3,8-disubstituted deuterorphyrins, but apparently not to pigments containing vinyl groups, e.g., protoporphyrin.

E. Isomeric bilirubins-IIIα and -XIIIα (20, 21) and their dimethyl esters (22, 23) by scrambling of bilirubin-IXα(8)

Asymmetric bilirubins exchange their two halves to yield a mixture of the possible isomers under a variety of conditions, including light, acid, and base (see MacDonagh, 1979, for leading references). Commercial bilirubin (8) may contain undesirable contaminations of these isomers. The equilibrium mixture is, on the other hand, a useful source of symmetric bilirubins, e.g., dimethyl esters 22 and 23. The time-consuming step is again the chromatographic separation. Monti and Manitto (1981) have devised an elegant route to the IIIα isomer (22), which takes advantage of the ready addition of nucleophiles to exo vinyl groups (Section III. J). By adding a hydrophilic group, 22 can be fished out of the isomer mixture by means of a simple solvent extraction since 22 is the only substance not containing such a

group. The scrambling reaction is typical for 10,23-dihydrobilindiones (bilirubins), but is of great potential for all bile pigments which are reversibly
convertible to bilirubins. A striking example is the preparation of a great
variety of pigments with unusual asymmetric substitution pattern by Stoll
and Gray (1977). A key step is the scrambling of a mixture of the dimethyl
ester of one 10,23-dihydrobilindione with the free dicarboxylic acid of another 10,23-dihydrobilindione, which again greatly simplifies the isolation
of the scrambling product, a monomethyl ester, owing to the large differences in polarity.

The following general procedure is taken from Stoll and Gray (1977).
The bilirubin (100 mg) is dissolved in dimethyl sulphoxide (20 ml). The
mixture is flushed for 10 min with purified nitrogen, and hydrochloric acid
(12 M, 1.25 ml) is added dropwise under nitrogen. The reaction is
quenched after 1 min by the addition of water (20 ml); this precipitates the
products, which are then extracted with chloroform and worked up.

F. Biliverdins-IIIα and -XIIα

Biliverdins-IIIα and -XIIIα (**24, 25**) are available by oxidation (Section III.
A) of the respective bilirubins (Section III. E).

G. Hydrogenation of vinyl groups — example: meso-bilirubin-IXα

The β-pyrrolic vinyl groups of 10,23-dihydrobilindiones (bilirubins) are
readily converted to ethyl groups by catalytic hydrogenation (Fischer and
Haberland, 1935). Care must be taken to avoid scrambling (Section III. E).
We obtained good results with hydrogenation over palladium on coal (10%)
in a moderately basic solvent system (0.1 N NaOH). The reaction can be

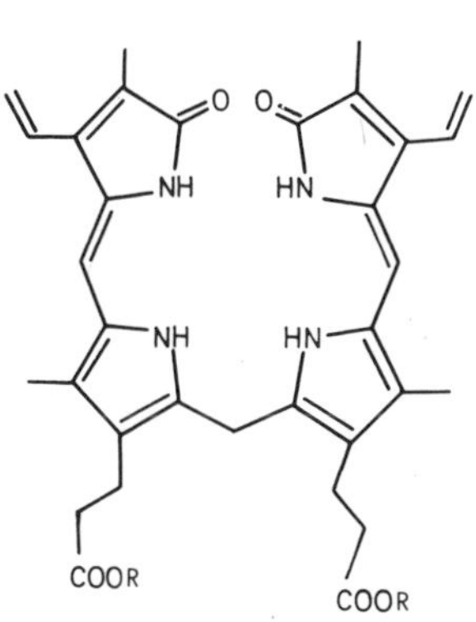

20: R = H
22: R = CH₃

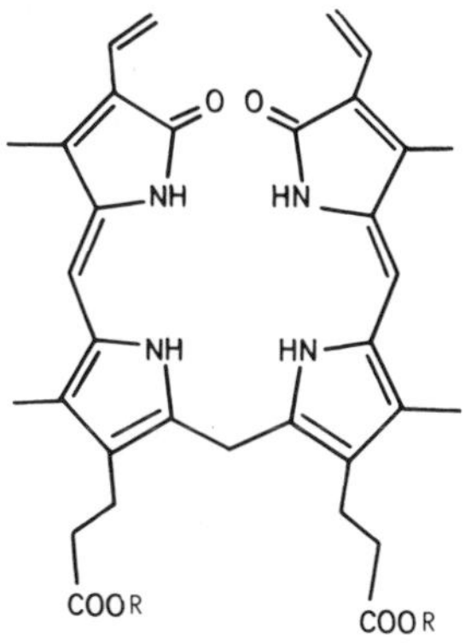

21: R = H
23: R = CH₃

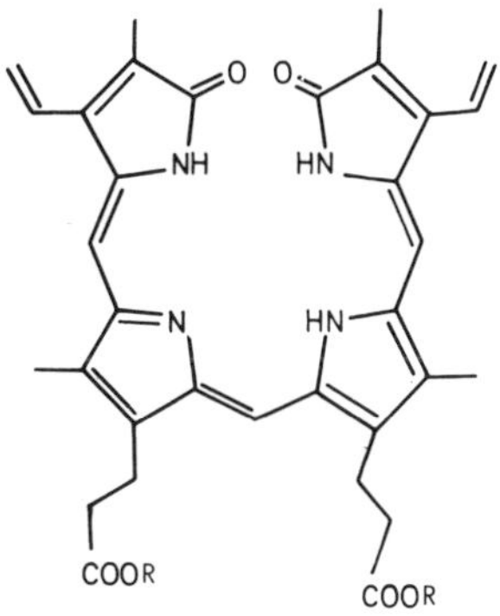

24 : R = H

25 : R = H   26 : R = H   27 : R = H

followed spectrophotometrically (shift of the absorption from 452 to 428nm, in methanol) and is complete within 15–30 min. No scrambling occurs under these conditions according to TLC analysis. The reaction can not be applied to bilindiones, e.g., biliverdin-IXα (7), because of the concurrent reduction of the tetrapyrrole π-system. For the preparation of mesobiliverdin-IXα (27), it is therefore better first to hydrogenate bilirubin-IXα (8) to the meso pigment *meso*-bilirubin-IXα (26), and oxidise the latter to the biliverdin 27.

H. Phorcabilin dimethylester

The central vinyl groups in biliverdins-IXβ,γ, and δ can react with the nitrogen atoms of the neighbouring pyrrole ring(s) to yield bridged bilindiones of restricted conformational freedom. Pigments of this type are useful for studying the influence of conformation on the properties of bile pigments. They were originally detected as natural pigments in some butterflies and caterpillars (Choussy and Barbier, 1975) and are obtainable in moderate yields from protohemin (17) via the respective biliverdins (Choussy and Barbier, 1975; Petrier, 1978; Petrier *et al.*, 1982). Below is an outline of the method of Petrier (1978) for the preparation of phorcabilin dimethyl ester (18) from biliverdin-IXγ dimethyl ester (15). Two further examples for conformationally restricted pigments are given in Sections III. I and II. D. Biliverdin-IXγ dimethyl ester (15) (100 mg) in dimethyl sulphoxide (100 ml) is heated for 1 h under nitrogen to 100°C. After workup, 18 is isolated by chromatography on silica plates with chloroform–acetone (8:2) and subsequent crystallisation from chloroform–pentane (1:25). The yield is 30–40%.

I. Isophorcabilin dimethyl ester

The synthesis of isophorcabilin dimethyl ester (**19**) starts from phorcabilin dimethyl ester (**18**, Section III. H), in which the remaining central vinyl group is cyclised by acid catalysis. According to the procedure of Petrier (1978), **18** is refluxed for $\frac{1}{2}$ h in methanolic sulphuric acid (20%). The products are isolated by chromatography on silica plates with chloroform – acetone (8:2) and subsequent crystallisation from chloroform – hexane (1:25).

J. Addition of nucleophiles to vinyl or ethylidene groups

Bilindiones (biliverdins) add nucleophiles in a dark reaction to C-10 (Section II. C). 10,23-Dihydrobilindiones (bilirubins) do not show this reactivity but can rather add nucleophiles in a photochemical reaction to vinyl substituents (Manitto and Monti, 1972). The reaction is regioselective to the exo vinyl groups, e.g., the ones at C-2 and/or C-18, and has been used in a clever way to prepare the bilirubin-IIIα (**20**) containing both vinyl groups in the endo position (C-3 and C-17) (Monti and Manitto, 1981). Although this reaction is not directly applicable to biliverdins, conversion of the latter to the corresponding rubin, addition to the vinyl group, and reoxidation is a possible means to that end. It should be noted, that inner vinyl groups in biliverdins are principally reactive to nucleophilic addition as evidenced by the intramolecular cyclisations discussed in Sections III. H and I, and that the quasi-vinylic substituent in phycocyanobilin (**32**) (namely, the 3-ethylidene group) also adds nucleophiles (Gossauer *et al.,* 1981a; Klein and Rüdiger, 1979).

For the addition of thiols to the 18-vinyl group of bilirubin (**8**), it is dissolved in chloroform with an excess of the thiol (e.g., 5% thioglycolate) and irradiated for 1 h with UV light. The product is isolated by thin layer chromatography on polyamide with methanol – 10% ammonia (9:1) in 45% yield (Manitto and Monti, 1972).

K. *E* to *Z*-isomerisation of bile pigments

E to *Z* isomerisation of bile pigments is dealt with in Section II. C.

IV. PHYCOBILIPROTEINS AND CHROMOPEPTIDES

A. Phycocyanin and allophycocyanin

Phycocyanin (**28a**) is a major light-harvesting pigment of blue-green and red algae. It has a different function and protein structure from that of phytochrome, but its chromophore structure differs from Pr only by the exchange of the 18-vinyl with an ethyl substituent; the chromophore–protein interactions are also very similar [see Rüdiger and Scheer (1983) for leading references]. Phycocyanin is accompanied by smaller amounts of allophycocyanin (**29a**), which has the same chromophore as does **28a** but is spectroscopically even more similar to Pr than is phycocyanin. Both pigments have a long history as models for pytochrome, and it was this spectroscopic similarity to phytochrome which led first to the classification of pytochrome as a biliprotein.

The content of phycocyanin in blue-green algae can amount to up to 50% by weight, and the algae are thus a very good source for large amounts of **28a**. If possible, one should select a species which does not also contain the red pigments, phycoerythrins, inasmuch as this simplifies considerably the isolation. Phycocyanin has been isolated from many algae by a variety of procedures (see, e.g., Scheer, 1981, for leading references), and only an outline of the one commonly used in our laboratory is given here. The cells are broken mechanically in a vibration-type cell mill with 0.25-mm glass beads, and a crude extract is prepared by centrifugation at ∼15,000 rpm. This is freed from the remaining (membrane-bound) chlorophyll by high speed centrifugation ($\geq 30,000$ rpm for ≥ 1 h), and the supernatant is then purified by ammonium sulphate fractionation or chromatography on DEAE–cellulose. The final separation from and purification of the accompanying allophycocyanin is most efficient on calcium phosphate gels (Cohen-Bazire *et al.*, 1977). The purity of the material is indicated — as in the case of phytochrome — by the ratio of the chromophore absorption at 620 nm (**28a**) and 650 nm (**29a**), respectively, to the protein absorption at 280 nm; the ratio should be ≥ 4. Although they both have a similar size apoprotein, **29a** contains only two chromophores, as compared to three in **28a**; the same absorption ratio is due to an increased extinction coefficient of the chromophores of **29a**. The amount of residual **28a** in preparations of **29a** can be estimated from the ratio of the absorption at 650 and 620 nm, which should be ≥ 2. Since this ratio depends strongly on the state of the allophycocyanin (e.g., aggregation, MacColl and Berns, 1981; MacColl, 1982), polyacrylamide electrophoresis (with and without SDS) is a better criterion for purity.

B. Subunit separation of phycocyanin

Phycocyanin and allopycocyanin each contain two subunits bearing either one (α of **28a** and **29a**, β of **29a**) or two chromophores (β of **28a**) (see Scheer, 1981, for leading references). The chromophores have identical molecular

28a : R = ethyl
29a : R = ethyl } different proteins
30a : R = vinyl

31a

structure, but different spectroscopic properties owing to different noncovalent interactions with their different protein environments. Separation of the subunits requires denaturing conditions; urea and SDS are the most common agents used. The individual subunits are generally separated by ion-exchange chromatography (see, e.g., Glazer and Fang, 1973; Gysy and Zuber, 1979), and are subsequently renatured. The choice of the procedure depends mainly on the most desired properties of the final products. In the context of this work, the integrity of the chromophore is most important. Modifications during the time in which the protein is denatured (and hence the chromophore unprotected), should therefore be minimised. Reproducibly good separations can be achieved by the method of Glazer and Fang (1973) using urea ($\geq 8\ M$) in formic acid, with subsequent renaturation over a desalting gel (e.g., Biogel P2 or equivalent). The acid stabilises the chromophore in the denatured state, and the gel provides for a rapid and complete separation. Problems may occur with the β subunit, which is eluted last from the column and also seems to be more hydrophobic, as indicated by its tendency to precipitate in low salt concentrations. The yield is about 65% for the α and 40% for the β subunit. Concentration of the isolated subunits is difficult with many techniques, but good results can be obtained with aquacide (Calbiochem); (this method was suggested to us by G. Bjoern).

Although the absorption spectra of the subunits appear smooth, there are indications from time-resolved fluorescence data (Hefferle et al., 1983) of the presence of more than one chromophore population in the α subunit. Since the α subunit contains only one chromophore, this could be due to some irreversible change of the chromophore or its neighbouring amino acids.

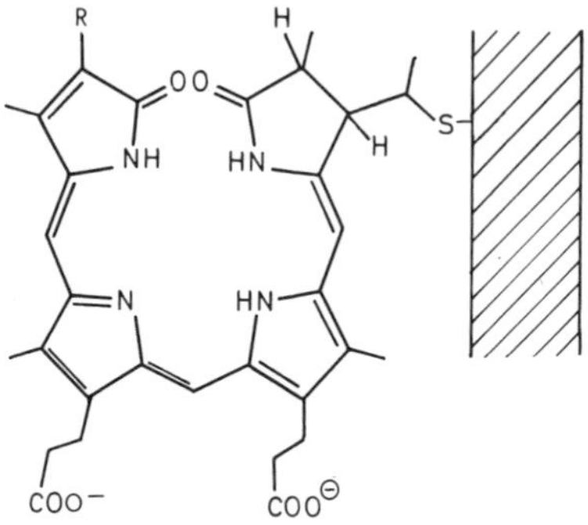

28b: R = ethyl
29b: R = ethyl   different
30b: R = vinyl   peptide chains

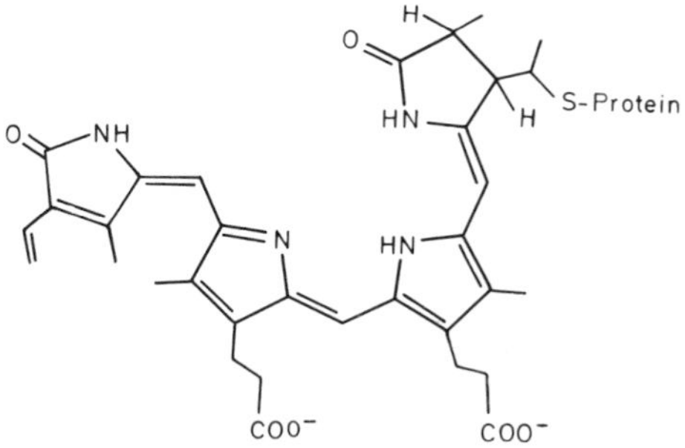

31b

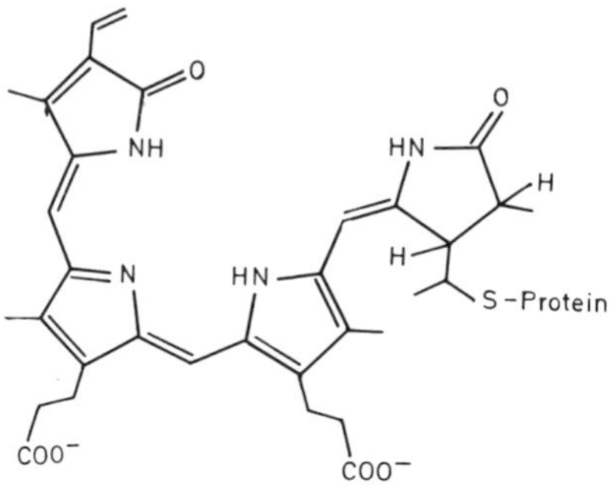

31c

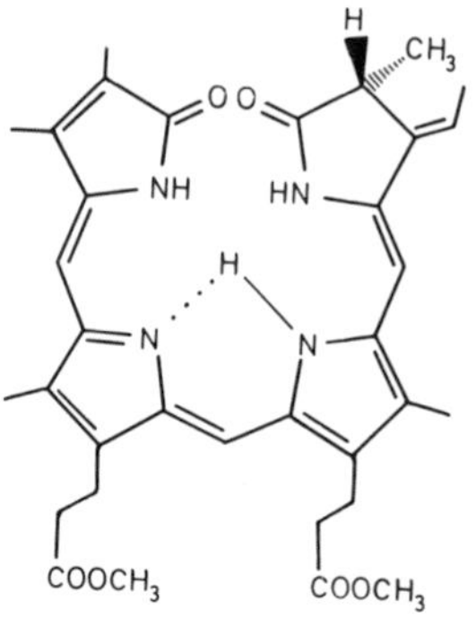

32 : R = ethyl
33 : R = vinyl

C. Chromopeptides of phycocyanin

Phycocyanins (28b) contain three chromophores with different peptide sequences in the chromophore region. All chromophores are bound to the protein via a thioether bond to a cystein residue (Klein *et al.*, 1977; Köst-Reyes and Köst, 1979; Zuber *et al.*, 1980; Glazer *et al.*, 1979; Lagarias *et al.*, 1979; Lagarias and Rapoport, 1980) and possibly sometimes a second ester(?) bond (see Scheer, 1981, for leading references). The molecular structure of all chromophores is identical, and there is also some homology in the peptide sequence around the chromophore of the α and one of the β subunits. Proteolysis generally yields more than three different chromopeptides. Digestion is possible with the common neutral proteases, such as trypsin, but acidic proteolysis with pepsin is advantageous because of the stabilisation of the chromophores at low pH. It has nonetheless been a good rule in the Munich laboratory never to prepare a stock of chromopeptides, but rather to prepare the necessary amount freshly whenever needed.

The chromophores remain bound to the peptide chain during digestion, but the noncovalent chromophore–protein interactions are uncoupled. The UV–visible spectroscopic properties of the different chromopeptides are therefore very similar, and the crude proteolysis mixture may already be useful for many purposes.

Several separation techniques have been reported for the chromopeptides. The one outlined here (Thümmler and Rüdiger, 1983) relies again on the increased stability of the chromophores at low pH. Phycocyanin (**28a**) (100 mg) is dissolved in buffer and brought to pH 1.5 with hydrochloric acid. It is digested with three subsequent portions of pepsin (10 mg) for 1 h each. The chromopeptides can be separated from the colourless peptides and chromopeptides with violet (i.e., oxidised) chromophores by chromatography on Biogel P10 (BioRad) with aqueous formic acid. They are then separated from each other by isoelectric focusing on Sephadex G-100 (Pharmacia). The final separation from the ampholyte introduced during the last step is possible on silica plates.

D. Chromopeptides of phytochrome Pr

The chromophores of phycocyanin and phytochrome Pr (**30**) are so similar, that their peptides can be prepared by essentially the same methods except for the lower yields obtained due to the larger size of the apoprotein.

E. Chromopeptides of phytochrome Pfr

The isolation of phytochrome Pfr chromopeptides (**31a**) is complicated by the facile photoreversion of the 15E-configurated chromophore **31b** to the thermodynamically more stable 15Z-configurated chromophore **30** of Pr. The key to the isolation (Thümmler and Rüdiger, 1983) of rather large amounts of **31b** for ^{1}H-NMR studies was the careful speeding up of all steps and the maintenance of a low (but not too low) pH throughout the entire procedure. Since the Pfr peptide **31b** reverts photochemically to the Pr peptide **30b**, the entire isolation procedure must be carried out under a dim safelight or, whenever possible, in the dark. Small phytochrome is first converted to ~80% of the Pfr form by irradiation with red light. It is digested within 1 h at 37°C in aqueous hydrochloric acid (pH 1.5) with a large amount of pepsin (1:1 w/w). Separation from colourless peptides was achieved on a Biogel P10 column (BioRad) with aqueous hydrochloric and then formic acids. The Pr peptide remaining on the column during this step can be eluted with aqueous pyridine in 30% yield. The Pfr peptide **31b** is further purified on silica, which is first washed with 1% formic acid. Compound **31b** is then released by elution with 30% formic acid. The yield is 30% with respect to the chromophore.

V. CHROMOPHORE CLEAVAGE REACTIONS OF BILIPROTEINS

Thioether bonds are rather stable. Although the thioether bond between the chromophores and the peptide chains in biliproteins is somewhat activated owing to the presence of the α,β double bond, there is currently no method available which can cleave the chromophore from the protein and leave the latter intact. A variety of chromophore cleavage reactions, leading to a variety of products, are described in the literature (for leading references, see, e.g., Rüdiger, 1979; Scheer, 1981). The nomenclature in the older literature is somewhat confusing, because many of the products having different structures and properties have been given the same names (e.g., phycocyanobilin for the phycocyanin-derived free bile pigments as well as for the protein-bound chromophores) and are only characterised by an index referring to their absorption maxima. Only some of the cleavage products have hitherto been fully characterised (e.g., Gossauer *et al.,* 1981a). The best character-ised ones are derived from a thioether elimination reaction, which yields 3-ethylidenebilindiones with the newly formed double bond in conjugation with the main π system. The 3-ethylidenebilindione **32** derived from phy-cocyanin (**28a**) is thus included as *the* phycocyanobilin in the IUPAC no-menclature, and the same is generally accepted for other biliprotein-derived free chromophores.

A. Boiling methanol — example: phycocyanobilin dimethyl ester

Phycocyanobilin (**32**) is cleaved from the protein in refluxing methanol. The mechanism of this reaction is not fully understood, but it is accelerated by the use of alcohols with a higher boiling point (Fu *et al.,* 1979). This thermal cleavage reaction can be applied to all common phycobiliproteins, including phycoerythrin, but not to phytochrome. The procedure given below (H. P. Kost, personal communication) can be used to obtain large amounts of phycocyanobilin from dry blue-green algae containing no phy-coerythrin. A good source is *Spirulina geitleri,* which is commercially avail-able in spray-dried form from the SOSA Texcoco Corporation, Mexico. The method involves the extraction of the chlorophylls with hot methanol, chromophore cleavage in refluxing methanol, esterification to the dimethyl ester, and a final chromatographic purification of **32**. The entire isolation should be carried out in dim light, and exposure to oxygen should be kept to the minimum. The dry, powdered material (40 g) is extracted with 100-ml portions of hot methanol until the extracts remain colourless; the residue is then refluxed in 400 ml of methanol for 4 h under nitrogen. The mixture is filtered to yield a blueish-green filtrate, which may be stored in the refrigerator. The residue is again refluxed with methanol, and the procedure is repeated three to four times until the extract is only lightly coloured. The combined extracts are then evaporated to dryness in a rotary evaporator, and the residue is treated overnight at 4 °C under nitrogen with boron trifluoride in methanol (5% w/v, 25 ml). The mixture is partitioned between methylene chloride and water, and the organic phase is washed, dried, and evaporated to dryness. Chromatography on silica with carbon tetrachloride – acetone (9:1) yields the blue **32** (about 50 mg) as the first main band, preceding a series of green, violet, and red by-products.

B. Hydrogen bromide in trifluoroacetic acid — example: phycocyanobilin dimethyl ester

Treatment with hydrogen bromide in trifluoroacetic acid is the only known method which is useful for the cleavage of the phytochrome Pr chromophore, although the elimination is accompanied by addition reactions to the 18-vinyl group of the phytochromobilin **33** (Rüdiger *et al.,* 1980). The method is therefore described for phycocyanin, as applied first by Schram and Kroes (1971). It gives very good yields even with small sample sizes. Its only drawback is the initiation of acid-catalysed side reactions. Although of no importance for the preparation of phycocyanobilin dimethyl ester (**32**), this may lead to complications with other biliproteins (e.g., epimerization at C-16 in phycoerythrobilin, Scheer and Bubenzer, unpublished).

Phycocyanobilin is suspended or dissolved in trifluoroacetic acid. After thorough deoxygenation with a stream of nitrogen, hydrogen bromide is bubbled through the solution for ½ h. The solvent is then evaporated by blowing nitrogen through it, and the residue is esterified and worked up as described in the preceding section to yield the dimethyl ester of phycocyanobilin (**32**).

VI. CHROMOPHORE DEGRADATION REACTIONS

A. Chromic acid and chromate degradation

Oxidative cleavage of tetrapyrroles is one of the classical degradation techniques for structure elucidation. Chromic acid (e.g., solutions of chromates in sulphuric acid) is the standard reagent, and Rüdiger (1969) has worked out its application to linear tetrapyrroles including the chromophores of biliproteins. The reaction cleaves the tetrapyrrole skeleton into four imides, which, at least in principle, retain the β-pyrrolic substituents and hydrogenation pattern of the four rings. The degradation of phycocyanobilin is given as an example in Scheme 1. Information on the methine and α-pyrrolic substituents is lost, and side reactions may occur with acid-labile substituents.

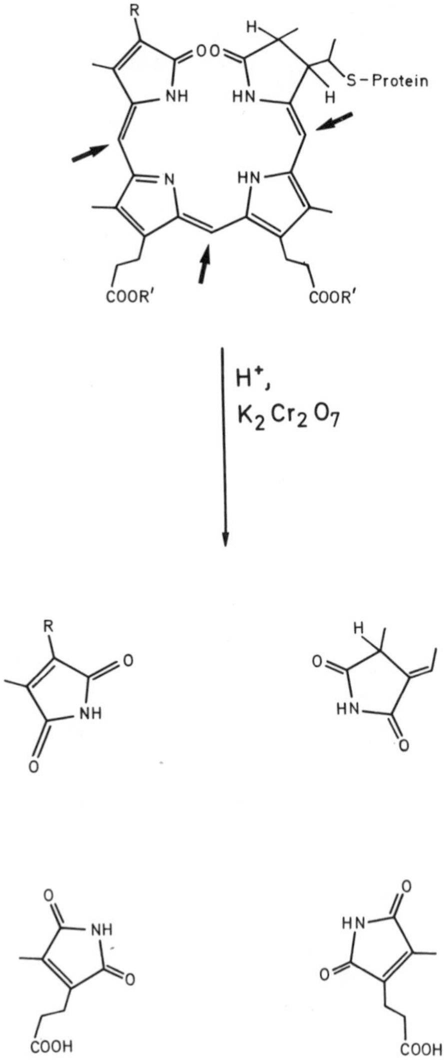

Scheme 1

The reaction requires only nmol amounts if the individual imides are separated and analysed by high performance thin layer chromatography and stained with the chlorine – benzidine technique (substitution of benzidine by tetramethylbenzidine is recommended for its decreased carcinogenicity). The identification of the imides is usually done by comparing their mobility with known imides, and by co-chromatography. Most of the relevant imides are accessible from the various natural and semisynthetic bile pigments (Sections II – V).

Several modifications of the original procedure have been important in biliprotein studies. One is chromate oxidation, which is carried out with the same reagents but under less acidic conditions (Rüdiger, 1969). It allows distinction between the inner (B,C) and outer rings (A,D), because the methine carbon atoms are retained in the cleavage products. Only the inner rings can yield 2,5-diformylpyrroles with the respective β-pyrrolic substituents of the parent tetrapyrrole. Another important modification is the reaction temperature (Klein et al., 1977). At 20°C, the thioether bonds between the biliprotein chromophores and the peptide chain are stable, and the succinimide derived from ring A remains attached to the protein. At 100°C, this linkage is broken to yield the ethylidenesuccinimide. This so-called hydrolytic cleavage has been further modified (Klein et al., 1977) by a two-step procedure. The thioether bond is already oxidised, albeit not cleaved, under nonhydrolytic (20°C) conditions, and the resulting sulphone can be cleaved with ammonia at room temperature. The overall reaction sequence is thus much milder than the original hydrolytic (100°C) procedure.

The bile pigment (~ 50 μg), or the corresponding amount of the denatured biliprotein or bilipeptide, is treated with a solution of potassium dichromate (1%) in sulphuric acid (1 M, 0.2 ml) and stored for 15 h in the dark. The mixture is extracted several times with ethyl acetate, which is evaporated in a stream of nitrogen. The residue is taken up in chloroform, analysed by thin layer chromatography or gas-liquid chromatography, and compared with standard imide mixtures prepared from suitable bile pigments.

B. Reduction and diazo reaction

The reduction and diazo reaction sequence has been devised as an even milder and more selective cleavage (Kufer et al., 1982b). It makes use of a classical analytical tool, the diazo reaction of bilirubin (Heirwegh et al., 1974). Although the biliprotein chromophores are not directly accessible to the reaction with diazonium salts, they can be first reduced to the reactive rubinoid pigments (phycorubins) with sodium borohydride (Kufer and Scheer, 1982). The reaction cleaves the molecule selectively between rings

B and C, and the entire sequence can be carried out at 4°C and pH 7. Under these conditions, a 1:1 mixture of 9-azopyrromethenones and 9-oxymethyl-pyrromethenones is obtained (see the general Scheme 2 and Scheme 3 as an example for phycocyanin). Even weak bonds, such as the suspected ester bond to one of the propionic acid side chains, should be stable under these conditions, and a separation of free and protein-bound degradation products is possible, using solvent extraction (Kufer *et al.*, 1982b). The identification of the products is possible by UV–visible spectroscopy and chromatography, and reference compounds can be obtained from bile pigments of known structure.

For the reduction to the phycorubin (Kufer and Scheer, 1982), the phycocyanin (**28a**) solution (50 μM) in phosphate buffer (50 mM) is denatured with urea (0.96 g per 1.25 ml to yield a final concentration of 8 M). The solution is cooled with ice and treated twice with sodium borohydride (10 mg/ml in the same buffer, 25 μl of this solution per ml of the phycocyanin solution) for 30 min. Excess sodium borohydride is removed by

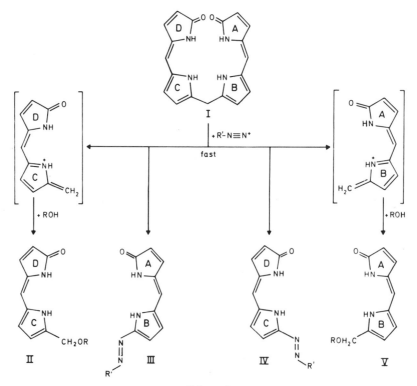

Scheme 2

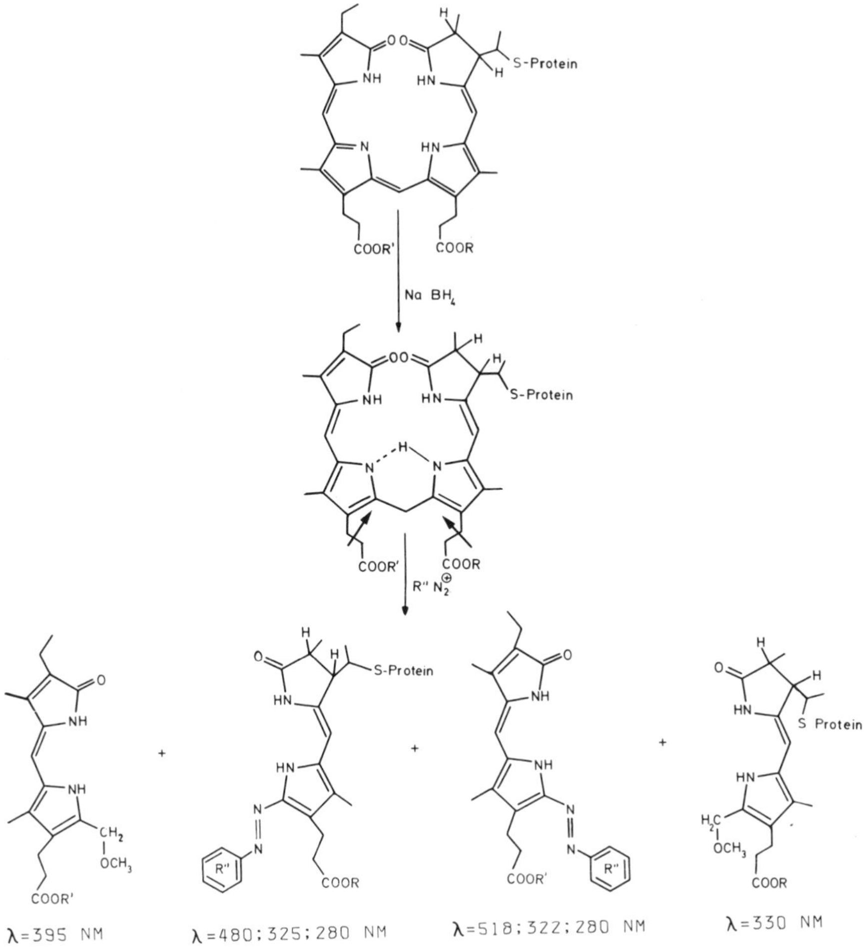

Scheme 3

filtration over a Biogel P2 (BioRad) column, pre-equilibrated and developed with the phosphate buffer containing urea (8 M). A volume of 2 ml of the resulting solution of phycorubin ($\sim$40 nmol) is coupled with diazotised ethyl anthranilate ($\sim$400 nmol) at 0°C for 15 min. It is then extracted twice with 1 ml of isobutanol. The phases are separated by centrifugation, and the isobutanol solution containing the nonprotein-bound azopigment(s) and oxymethylpyrromethenones is analysed by high performance thin layer chromatography and/or UV–visible spectroscopy (Kufer and Scheer, 1983).

C. Cleavage to formyltripyrrinones

One of the end rings in bilindiones (biliverdins) can be cleaved selectively with thallium trifluoroacetate (Eivazi and Smith, 1980). This reaction is even more facile in 2,3-dihydrobilindiones, where the saturated ring A is cleaved regioselectively (Krauss *et al.*, 1979). Since all biliprotein chromophores belong to this class of bile pigments, the reaction is suitable for their analysis by chemical degradation (Kufer *et al.*, 1982b; see Scheme 4). The reaction proceeds at pH $\geq$ 8.5 in the presence of zinc ions, probably via the free radical of the zinc complex (Krauss and Scheer, 1979). The products can be analysed by chromatography (preferentially after esterification with methanol) and/or spectroscopy, and reference compounds are again available by the degradation of suitable free bile pigments.

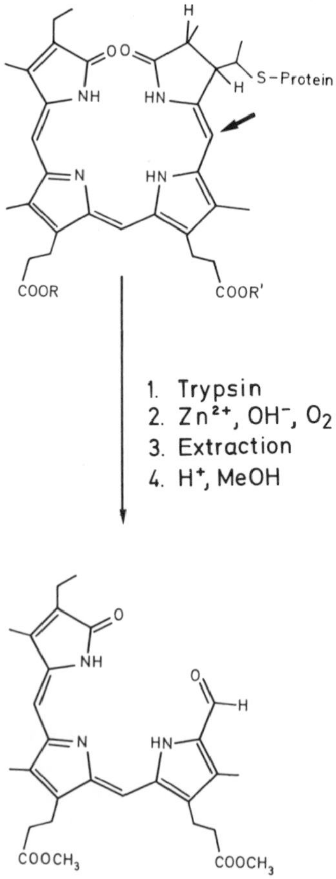

1. Trypsin
2. Zn^{2+}, OH^-, O_2
3. Extraction
4. H^+, MeOH

Scheme 4

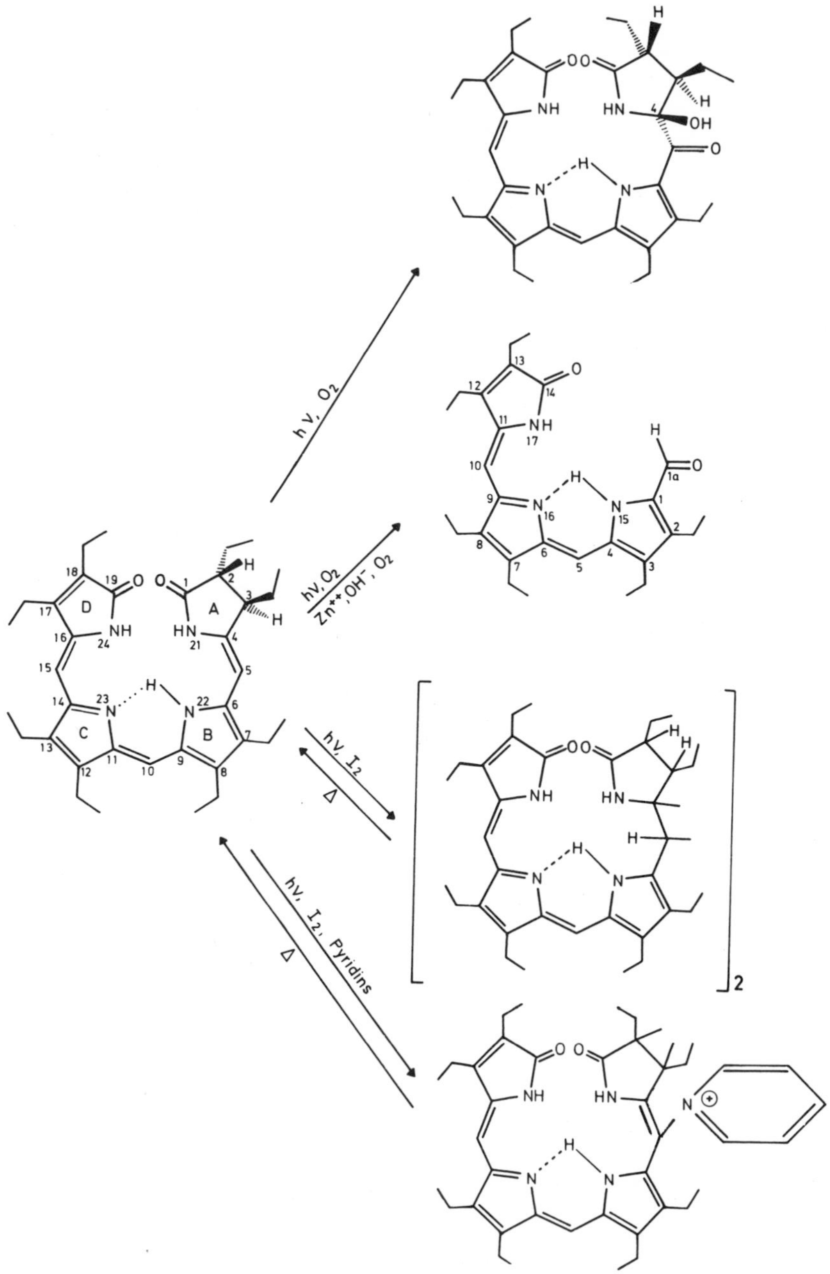

Scheme 5

The biliprotein is denatured with urea (Section VI. C) or preferentially degraded to bilipeptides (Section IV. C). The respective solution is adjusted with sodium hydroxide to pH 9 and treated with a fivefold molar excess of zinc acetate to yield a greenish solution ($\lambda_{max} \sim 725$ nm). The solution is kept for 15 min at ambient temperature, extracted by the procedure given in Section VI. B, and analysed after esterification (Section III. B).

D. Unspecific oxidation reactions

As mentioned in the introduction, bile pigments, and, in particular, 2,3-di-hydrobilindiones like the biliprotein chromophores, are readily photooxi-dised. While the chromophores of native biliproteins are stabilised surpris-ingly well, this is no longer true for the denatured or proteolytically degraded biliproteins. It is almost inevitable that at least part of these pigments become oxidised during any treatment involving denaturation (see Section I. B for precautions). A variety of such oxidation products has been identified with the 2,3-dihydrobilindione **4a** as a model (Scheme 5, see Scheer, 1981, for leading references). A product mixture is thus expected whenever the conditions are not carefully optimised for a specific type of reaction (see, e.g., Section VI. C). While it is highly impractical to analyse such mixtures, the presence of oxidation products can be checked for in a straightforward way because they all absorb at shorter wavelengths than do the starting mate-rials. The only requirement is that the spectrum be recorded under condi-tions where all of the products are in the same state. From a practical point of view, this is that of the protonated pigments which are formed at pH ≤ 1.5, and oxidation products then show up as one or more distinct shoulders or peaks in the short-wavelength side of the visible absorption band.

ACKNOWLEDGMENTS

The cited work of the author was supported by the Deutsche Forschungsge-meinschaft, Bonn.

REFERENCES

Bonnett, R. (1978). *In* "The Porphyrins" (D. Dolphin, ed.), Vol. 1, p. 1. Academic Press, New York.
Bonnett, R. and MacDonagh, A. F. (1974). *J. Chem. Soc., Perkin Trans. 1* p. 881.
Brown, S. B., Holroyd, J. A., Troxler, R. F. and Offner, G. D. (1981). *Biochem. J.* **194**, 137.
Cavaleiro, J. A. S. and Smith, K. M. (1973). *J. Chem. Soc., Perkin Trans. 1* p. 2149.
Choussy, M. and Barbier, M. (1975). *Helv. Chim. Acta* **58**, 2651.

Cohen-Bazire, G., Béguin, S., Rimon, S., Glazer, A. N. and Brown, D. M. (1977). *Arch. Microbiol.* **111**, 225.

Cole, W. J., Chapman, D. J. and Siegelman, H. W. (1967). *J. Am. Chem. Soc.* **89**, 3643.

Eivazi, F. and Smith, K. M. (1979). *J. Chem. Soc., Perkin Trans. 1* p. 544.

Falk, H. and Grubmayr, K. (1977). *Synthesis* p. 614.

Falk, H. and Müller, N. (1981). *Monatsh. Chem.* **112**, 791.

Falk, H. and Thirring, K. (1981). *Tetrahedron* **37**, 761.

Falk, H., Grubmayr, K. and Neufingerl, F. (1979). *Monatsh. Chem.* **110**, 1127.

Falk, H., Müller, N. and Schlederer, T. (1980). *Monatsh. Chem.* **111**, 159.

Fischer, H. and Haberland, H. W. (1935). *Hoppe-Seyler's Z. Physiol. Chem.* **232**, 236.

Fu, E., Friedman, L. and Siegelman, H. W. (1979). *Biochem. J.* **179**.

Glazer, A. N. and Fang, S. (1973). *J. Biol. Chem.* **248**, 663.

Glazer, A. N., Hixson, C. S. and De Lange, R. J. (1979). *Anal. Biochem.* **92**, 489.

Gossauer, A., Blacha-Puller, M., Zeisberg, R. and Wray, V. (1981a). *Leibigs Ann. Chem.* p. 342.

Gossauer, A., Hinze, R. P. and Kutschan, R. (1981b). *Chem. Ber.* **114**, 132.

Gysi, J. R. and Zuber, H. (1979). *Biochem. J.* **181**, 577.

Hefferle, P., Geiselhart, P., Mindl, T., Schneider, S., John, W. and Scheer, H. (1984). *Z. Naturforsch.* **39C**, 606.

Heirwegh, K. P. M., Fevery, J., Meuwissen, J. A. T. P., Compernolle, F., Desmet, V. and van Roy, F. P. (1974). *Methods Biochem. Anal.* **22**, 205.

IUPAC-IUB Joint Commission on Biochemical Nomenclature (JCBN) (1979). *Pure Appl. Chem.* **51**, 2251.

Klein, G. and Rüdiger, W. (1979). *Z. Naturforsch.* **34C**, 192.

Klein, G., Grombein, S. and Rüdiger, W. (1977). *Hoppe-Seyler's Z. Physiol. Chem.* **358**, 1077.

Köst, H.-P., Rüdiger, W. and Chapman, D. J. (1975). *Liebigs Ann. Chem.* p. 1582.

Köst-Reyes, E. and Köst, H.-P. (1979). *Eur. J. Biochem.* **102**, 83.

Krauss, C. and Scheer, H. (1979). *Tetrahedron Lett.* p. 3553.

Krauss, C., Bubenzer, C. and Scheer, H. (1979). *Photochem. Photobiol.* **30**, 473.

Kufer, W. and Scheer, H. (1979). *Hoppe-Seyler's Z. Physiol. Chem.* **360**, 935.

Kufer, W. and Scheer, H. (1982). *Z. Naturforsch., C: Biosci.* **37C**, 179.

Kufer, W. and Scheer, H. (1983). *Tetrahedron* **39**, 1187.

Kufer, W., Cmiel, E., Thümmler, F., Rüdiger, W., Schneider, S. and Scheer, H. (1982a). *Photochem. Photobiol.* **36**, 603.

Kufer, W., Krauss, C. and Scheer, H. (1982b). *Angew. Chem.* **94**, 455; *Angew Chem. Int. Ed. Engl.* **21**, 446; *Angew. Chem., Suppl.* p. 1050.

Lagarias, J.C. and Rapoport, H. (1980). *J. Am. Chem. Soc.* **102**, 4821.

Lagarias, J. C., Glazer, A. N. and Rapoport, H. (1979). *J. Am. Chem. Soc.* **101**, 5030.

MacColl, R. (1982). *Photochem. Photobiol.* **35**, 899.

MacColl, R. and Berns, D. S. (1981). *Isr. J. Chem.* **21**, 296.

MacDonagh, A. F. (1979). *In* "The Porphyrins" (D. Dolphin, ed.), Vol. 6, p. 294. Academic Press, New York.

MacDonagh, A. F. and Palma, L. A. (1980). *Biochem. J.* **189**, 193.

MacDonagh, A. F., Palma, L.A., Trull, F. R. and Lightner, D. A. (1982). *J. Am. Chem. Soc.* **104**, 6865.

McKillop, A., Hunt, J. D., Zelesko, M. J., Fowler, J. S., Taylor, E. C., McGillivray, G. and Kienzle, F. (1971). *J. Am. Chem. Soc.* **93**, 4841.

Manitto, P. and Monti, D. (1972). *Experientia* **28**, 379.

Manitto, P. and Monti, D. (1979a). *Experientia* **35**, 1.

Manitto, P. and Monti, D. (1979b). *Experientia* **35**, 1418.

Monti, D. and Manitto, P. (1981). *Synth. Commun.* **11,** 811.

O'Carra, P. (1975). *In* "Porphyrins and Metalloporphyrins" (K. M. Smith, ed.), pp. 123–157. Elsevier, Amsterdam.

O'Carra, P. and Colleran, E. (1970). *J. Chromatogr.* **50,** 458.

Petrier, C., Dupuy, C., Jardon, P. and Gautron, R. (1982). *Nouv. J. Chim.* **6,** 495.

Petrier, J. (1978). Ph.D. Thesis, Université Scientifique et Médicale de Grenoble.

Rüdiger, W. (1969). *Hoppe-Seyler's Z. Physiol. Chem.* **350,** 1291.

Rüdiger, W. (1979). *Ber. Dtsch. Bot. Ges.* **92,** 413.

Rüdiger, W. and Scheer, H. (1983). *In* W. Shropshire, Jr., and H. Mohr (eds.) "Encyclopedia of Plant Physiology," **16A,** 119. Springer-Verlag, Berlin and New York.

Rüdiger, W., Brandlmmeier, T., Blos, I., Gossauer, A. and Weller, H.-P. (1980). *Z. Naturforsch., C: Biosci.* **35C,** 763.

Scheer, H. (1976). *Z. Naturforsch., C: Biosci.* **31C,** 413.

Scheer, H. (1981). *Angew. Chem.* **93,** 230; *Angew. Chem., Int. Ed. Engl.* **20,** 241.

Schram, B. L. and Kores, H. H. (1971). *Eur. J. Biochem.* **19,** 581.

Smith, K. M., Sharkus, L. C. and Dallas, J. L. (1980). *Biochem. Biophys. Res. Commun.* **97,** 1370.

Stoll, M. S. and Gray, C. H. (1977). *Biochem. J.* **163,** 59.

Thümmler, F. and Rüdiger, W. (1983). *Isr. J. Chem.* **23,** 195.

Thümmler, F., Rüdiger, W., Cmiel, E. and Schneider, S. (1983). *Z. Naturforsch., Biosci. C:* **38C** 359.

Whitlock, H. W., Jr., Hanauer, R., Oester, M. Y. and Bower, B. K. (1969). *J. Am. Chem. Soc.* **91,** 7485.

Zuber, H. (1978). *Ber. Dtsch. Bot. Ges.* **91,** 459.

11

Phytochrome in Membranes

Stanley J. Roux

I. INTRODUCTION

In the last 5 years, over 100 articles have been published on the topic of phytochrome and membranes. An implicit hypothesis behind most of these articles is that understanding how phytochrome affects plant cell membranes will greatly clarify mechanisms of photomorphogenesis. Although data on specific membrane alterations attributable to phytochrome are plentiful (for review, see Marmé, 1977), convincing evidence that any one or group of these alterations is critical for any specific morphogenetic event is lacking. One can confidently hope for significant progress on this point in the coming years, since there are a variety of promising techniques available and currently being applied to clarify this and other basic issues on phytochrome – membrane interactions.

The methods presented here have already been employed to obtain useful information on phytochrome in membranes at four different levels of organization: tissue, cell, organelle, and the lipid bilayer. Because of the breadth of the topic and limitations of space, the methods described are a minimal representation of the many possible approaches. The two main criteria for selecting the described techniques, other than their relevance to the topic, were the experience of the author in employing them and the simplicity of the technique and probable availability of the materials and instruments needed. No judgment is implied as to the relative merit of these approaches compared to others. In addition to methods described in detail, some other techniques will be referred to and discussed briefly.

TECHNIQUES IN PHOTOMORPHOGENESIS
0-12-652990-6

II. METHODS FOR EXPERIMENTS AT THE MULTICELLULAR LEVEL: Rb⁺ FLUX

A. Introduction

Historically, data obtained from experiments on intact plant tissue in the middle 1960s first drew attention to the potential role of phytochrome as a membrane agent. The observations of Fondeville *et al.* (1966) on the influence of R and FR light on the nyctinastic movements of *Mimosa* leaflets and Tanada's (1968) report on photoreversible adhesion of barley root tips to negatively charged glass surfaces were most influential in this respect. Both types of experiments have been repeated successfully in many other laboratories and have inspired a variety of follow-up experiments. Galston and Satter and colleagues exploited the system of light-induced turgor changes in pulvini to document phytochrome-controlled K^+ fluxes *in situ* (cf. review, Satter, 1979). Jaffe (1968) reported phytochrome-controlled surface potential changes in root tips, and Tanada utilized his root-tip adhesion assay to study interactions of phytochrome with hormones in the control of root-tip surface potential (Tanada, 1972, 1973a,b).

Many other laboratories have gained important insights into phytochrome control of membrane functions from *in vivo* studies, using a variety of different approaches. For example, two different laboratories (Pike and Richardson, 1979; Brownlee and Kendrick, 1979) have explored the hypothesis that phytochrome can modulate K^+ fluxes by employing radiotracer techniques with the $^{86}Rb^+$ isotope. The simplicity of the methods used and the ease with which the isotope assay can be manipulated recommend this experimental approach for a detailed description here.

B. Materials and methods

Small sections of tissue are transferred to a $^{86}Rb^+$-containing solution, irradiated with actinic R or FR light, incubated for a short time and then rinsed, dried, weighed, and dissolved in a scintillation fluid for determination of radioactivity by liquid scintillation counting. Key parameters of the experiments are highlighted and discussed below.

(1) Choice of isotope

$^{86}Rb^+$ has a much longer half-life than $^{45}K^+$, and is therefore more convenient to use. Unfortunately, one can not assume *a priori* that $^{86}Rb^+$ flux accurately mirrors K^+ flux (Jacoby and Nissen, 1977).

(2) Tissue choice

By choosing tissue with a high phytochrome content, such as apical tip and hook regions of etiolated dicotyledonous seedlings and coleoptile tips of etiolated monocotyledonous seedlings, the probability of detecting phytochrome-modulated ion fluxes may be increased. Tissue slices from these regions of pea, bean, and oat seedlings have been used in the published studies on $^{86}Rb^+$ fluxes. As there is no clear indication in the literature that phytochrome content is quantitatively correlated with response in the case of ion uptake, it is possible that tissues with far lower phytochrome levels also show significant responses.

(3) Tissue preparation

Small tissue slices, 2–10 mm in length, are commonly used. With pea it may be advisable to peel off the cuticle to allow shorter uptake time. In general, the lower the ratio of cut (i.e., damaged) surface area to intact surface area, the less chance there is that wound reactions will complicate data interpretation. For each experiment, 20–30 tissue segments (∼0.2–0.4 g) are treated together. For ease of handling, it is convenient to enclose the segments in a small bag of nylon mesh (Pike and Richardson, 1979).

(4) Preincubation in unlabeled buffer

The freshly sliced tissue should be placed in an aerated "holding" buffer containing ∼0.5 mM CaCl$_2$. Epstein (1961) has shown that Ca^{2+} is required to reduce nonspecific adsorption of monovalent cations and to permit selective uptake of those cations which are to be detected. Brownlee and Kendrick (1979) indicated that mung bean sections required 3 h in a holding solution before the considerable K$^+$ leakage, initiated when the tissue was sliced, is slowed to a low rate. For these long pretreatments it is advisable to buffer the solution between pH 6 and 7, using MOPS (pK_a 7.2) or PIPES (pK_a 6.8) at 1 mM. The temperature of the preincubation should be the same as that of the subsequent isotope uptake solution. After the preincubation period, the segments can be transferred together into the isotope uptake solution.

(5) Isotope uptake conditions

Epstein et al. (1963) recommend that the isotope absorption period be as brief as possible and that the volume of uptake buffer be large so that the amount of cation absorbed by the tissue will not be a significant fraction of the total solute initially present. These conditions guarantee that the uptake solution will not be depleted of any salts and therefore favour the maintenance of constant uptake rates during the absorption period. The concentration of the substrate ion (Rb^+) should not be allowed to drop by more than 5% during the absorption period. A typical concentration of $^{86}Rb^+$ is 0.1 μCi/ml, in volumes that vary, depending on the amount of tissue used, from about 10 to 200 ml.

If uptake periods are short (10 min or less), the use of a buffer is optional; however, it is usually convenient to use the same buffer and pH as in the preincubation solution. Ca^{2+} at 0.5 mM should be present, and Brownlee and Kendrick (1979) found that addition of 25–50 mM KCl improved $^{86}Rb^+$ uptake significantly.

(6) Irradiation treatments

Use of narrow bandpass interference filters for R and FR light allows less ambiguous data interpretation than broad-band sources. Normally, irradiation treatments should be kept as brief as compatible with saturation of the photoconversion of Pr to Pfr and Pfr to Pr. Treatments can be given immediately after the tissue is introduced into the isotope solution.

(7) Desorbing (wash) treatment

The wash solution should be identical to the isotope uptake solution, except that it does not contain labelled Rb^+. The wash may consist of rapidly dipping the bag of tissue pieces through five successive unlabelled solutions for a total desorbing time of 1 min, or may consist of much longer washes, depending on the kind of information required from the experiment. Brownlee and Kendrick (1979) demonstrated that the removal of isotope from the free space of tissue that has taken up $^{86}Rb^+$ is rapid, but that it takes longer washes to remove $^{86}Rb^+$ that has been compartmented into cytoplasmic or vacuolar space.

(8) Radioactivity counting

The washed tissue segments are placed in a preweighed scintillation vial, dried, and weighed. The $^{86}Rb^+$ can then be counted in a mixture of 1 ml of distilled H_2O plus 10 ml of NE 260 liquid scintillation fluid (Nuclear Enterprises Ltd., Edinburgh, U.K.). There should be little variation in quenching between samples, so that the results may be expressed in cpm/g dry weight tissue.

C. Comments on expected results

Based on the data given in Pike and Richardson (1979) and Brownlee and Kendrick (1979), R light should induce a small but statistically significant difference in the $^{86}Rb^+$ uptake rate of some tissues. If the effect of R light is reversible by FR, it would be attributable to the mediation of phytochrome.

By testing the effect of changes in the experimental procedures (e.g., longer or shorter pre-incubation periods, longer or shorter washes), or of certain inhibitors (e.g., carbonyl cyanide *m*-chlorophenylhydrazone), it is possible to study how phytochrome affects cellular membranes.

Since changes in $^{86}Rb^+$ flux are measured, at the earliest, 10 min after actinic irradiations, judgment must be reserved on whether this result is primary or follows secondarily from a prior effect of phytochrome photoconversion. The techniques described in the following section can be used to reveal the short-term kinetics of phytochrome-induced ion fluxes and the fluxes which occur during the actinic irradiations. These methods have been used successfully with tissue sections and cell suspensions (Hale and Roux, 1980).

III. METHODS FOR EXPERIMENTS AT THE CELLULAR LEVEL

A. Introduction

Four different approaches have yielded valuable information about phytochrome in membranes at the cellular level. One approach, used by Haupt's laboratory in Erlangen, West Germany, uses polarized microbeams of R and FR light to study phytochrome control of chloroplast orientation in *Mougeotia*. One of the more important conclusions from their data was that the chromophores of both Pr and Pfr were preferentially oriented in different planes, indicating that at least some fraction of both forms of phytochrome was membrane associated. Another approach used techniques of microelectrophysiology to implant electrodes in cells for the measurement of photoreversible cell membrane potential changes in response to R and FR light. Several laboratories have obtained useful data from this approach, which is described in detail by Racusen (1976). A third approach, demonstrated by Dreyer and Weisenseel (1979), used autoradiographical techniques to study phytochrome-controlled $^{45}Ca^{2+}$ fluxes in single cells of a filament of *Mougeotia*. Finally, with an approach that has only begun to be exploited, Pratt's group has shown phytochrome–membrane associations in cells by immunocytochemistry (Pratt *et al.*, 1976; Chapter 8, this volume).

The method presented here has the advantage of allowing rapid Ca^{2+} fluxes to be monitored continuously within seconds after, or even during, irradiation treatments with R and FR light. The method uses the metallochromic indicator murexide, which undergoes a rapid absorbance change at 540 nm when the concentration of Ca^{2+} in the solution changes (Scarpa, 1972). Since murexide is not taken up by live cells and does not bind to them, the method permits changes in [Ca^{2+}] to be measured only in the solution surrounding the test cells. Under the conditions described here, the dye is selectively sensitive to Ca^{2+} and can detect changes in its concentration as small as 1 μM (Ohnishi, 1978). Arsenazo III, another Ca^{2+} indicator dye, is much more sensitive than murexide, but because its absorbance changes are in the red region of the spectrum it is not suitable for phytochrome studies because the measuring beams required to detect its ΔA would be actinic.

B. Materials and methods

The method involves placing the test cells in a murexide-containing buffer in a spectrophotometer cuvette, and then monitoring the $\Delta A540$ nm of the solution in a dual-wavelength spectrophotometer, using 507 nm as the reference beam. When the A540 – A507 nm equilibrates at a constant value, the cells can be irradiated with various combinations of actinic R and FR light and the influence of these irradiations on the solution Ca^{2+} concentration (i.e., the $\Delta 540$ nm) can be monitored. Key parameters of the experiment are discussed below.

(1) Choice of cell source

Oat coleoptile cells are described here because our laboratory has extensive experience in applying the murexide method to measure Ca^{2+} fluxes from these cells and because they show phytochrome modulation of their Ca^{2+} fluxes. Other cell sources rich in phytochrome might also be suitable. Some cell sources not rich in phytochrome have failed to show photoreversible Ca^{2+} fluxes (Hale and Roux, 1980).

Although tissue slices from coleoptile tips may be used, the presence of cell walls, which have a high Ca^{2+}-binding capacity (Rains, 1976), could depress the apparent magnitude of Ca^{2+} flux from the cells by acting as a Ca^{2+}-buffer zone between the cells and the external medium. Higher apparent Ca^{2+} fluxes have been measured from coleoptile protoplasts than from excised coleoptile tips (Hale and Roux, 1980); thus, the method described below employs coleoptile protoplasts.

(2) Preparation of cells

All manipulations should be made under dim green safelight. Four-day-old etiolated oat seedlings are harvested near the base of the coleoptile, sterilized in 70% ethanol for 2 min, then rinsed in sterile, distilled water. The top 20 mm of the coleoptiles are excised and floated on a sterile solution of Buffer A (0.6 M mannitol, 10 mM PIPES, titrated to pH 6.8 with NaOH) until use. Three-gramme lots of the seedling tips are digested for 5 – 7 h in the dark at 30°C in a solution containing 2% w/v cellulysin and 1% w/v macerase (Calbiochem), 0.6 M mannitol, and 10 mM PIPES, pH 5.8. Periodic, gentle agitation of the digest improves the efficiency of the process. Protoplasts are harvested by filtering through 50 μ-mesh nylon screening and then washing twice in Buffer A by low-speed centrifugations (1000 $\times$ g, 5 min), gently resuspending the pellet each time by drawing it up through a wide-bore pipette. Protoplast viability may be determined by observing cyclosis and/or Evans Blue dye exclusion. Protoplast counts can be made with a haemocytometer. Optimal yields range from $1-3 \times 10^5$ protoplasts from 3 g of etiolated coleoptile tips and $1-3 \times 10^6$ protoplasts from 3 g of greened leaves.

If a 1-ml, 1-cm path-length cuvette is used, and the cells are concentrated just below the measuring beam as described below, about 10^5 cells should yield sufficiently large Ca^{2+} fluxes to be detectable by murexide (lower limit = $\Delta[Ca^{2+}]$ of about 1 μM = A540 nm cm^{-1} of ~0.0002 A). To provide interpretable results, the protoplast preparation must not be contaminated with bacteria, and the protoplasts themselves should be relatively turgid and stable against spontaneous breakage.

We have not yet developed a protoplast isolation procedure from oat coleoptile tips which will consistently produce 10^5 protoplasts that are viable and stable against spontaneous breakage. This difficulty is not surprising considering how densely packed the apex tissue is and how poorly the wall-digestive enzymes would penetrate such tissue. Consequently, we have begun to examine othe phytochrome-containing plant material, which is not so resistant to wall-digestive enzymes. In preliminary results we have found that large numbers of protoplasts may be obtained easily from the primary leaves of 7-day-old etiolated oat seedlings and that these protoplasts show photoreversible Ca^{2+} fluxes that are qualitatively the same as those reported from the coleoptile tip protoplasts (R. L. Biro and S. J. Roux, unpublished results). Tissue culture cells (Tanada, 1977) may also prove to be excellent experimental material for this purpose.

(3) Composition of murexide solution for measurements
For Ca^{2+}-flux measurements, the cells are suspended in the same buffer as that used to isolate them (Buffer A) plus 25 μM murexide. The 10-mM PIPES keeps the solution buffered against pH changes that could alter the extinction coefficient of Ca^{2+}-murexide complexes at 540 nm. Two mM Ca^{2+} can also be added to the buffer, but the cells must be allowed to equilibrate in Ca^{2+} before the actinic irradiations are started (usually about 0.5 h). If no Ca^{2+} is used in the measuring buffer, the cells should be suspended in Buffer A plus 2 mM Ca^{2+} for about 0.5 h before they are washed and introduced into the murexide solution. Our results indicate

Table I. Summary of organelle reactions controlled by phytochrome *in Vitro.*

Tissue source	Organelle	Photoreversible function	Phytochrome source	Reference
Pea	Mitochondria	NADP reduction	Endogenous	Manabe and Furuya (1974)
Barley, wheat	Etioplast	Regulate levels of biologically active GA	Endogenous	Cooke *et al.* (1975); Evans and Smith (1976); Hilton and Smith (1980)
Oat	Mitochondria	NADH oxidation	Exogenous	Cedel and Roux (1980b)
Oat	Mitochondria	Ca^{2+} fluxes	Endogenous-exogenous	Roux *et al.* (1981)
Zucchini	Microsomes	Peroxidase activity	Endogenous	Penel *et al.* (1976)
Oat	Mitochondria, etioplasts	Permeability of organelle membranes to metabolites	Endogenous	Schmidt and Hampp (1977)

that protoplasts must have a prior incubation in a Ca^{2+}-containing measuring buffer in order to show R–FR reversible Ca^{2+} fluxes (Hale and Roux, 1980).

(4) Positioning of cells in cuvette relative to the measuring beam
The protoplasts should be introduced into the 1-ml cuvette in a volume of 0.5 ml or less. The vast majority will settle to the bottom of the cuvette within 30 min. At this point the cuvette should be raised about 10 mm so that the settled cells are just below where the measuring beams cross through the cuvette. Alternatively, a Ficoll (Sigma Co.) cushion, consisting of 7% w/v Ficoll in Buffer A plus 25 μM murexide, may be placed in the bottom of the cuvette for the cells to settle on. By concentrating the protoplasts just below the measuring beams, two problems are avoided. First, the signal is not disturbed by protoplasts falling through the beam path, and, second, changes in $\Delta 540$ nm due to photoconversion of cellular phytochrome by R and FR light or to the reduction of *c*-type cytochromes do not occur (Scott and Jeacocke, 1980). Clear interpretation of the data requires that all the $\Delta A540$ nm measured be solely attributable to the formation of Ca^{2+}–murexide complexes; the protoplasts and their endogenous phytochrome and cytochromes should not be in the path of the beam.

(5) Irradiation treatments
The test cells can show photoreversible Ca^{2+} fluxes whether they are irradiated inside or outside of the cell compartment of the spectrophotometer (Hale and Roux, 1980). Leaving the cuvette inside the spectrophotometer for the irradiation treatments allows faster monitoring of the Ca^{2+} fluxes; to

do this, the photomultiplier must either be protected with a green interference filter or the voltage to the photomultiplier must be shut off during the light treatments. To measure continuously small changes in the absorbance of murexide during the R – FR irradiations of cells, it is necessary to block the photomultiplier with a green interference filter (high transmittance > 500 nm; none > 610 nm), and to minimize scatter and spreading of the actinic beam by guiding it to be flush against the window of the cuvette through a light pipe, or focusing it narrowly on the cells through lenses.

C. Comment on expected results

Hale and Roux (1980) found that R light induced an apparent efflux of Ca^{2+} (i.e., decrease in the A540 – A507 nm of the murexide solution) from oat coleoptile protoplasts and that FR light could reverse this flux. To detect even minimal changes (ΔA540 nm $\geq$ 0.0005), at least 10^5 functioning cells with a high phytochrome titre were required. Cell quantities $> 2 \times 10^5$ consistently gave changes in A540 – A507 nm > 0.001 A cm^{-1}. Depending on as yet incompletely characterized variables in the experiment (such as the $[Ca^{2+}]$ of the measuring medium, total cell number, age, and condition of the protoplasts), the R light induced efflux may last less than 30 sec (i.e., less than the duration of the irradiation), or may continue for 10 min or more after the end of the irradiation period. Although murexide undergoes a ΔA in response to binding a number of other cations besides Ca^{2+}, it has a much lower affinity for Ca^{2+} and therefore dissociates from it more readily; when binding other cations, murexide undergoes essentially irreversible colour changes (Ohnishi, 1978). Thus, to measure irreversible changes in $[Ca^{2+}]$ with murexide, such as when protoplasts are lysed, the portion of the ΔA of the murexide due to Ca^{2+} must be estimated by back-titrating with an excess of the specific Ca^{2+} chelator EGTA. Any colour change which cannot be reversed by EGTA must be attributed to some process other than the binding of Ca^{2+} by murexide.

IV. METHODS FOR EXPERIMENTS AT THE ORGANELLE LEVEL

A. Introduction

Available evidence suggests that phytochrome is present mainly as a soluble cytosolic protein in the dark-grown plant cell, but that upon being activated by light it associates with some subcellular structures (cf. Mackenzie et al., 1975). If this association is a prerequisite for phytochrome-mediated morphogenesis, then the question of with what structures phytochrome associates takes on obvious significance. Attempts to answer this question directly by immunocytochemical localisation methods have provided no definite answers, although it has been suggested that Pfr probably associates with more than one type of subcellular structure (Coleman and Pratt, 1974). Another approach has been to purify various subcellular membrane fractions and study whether they exhibit any R–FR reversible functions; membranes studied include mitochondria, etioplasts, plasma membranes, microsomes, and endoplasmic reticulum (Marmé, 1977). Each study argues for a meaningful association of phytochrome with the tested membrane. The follow-up studies necessary to substantiate these claims are still in progress.

Several other laboratories, most notably those of Marmé, Pratt, Quail, Smith, and Yu, have examined in detail the phenomenon of phytochrome "pelletability": the observation that phytochrome from irradiated cells fractionates mainly with the "pelletable" component of the cell. The induction of pelletability is an R–FR reversible event that occurs *in vivo* within seconds after irradiation (Quail and Briggs, 1978). A large percentage of the total phytochrome is associated with particulate fractions from irradiated tissue when Mg^{2+} is present in the homogenization buffer. Less association occurs in the absence of Mg^{2+}, but this association is tight (Watson and Smith, 1982) and can induce photoreversible changes in the mitochondrial fraction of the particulates (Roux *et al.*, 1981). Since the organelles in crude homogenates which contain ≥ 5 mM Mg^{2+} are highly aggregated and difficult to dissociate, little or no Mg^{2+} should be included in the homogenization buffer if one wishes to subfractionate phytochrome-containing particulates from irradiated tissue. A good analysis of pelletability experiments, their utility, and their pitfalls may be found in Pratt (1979).

As indicated previously, it is possible that phytochrome is modified *in vivo* following its conversion to Pfr, and that such a modification could facilitate the interaction of phytochrome with cell membranes. Phytochrome isolated as Pr from unirradiated plants, however, can photoreversibly alter the ion permeability of planar lipid membranes (Roux and Yguerabide, 1973). This indicates that phytochrome unmodified by light *in vivo* has at least some of the structural requisites needed to affect artificial membranes. Whether it can function in cellular membranes may be tested *in vitro* by examining whether phytochrome purified from dark-grown seedlings can specifically bind to and affect the functions of a well-characterized membrane preparation.

Tests for binding "specificity" involve multiple criteria which may vary, depending on the kind of ligand and binding site involved. Cedel and Roux (1980a) describe in detail a method for determining the amount of phytochrome which binds to a purified organelle and show how this method can be exploited to test whether the binding meets a number of minimal criteria for specificity. The test for phytochrome involvement is generally accepted to be photoreversibility of the effect of R light by FR light. The rest of this section will focus on experiments designed to examine whether phytochrome (exogenous or endogenous) can photoreversibly alter some organelle function. Selected data from the papers noted in Table I will be used to illustrate the most important technique considerations in this section.

B. Materials and methods

(1) Organelle choice

There are no data in the literature to suggest that phytochrome preferentially reacts with any one cellular organelle, and, as reviewed by Quail (1979), there are methods available for purifying every well-characterized organelle in plant cells. Plastids and mitochondria may be prepared more readily in bulk quantities and characterised more easily as to their purity, intactness, and function, but the methods for preparing the other major organelles are also reliable. The results reported for plastids and mitochondria in Table I should not focus undue attention on these organelles, and other organelles should be considered equally appropriate subjects for study.

(2) Tissue choice

(a) Age

Tissues that are rapidly undergoing development, such as the phytochrome-rich regions of cereal coleoptiles and of dicotyledonous seedling epicotyls and hypocotyls, are also rapidly changing the content and functional potentials of their organelles (Hampp and Wellburn, 1979; Quail, 1979). For example, mitochondria from 3-day-old corn seedlings show an R-light response that is absent in mitochondria from 7-day-old seedlings (Gross *et al.*, 1979). Awareness of this organelle aging factor will help to achieve reproducibility of results. Note that "age" here refers to physiological age; a 4-day-old seedling grown at 28°C is physiologically older than a 4-day-old seedling grown at 20°C.

(b) Tissue type

Tissue source will determine the morphological appearance and membrane function of the organelles derived from it (Quail, 1979). The variability of structure and function of isolated organelles may often be correlated with a similar variability *in vivo* (Malone *et al.*, 1974). In oat coleoptiles, phloem mitochondria differ morphologically from parenchyma mitochondria, and parenchyma mitochondria located in cells near the apical tip of a coleoptile differ from those from cells 20 mm from the tip (Slocum, 1981). Thus it is practically impossible to avoid obtaining a mixture of mitochondrial types at the end of a purification. One can reduce this heterogeneity and ultimately clarify data interpretation and reproducibility by minimizing the heterogeneity of the tissue source. For example, harvesting etiolated oat seedlings near the base of the shoot yields a very heterogeneous mixture of mesocotyl, primary leaf, and coleoptile tissue; harvesting them above the node yields mostly coleoptile tissue with some primary leaf and no mesocotyl. Mitochondria from these two different tissue sources yield different responses to R and FR irradiation when there is no exogenous phytochrome in the reaction mixture.

(3) Exposure of tissue to light prior to extraction

Irradiation *in vivo* could initiate two separate processes simultaneously: (1) The conversion of Pr to Pfr and (2) The chemical modification of either phytochrome (e.g., phosphorylation or methylation) or the organelle receptor sites or both. Whereas process 1 would certainly proceed as well as *in vitro* as *in vivo,* process 2 might require intact cells to progress to completion.

From the work of Hampp and Wellburn (1979) and Cedel and Roux (1980a) it seems clear that mitochondria isolated from R-irradiated tissue differ functionally from those isolated from unirradiated tissue. In contrast, none of the references in Table I on phytochrome-controlled GA levels in etioplast preparations, nor the study of Manabe and Furuya (1974) with pea mitochondria, noted any requirement for an *in vivo* irradiation, other than the green-light illumination used for harvesting, to obtain R – FR responsive organelles. An actinic irradiation of the source tissue *in vivo* may or may not be an important feature of the protocol followed; however, an awareness of its potential influence on the *in vitro* results is warranted. The *caveat* of Smith (1975) that no visible light, green or otherwise, is truly "safe," makes it important to note how long plants were exposed to green light and at what irradiance level prior to their homogenization.

(4) Use of exogenous phytochrome

It is clear from Table I that most organelle experiments reported do not require exogenous phytochrome for the induction of photoreversible responses. Methods for the use of exogenous phytochrome in *in vitro* experiments are described in Section V.

(5) Quantity and quality of organelles required

Whenever possible, the intactness and normal function of the organelle being tested should be assayed before the phytochrome responses are examined. Mitochondria are ideal because the biochemical assays for intactness and ADP:O ratios may be easily and rapidly carried out (see Cedel and Roux, 1980b). The magnitude of any actinic light response should be relatable to the quantity (as assayed, for example, by protein content) and quality (relative purity, intactness, metabolic rate) of the organelles in the preparation. In the case of mitochondria, the higher the percentage of these organelles with intact outer membranes, the less mitochondria are needed to give a photoreversible response (Roux *et al.,* 1981). The "quality" of an organelle preparation when assayed for phytochrome responses is often inversely related to the length of time intervening between the homogenization of the tissue and the start of the assay. In the experience of this author's laboratory, mitochondria rarely function normally 14 h after tissue homogenization.

(6) Identification of organelle responding to actinic light
Quail (1979) discusses the problem of evaluating the purity of a given organelle preparation. With currently available methods, it is unlikely that any membrane preparation can be made more than 80–90% pure. This can lead to difficulties in determining which membrane produces the photoreversible response in an organelle preparation. To attribute a photoreversible response to an organelle known to be present in significant quantity in the preparation, at least two criteria should be met. The response being measured should be a proven (or at least a probable) biochemical function of that organelle under the *in vitro* conditions used. This criterion may appear to be self-evident, but it has not always been met, as discussed by Cedel and Roux (1980b). A second criterion, which is more difficult to evaluate, is that there should be clear evidence of phytochrome association with that organelle under the conditions of the experiment. Indirect evidence may be used to infer which organelle is the one affected by the actinic light (e.g., Hilton and Smith, 1980). Whenever possible, however, direct immunocytochemical evidence for the association of phytochrome with the organelle is preferred. Methods for this are described for phytochrome bound to mitochondria (Roux *et al.,* 1981).

V. METHODS FOR EXPERIMENTS AT THE LIPID BILAYER LEVEL

A. Introduction

The "lipid bilayer" refers to an artificial membrane made up solely of lipids which, at best, only partially represent the lipid composition of cellular membranes. Model membranes greatly simplify the situation in the intact cell and can, under specified conditions, provide data that accurately reflect native membrane interactions (Racker, 1976).

Phytochrome has been implicated to have membrane functions, and, as noted in Sections II–IV above, the facilitation of ion movement across membranes would certainly be one membrane function suggested by *in vivo* data. To date, the studies on phytochrome–model–membrane interactions have examined mainly this ion-permeability function. Two kinds of artificial membrane systems have been used in these studies: planar lipid bilayers, also called "black lipid membranes" or BLMs (Roux and Yguerabide, 1973; Roux, 1974), and liposomes (Georgevich and Roux, 1982). The rest of this section will discuss general features of these studies and will include both the technical aspects of model membrane production and the handling of exogenous phytochrome in model membrane experiments.

B. Materials and methods

(1) Choice of model membrane system

In both BLMs and liposomes, Pfr induces a greater permeability to ions than Pr. Both types of membranes are easy to make. The experimental setup for monitoring ion fluxes is simpler for liposomes than for BLMs, and the ease of harvesting membrane-bound phytochrome after the reaction is far greater with liposomes than with BLMs. The BLM would be the system of choice for accurately monitoring very rapid changes in ion permeability induced by

Table II. Summary of phytochrome-model membrane studies.

Bilayer type	Lipid Source	Pfr-enhanced ion permeability?	Reference
BLM	Oxidised cholesterol	Yes	Roux and Yguerabide (1973)
BLM	PC[a]	No	Roux (1974)
	PC–cholesterol	Yes	
Liposome[b]	Oat membranes	Yes	Georgevich and Roux (1982)
	PC–sitosterol	Yes	Georgevich and Roux (1982)
	Soybean lecithin (80% pure, Sigma)	No	Georgevich and Roux (1982)
	PC–DP[c]	Yes	Georgevich and Roux (1982)

[a] PC, dipalmatoylphosphatidylcholene (Sigma).

[b] Ca–EDTA liposomes (Papahadjopoulos *et al.*, 1975) were the only variety that consistently showed Pfr-enhanced ion permeability.

[c] DP, dicetyl phosphate (Sigma).

phytochrome after R or FR light; for most other purposes the liposome system is better suited. The discussion below will highlight liposome experiments because of their greater feasibility.

(2) Production of lipid bilayers

The reviews by Finkelstein (1974) and Szoka and Papahadjopoulos (1980), together with the brief description below, provide a guide for the formation of stable lipid bilayers.

Unilamellar liposomes afford simpler data interpretation than multilamellar liposomes; large ($> 1 \mu$m diameter) unilamellar liposomes have a degree of surface curvature that more closely approximates that of cellular membranes than does the severe surface curvature found on small unilamellar liposomes ($< 0.05 \mu$m diameter). Large unilamellar liposomes are formed easily by the method of Papahadjopoulos *et al.* (1975). This method first requires making small unilamellar liposomes by sonicating a lipid, which has been dispersed in a test tube, in a 100-w bath sonicator for about 1 h. If the ambient temperature is maintained at 4°C, the temperature of the bath will rise to only 30–40°C during the sonication. For the lipids used in Georgevich's (1980) liposome studies (cf. Table II), this temperature was above the phase transition temperatures of the dispersed lipids, a factor important for the formation of well-sealed liposomes (Szoka and Papahadjopoulos, 1980). The liposomes formed by this sonication procedure average 300 Å in diameter. To form large diameter unilamellar liposomes, $CaCl_2$ (final concentration 10 mM) is added to the sonicated liposomes, causing them to fuse and flocculate. The lipid precipitate is then incubated at 40°C for 1 h, then treated with 15 mM EDTA. This induces the spontaneous formation of large unilamellar liposomes, with diameters up to 50 μm. Encapsulated in the lumen of these liposomes would be the same buffer in which the lipids were originally sonicated, as discussed in Section V. B. 4 and in Georgevich and Roux (1982).

Whether or not phytochrome will induce photoreversible ion fluxes across model membranes depends to a significant extent on the lipid composition of the bilayer (Georgevich and Roux, 1982). Table II summarizes some of the lipids that have been used in previous studies. This list is not exclusive and our expectation is that other lipid combinations would also support phytochrome responses. Lipids may be extracted from plant membranes by standard methods (e.g., Bligh and Dyer, 1959), and, as noted in Table II, these can form liposomes which are useful for phytochrome studies. However, to standardize lipid sources and to simplify and specify the exact lipid composition of the liposomes, it is usually more convenient to use commercially prepared lipids.

(3) Quality and quantity of phytochrome reacted with liposomes
One of the main advantages of a model system is that most or all of the components in it are defined. The greater the purity of the phytochrome preparation used for model membrane studies, the more defined the system will be and the more confident can be the attributions of permeability effects to photochrome alone. On the other hand, when the effects of R or FR light or of Pr and Pfr on the membranes are being compared, photoreversible or differential effects may be reasonably attributed to phytochrome, even when it represents less than 50% of the total protein added, as long as no other R- or FR-absorbing pigment is present in the preparation.

Degraded "small" (60,000 daltons) phytochrome will photoreversibly alter the permeability of model membranes as effectively as "large" (400,000 daltons) phytochrome, whereas spectrally denatured phytochrome will have no specific effects on membrane permeability (Georgevich and Roux, 1982). A shift in the Pr absorbance peak to lower wavelengths is one of the more sensitive indicators of spectral denaturation. The absorbance peak of spectrally native Pr is near 667 nm. If during sample handing this shifts to below about 662 nm (usually also with some loss of photoreversibility), the competence of phytochrome to affect membrane ion fluxes in either BLMs (Roux and Yguerabide, 1973) or in liposomes (Georgevich and Roux, 1982) is lost. This principle also holds true for the effects of exogenous phytochrome on purified mitochondria (S. J. Roux and T. E. Cedel, unpublished). Phytochrome labeled with ^{125}I, which retains its native spectral properties (Georgevich *et al.*, 1977), can modulate the ion permeability of liposomes as well as unlabeled phytochrome.

To minimize the possibility of phytochrome denaturation, phytochrome can be treated with bilayers after they are formed instead of mixing phytochrome with the lipids before they form bilayers. Final concentrations of phytochrome in the reaction medium of between 10^{-6} and 10^{-7} M (based on a molecular weight of 120,000) almost invariably suffice to give differential Pr/Pfr responses. Higher concentrations often lead to general membrane leakiness and a loss of any detectable differential effects of Pr and Pfr on the membrane.

(4) Measurement of the efflux of liposome-encapsulated electrolytes
Ion influx or efflux may be monitored, but the probability of liposomes bursting is less if efflux is promoted during the experiment. The procedure is to form the large diameter (Ca–EDTA treated) liposomes in the medium containing the species whose efflux is to be studied and then to pass the mixture through a molecular-sieve column, such as Sephadex G-50 (coarse), to remove untrapped solutes. To promote efflux, the composition of the column buffer is selected to give it a slightly higher osmolarity than that of

the solution trapped inside the liposomes. The purified liposomes are then mixed with the phytochrome as Pr or Pfr, after which the efflux of one of the entrapped substances is monitored.

Radiotracers have also been used to monitor the efflux of ions from the liposomes (Georgevich, 1980). The method was to place ^{86}RbCl-loaded liposomes with phytochrome into a dialysis bag at time zero, allow the mixture to dialyze into $20-50 \times$ the bag volume, and monitor the appearance of ^{86}Rb in the dialyzate over time. Often within 20 min after the reaction of liposomes with phytochrome is initiated, liposomes begin fusing. This process tends to blur the differences between the efflux rates promoted by Pr and Pfr; consequently, to detect any differences, most of the efflux sampling must be completed within the first 15 min. It is technically difficult to obtain enough data points during this time period, using the radiotracer methodology.

A more convenient way to accomplish this rapid sampling is to monitor continuously the electrolyte leakage from the liposomes with a conductivity meter. With this method, desalted liposomes are placed in a temperature-controlled (water-jacketed) chamber. A standard conductivity probe, connected to the conductivity meter, is placed into the chamber and the efflux of electrolytes is monitored until temperature equilibration is reached and the efflux rate becomes constant, usually within 10 min. The probe is then removed for the few seconds necessary to introduce phytochrome (or ionophore, or detergent, etc.) into the chamber. This step is followed by gentle mixing of the liposome suspension and immediate resumption of recording.

The measuring and recording apparatus used by Georgevich (1980) for this method consisted of a Radiometer CDM 3 (alternating current) conductivity meter with a CDC 304 probe connected to a variable range chart recorder through a homemade continuously variable offset-voltage generator. Using this arrangement, the signal generated by the conductivity meter could be amplified up to 1000 times, allowing the measurement of low-level changes of the conductivity in the liposome suspension. These changes were indicated by changes in the slope of the line drawn by the recorder.

(5) Monitoring the efflux of other entrapped substances from the liposomes
The significance of any change in the rate of electrolyte efflux induced by Pr or Pfr can only be gauged if the rate of general release of all entrapped substances induced by the reaction is also known. To do this, one need only entrap some large macromolecule in the liposomes and follow its release after the vesicles are treated with phytochrome. Trypsin is a sufficiently large macromolecule to serve this purpose.

The following method was used by Georgevich (1980) to encapsulate trypsin in liposomes and then to monitor its release following phytochrome reaction with these liposomes. Trypsin is trapped in liposomes by dispersing 30 mg of lipid and 1 mg of trypsin in 1 ml of 10 mM Tris-HCl at pH 8.5. The temperature of the sonicating bath is kept below 30°C throughout the 0.5 h sonication. A 5-min centrifugation is used to clarify the sample before it is passed through a Bio Gel A 1.5-m column (280 × 3 mm), equilibrated in 30 mM Tris-HCl at pH 8.5, to remove untrapped trypsin from the liposomes. The assay for trypsin leakage from the liposomes is performed at 20°C in a 1-ml, 1-cm path-length cuvette which is positioned as close as possible to the photomultiplier tube to minimize scatter effects. The leakage rate is measured by recording the appearance of hydrolysis products of an N-benzyl-L-arginine ethyl ester (BAEE) at A_{235} nm in the liposome suspension. The assay is initiated by the addition of 3 μg of BAEE of 1 ml of liposomes in the sample cuvette. Three min later, 7 μg/ml of phytochrome is added as Pr or Pfr and the recording of A_{253} is continued. Five minutes later, Triton X-100 is added to a final concentration of 0.1% to both the sample and reference cuvettes. This bursts the liposomes and ensures that the entrapped trypsin is still active.

C. Comment on expected results

Table II summarises the types of model bilayers which have shown a significantly increased ion permeability after reaction with Pfr. In the planar bilayer system, photoreversible conductance changes could be detected; in the liposomes studies, the ion efflux rate established by Pr or Pfr were only marginally affected by actinic light, although the rate of ion efflux from liposomes after they reacted with Pfr was consistently many times faster than the rate after reaction with Pr.

The poor photoreversibility of the permeability effects of Pfr after it reacts with liposomes suggests that Pfr may lose its spectral photoreversibility after reacting with liposomes. Georgevich (1980) proposed the following explanation for these observations. In the liposome experiment, most of the Pfr added to the reaction system comes into contact with a bilayer within minutes after addition to the liposome suspension. In contrast, in the planar bilayer experiments, most of the Pfr remains free in the buffer reservoir, untreated with a bilayer, and thus retains its spectral photoreversibility. These observations lead to a hypothesis to account for the reason why the permeability effects of phytochrome on BLMs were strongly photoreversible (though usually through only one R–FR cycle) whereas they were only poorly so with liposomes: the continuance of Pfr permeability effects requires that the membrane-bound Pfr be replaceable with previously unbound Pfr. If the reservoir of unbound Pfr is converted to Pr, the enhanced ion permeability of Pfr-treated bilayers is rapidly lost—i.e., shows photoreversibility, as in BLMs. If there is no unbound Pfr or Pr in the reservoir as in liposomes, the Pfr-enhanced permeability will be negligibly affected by light, but will gradually be lost in darkness. This hypothesis has not been rigorously tested in model membrane experiments, although it is consistent with all the observations made so far.

VI. CONCLUDING DISCUSSION

It is remarkable that at every level examined — tissue, cell, organelle, and lipid bilayer — phytochrome has been clearly demonstrated to alter rapidly the ion permeability of membranes shortly after actinic irradiation. It appears unlikely that this "coincidence" is without physiological significance. On the other hand, until such ion permeability changes can be causally linked with photoreversible macroscopic events, such as the rapid growth changes demonstrated by Morgan and Smith (1978), it would be premature to conclude that this ability to alter membrane ion permeability is a primary function of phytochrome.

It is currently fashionable to focus on Ca^{2+} ions as a key instigator of major physiological responses in animals and plants. If phytochrome turns out to be an important regulator of membrane permeability to this cation in most plants, as it appears to be in oats and some green algae, then it will be reasonable to designate this function as a primary one and to let this serve as a working hypothesis. Given the tight coupling of various membrane transport systems (Clarkson, 1977), it is almost certain that the permeability of membranes to other ions (and to larger substances) is affected simultaneously with the changes in Ca^{2+} flux. These permeability changes and concomitant changes in other membrane functions will, in the long run, have to be integrated into the picture describing the primary effects of phytochrome in membranes.

REFERENCES

Bligh, E. G. and Dyer, W. J. (1959). *Can. J. Biochem. Physiol.* **37**, 911–917.
Brownlee, C. and Kendrick, R. E. (1979). *Plant Physiol.* **64**, 206–210.
Cedel, T. E. and Roux, S. J. (1980a). *Plant Physiol.* **66**, 696–703.
Cedel, T. E. and Roux, S. J. (1980b). *Plant Physiol.* **66**, 704–709.
Clarkson, D. T. (1977). *In* "The Molecular Biology of Plant Cells" (H. Smith, ed.), pp. 24–63. Univ. of California Press, Berkeley.
Coleman, R. A. and Pratt, L. H. (1974). *J. Histochem. Cytochem.* **22**, 1039–1047.
Cooke, R. J., Saunders, P. F. and Kendrick, R. E. (1975). *Planta* **124**, 319–328.
Dreyer, E. M. and Weisenseel, M. H. (1979). *Planta* **146**, 31–39.

Epstein, E. (1961). *Plant Physiol.* **36**, 437–444.

Epstein, E., Schmid, W. E. and Rains, D. W. (1963). *Plant Cell Physiol.* **4**, 79–84.

Evans, A. and Smith, H. (1976). *Proc. Natl. Acad. Sci. U.S.A.* **73**, 138–142.

Finkelstein, A. (1974). *In* "Methods in Enzymology" (S. Fleischer and L. Packer, eds.), Vol. 32, Part B, pp. 489–501. Academic Press, New York.

Fondeville, J. C., Borthwick, H. A. and Hendricks, S. B. (1966). *Planta* **69**, 357–364.

Georgevich, G. (1980). Ph.D. Thesis, University of Pittsburgh, Pittsburgh, Pennsylvania.

Georgevich, G. and Roux, S. J. (1982). *Photochem. Photobiol.* **36**, 663–671.

Georgevich, G., Cedel, T. E. and Roux, S. J. (1977). *Proc. Natl. Acad. Sci. U.S.A.* **74**, 4439–4443.

Gross, J., Ayadi, A. and Marmé, D. (1979). *Photochem. Photobiol.* **30**, 615–621.

Hale, C. C., II and Roux, S. J. (1980). *Plant Physiol.* **65**, 658–662.

Hampp, R. and Wellburn, A. R. (1979). *Planta* **147**, 229–235.

Hilton, J. and Smith, H. (1980). *Planta* **149**, 312–318.

Jacoby, B. and Nissen, P. (1977). *Physiol. Plant.* **40**, 42–44.

Jaffe, M. J. (1968). *Science* **162**, 1016–1017.

Mackenzie, J. M., Jr., Coleman, R. A., Briggs, W. R. and Pratt, L. H. (1975). *Proc. Natl. Acad. Sci. U.S.A.* **72**, 799–803.

Malone, C., Koeppe, D. E. and Miller, R. J. (1974). *Plant Physiol.* **53**, 918–927.

Manabe, K. and Furuya, M. (1974). *Plant Physiol.* **53**, 343–347.

Marmé, D. (1977). *Annu. Rev. Plant Physiol.* **28**, 173–222.

Morgan, D. C. and Smith, H. (1978). *Nature (London)* **273**, 534–536.

Ohnishi, S. T. (1978). *Anal. Biochem.* **85**, 165–179.

Papahadjopoulos, D., Vail, W. J., Jacobson, K. and Roste, G. (1975). *Biochim. Biophys. Acta* **394**, 483–491.

Penel, C., Greppin, H. and Boissard, J. (1976). *Plant Sci. Lett.* **6**, 117–121.

Pike, C. S. and Richardson, A. E. (1979). *Plant Physiol.* **63**, 139–141.

Pratt, L H. (1979). *Photochem. Photobiol. Rev.* **4**, 59–124.

Pratt, L. H., Coleman, R. A. and Mackenzie, J. M., Jr. (1976). *In* "Light and Plant Development" (H. Smith, ed.), pp. 75–94. Butterworth, London.

Quail, P. (1979). *Annu. Rev. Plant Physiol.* **30**, 425–484.

Quail, P. and Briggs, W. R. (1978). *Plant Physiol.* **62**, 773–778.

Racker, E. (1976). "A New Look at Mechanisms of Bioenergetics." Academic Press, New York.

Racusen, R. H. (1976). *Planta* **132**, 25–29.

Rains, D. W. (1976). *In* "Plant Biochemistry" (J. Bonner and J. E. Varner, eds.), pp. 561–598. Academic Press, New York.

Roux, S. J. (1974). *In* "Proceedings of the Annual European Symposium on Photomorphogenesis" (J. de Greef, ed.), pp. 1–3. Antwerpen Univ. Press, Antwerpen.

Roux, S. J. and Yguerabide, J. (1973). *Proc. Natl. Acad. Sci. U.S.A.* **70**, 762–764.

Roux, S. J., McEntire, K., Slocum, R. D., Cedel, T. E. and Hale, C. C., II (1981). *Proc. Natl. Acad. Sci. U.S.A.* **78**, 283–287.

Satter, R. (1979). *Encycl. Plant Physiol., New Ser.* **7**, 442–484.

Scarpa, A. (1972). *In* "Methods in Enzymology" (A. San Pietro, ed.), Vol. 24, pp. 343–351. Academic Press, New York.

Schmidt, H.-W. and Hampp, R. (1977). *Z. Pflanzenphysiol.* **82**, 428–434.

Scott, I. G. and Jeacocke, R. E. (1980). *FEBS Lett.* **109**, 93–98.

Slocum, R. D. (1981). Ph.D. Thesis, University of Texas at Austin.

Smith, H. (1975). "Phytochrome and Photomorphogenesis." McGraw-Hill, New York.

Szoka, F., Jr. and Papahadjopoulos, D. (1980). *Annu. Rev. Biophys. Bioeng.* **9**, 467–508.

Tanada, T. (1968). *Proc. Natl. Acad. Sci. U.S.A.* **59,** 378–380.
Tanada, T. (1972). *Nature (London)* **236,** 460–461.
Tanada, T. (1973a). *Plant Physiol.* **51,** 150–153.
Tanada, T. (1973b). *Plant Physiol.* **51,** 154–157.
Tanada, T. (1977). *Planta* **134,** 57–59.
Watson, P. J. and Smith, H. (1982). *Planta* **154,** 121–127.

12

Blue-Light Photoreceptor

Dieter Dörnemann and Horst Senger

I. INTRODUCTION

A great variety of blue-light-induced phenomena in plants and microorganisms has been investigated since the early 1930s. Several reviews have summarised the remarkable diversity of the effects (Gressel, 1979; Presti and Delbrück, 1978; Schmidt, 1980, 1981; Senger and Briggs, 1981; Senger, 1982; Kowallik, 1982; Rau, 1976).

Blue light triggers many photomorphogenetic processes. Examples include protonemal development of ferns (Furuya *et al.,* 1980), chloroplast development in algae (Brinkmann and Senger, 1980; Hase, 1980; Senger, 1981; Schiff, 1980), roots (Richter *et al.,* 1980; Björn, 1967, 1980) and leaves of higher plants (Akoyunoglou *et al.,* 1980; Lichtenthaler *et al.,* 1980), fruiting of mushrooms (Eger-Hummel, 1980), photoperiodic responses (Clauss, 1979; Tsuchiya and Ishiguri, 1981; Edmunds, 1980), metabolic effects such as respiration enhancement (Kowallik, 1982), carotenogenesis (Rau, 1976, 1980), chlorophyll biosynthesis (Jeffrey and Vesk, 1981; Kamiya *et al.,* 1981), anthocyanin formation (Drumm-Herrel and Mohr, 1982), and enzyme regulation (Ruyters, 1982). Other responses which are modulated by blue light include phototropism (Shropshire, 1980; Lipson, 1980), photomovement (Nultsch, 1980; Zurzycki, 1980), and stomatal opening (Zeiger *et al.,* 1982; Ogawa, 1981; Donkin and Martin, 1981; Sharkey and Raschke, 1981).

Action spectra for many of these effects have been determined (Rau, 1980; Munoz and Butler, 1975; Schmidt, 1981; Presti and Delbrück 1978) and demonstrate a striking similarity with the main peak around 450 nm, two adjacent peaks at shorter and longer wavelengths, and another dominant peak around 370 nm. The similarity of these action spectra to the absorption spectra of carotenoproteins and flavoproteins stimulated intense dis-

TECHNIQUES IN PHOTOMORPHOGENESIS
0-12-652990-6

cussion about which one of these pigment groups would represent the photoreceptor for blue-light effects (Shropshire, 1980; Song, 1980a,b; De Fabo, 1980). Although these two pigment groups are still the favourite candidates for the photoreceptors, some action spectra extending into the green region of the spectrum raised the question of whether other pigments, such as haemoproteins (Schneider and Bogorad, 1978; Senger *et al.*, 1980), might be involved in photoregulatory processes which are different from photosynthesis and the phytochrome system.

Another well-investigated blue-light absorbing photoreceptor is rhodopsin, the vision pigment, which has retinal as the chromophoric group. Rhodopsins have also been shown to be the photoreceptor driving the proton pump in *Halobacterium* (Oesterhelt and Stoeckenius, 1971; Hildebrand, 1977). Elucidation of the extent to which the rhodopsins participate as photoreceptors in blue-light effects is impeded by several unanswered questions: what is the concentration of the photoreceptors, where are they located, are they oriented, to what extent are screening effects involved, do some of the action spectra represent accessory pigments, and are there photochromic effects? To answer some of the open questions and to reveal the true nature of the photoreceptor(s), isolation and identification of the relevant pigment groups and the knowledge of their specific reactions is of great importance.

II. CAROTENOPROTEINS

Carotenoids are found in almost every living system. They have been discussed as candidates for the blue-light photoreceptor since the early 1930s (Castle, 1935; Bünning, 1937). In photosynthetic organisms, there is no indication that they occur outside the plastids. In fungi, the location of carotenoids acting in photomorphogenetic processes is not known.

Physiologically active carotenoids all seem to be protein bound by noncovalent bonds. Their noncovalent binding to proteins and the resulting instability of the complexes has revealed their nature *in situ* as being carotenoproteins. The complex between carotenoid and protein is shown in at least some cases to have mutually stabilising effect against photooxidation of the pigment and denaturation of the protein (Ke, 1971). The embedding of the complexes into membranes explains energy transfer, location of photoreception, and subsequent reactions as well as their anisotropic behaviour. There is no doubt that binding to protein does alter the absorption and thus the possible action spectra of carotenoids. Red shifts in the absorption spectra of the isolated complexes versus the isolated carotenoids have been reported for extracts of higher plants (Vernon *et al.*, 1969) and for preparations from vertebrates (Cheesman *et al.,* 1967). Blue shifts have been demonstrated convincingly in carotenoid extracts upon increasing water content (Hager, 1970). They have also been found in preparations of carotenoproteins from *Rhodospirillum rubrum* (Izawa *et al.,* 1963; Vernon and Garcia, 1967; Fujimori, 1969) and again from invertebrates (Buchwald and Jencks, 1968).

Arguments for the identity of carotenoproteins with a blue-light photoreceptor are mainly based on their wide distribution and the similarity of their absorption spectra with the action spectra of blue light effects (cf. Schmidt, 1980). The dilemma that the 370-nm peak of most action spectra is not present in the absorption spectra of carotenoids (cf. De Fabo, 1980) might be explained by the various absorption changes in carotenoprotein complexes versus carotenoids (Fujimori, 1969), appearance of the 370-nm peak in a polar environment (Hager, 1970), or changes caused by its cis configuration.

Arguments that mutants lacking carotenoids (Presti *et al.,* 1977) or plants in which carotenoid biosynthesis is blocked (Bara and Galston, 1968) show no loss in their photoresponses have been questioned since traces of carotenoids could not be ruled out (cf. De Fabo, 1980). The argument that the lifetime of the first excited singlet state of carotenoids is too short to use the excitation energy directly for the primary photoprocess (cf. Song, 1980a; Bensasson, 1980) might be overcome by resonance interactions of more than one carotenoid molecule and an acceptor molecule (Song *et al.,* 1976; Koka and Song, 1977). Inasmuch as it has been suggested that blue light may act via induction of permeability changes (Kowallik, 1982; Laudenbach and Pirson, 1969; Hase, 1980), it is reasonable to speculate that carotenoids, tightly bound to the membrane, undergo cis–trans isomerisation expressed in light-induced absorbance changes (LIAC) and permeability changes. Blue-light-induced cis–trans isomerisation of ζ-carotene has been shown to be accompanied by LIACs (Fong and Schiff, 1978).

A. Isolation and identification of chromophores

The isolation and characterisation of the main carotenoids of plants is routinely performed in many laboratories. Due to modifications in the carbon skeleton (nor-, seco-, apocarotenoids), changes in the hydrogenation level (dehydro- or hydrocarotenoids), the degree of oxygenation (xanthophylls) and their configuration (cis and trans isomers), a great variability among carotenoids has been found. Therefore it is useful to give a brief description of some of the methods which result in these changes.

To extract carotenoids from higher plant tissues, solvents such as acetone, methanol, or ethanol are preferable (Davies, 1976). Algae are sometimes resistant to milder extraction methods so that even hot methanol must be applied (cf. Davies, 1976). General methods for extracting algae are described by Hager and Stransky (1970a–e). For fungi and yeasts, the method of Davies *et al.* (1974) may be used. A number of carotenoids are light sensitive (cis–trans photoisomerisation) and thermolabile, so exposure to sunlight, UV light, and heat must be avoided. Protection from oxygen, especially during thin layer chromatography (TLC) on acidic layers (Britton and Goodwin, 1971), as well as protection from acids is necessary because oxidative decomposition, cis–trans isomerisation, and isomerisation of 5,6-epoxides to furanoid 5,8-epoxides may take place.

Many chromatographic methods may be used to separate carotenoids. A compilation is given by Davies (1976). TLC solvent systems and procedures are described by Stahl (1967), Hager and Mayer-Bertenrath (1966), Britton and Goodwin (1971), and Taylor and Davies (1974). Gas–liquid chromatography of over 70 perhydro derivatives of carotenoids is described by Taylor and Davies (1975). The separation of carotenoids by high performance liquid chromatography (HPLC) is described by Stewart and Leuenberger (1976), Hagibrahim *et al.* (1978), Fiksdahl *et al.* (1978), and Braumann and Grimme (1981). The great advantage of HPLC is the high sensitivity (1–5 pmol of carotenoids) and the high reproducibility of this protective method. Further identification may be achieved by electronic absorption spectroscopy.

A great number of absorption data in different solvents for a variety of carotenoids as well as an extended description of carotenoid analysis, is given by Davies (1976). When using spectroscopy, one should pay attention to the fact that increasing water content in the solvent creates an absorption maximum around 370 nm at the expense of longer wavelengths (Hager, 1970). The identification of cis–trans isomers and their spectroscopic identification may be carried out using iodine-catalysed photoisomerisation (Zechmeister, 1962).

B. Isolation and characterisation of holochromes

Reports on isolated carotenoproteins date back as far as 1940 (Menke, 1940). Isolations have mainly focused on invertebrates and photosynthetic membranes. Attempts have rarely been made to isolate carotenoproteins with the aim of studying the blue-light photoreceptors.

If carotenoproteins are to be isolated from photosynthetic organisms, the first step is always the removal of the chlorophylls. This can be achieved from chloroplast preparations using methanol (Menke, 1940) and from total leaves or broken chloroplasts with ammoniacal acetone extraction (Nishimura and Takamatsu, 1957, 1960), or sequential acetone extraction (Ji *et al.*, 1968a,b). The extraction of material from higher plants, the breaking of the cells, the fractionation and the subsequent purifications are summarised by Ke (1971). Generally, carotenoproteins must be solubilised with detergents. All these procedures yield red β-carotenoprotein complexes with absorption peaks close to 538, 498, and 460 nm. The β-carotenoproteins are part of the photosynthetic apparatus and are not likely to act as specific blue-light photoreceptors.

It is now possible to isolate and purify chloroplast envelopes (Douce and Joyard, 1982) and to demonstrate their high carotenoid content. These envelopes promise to be a better starting material for the extraction of carotenoprotein complexes which may possibly act as photoreceptors. With a few exceptions, the isolations of pigment–protein complexes from photosynthetic membranes neglect the carotenoproteins. In the case of the marine diatoms a peridinin–chlorophyll complex could be isolated without solubilising the complex with detergent (Haxo et al., 1976; Koka and Song, 1977). By revealing the arrangement of this pigment complex, the mode of energy transfer could be studied (Song, 1980b).

Several authors have described the isolation of carotenoprotein complexes from Rhodospirillum rubrum (Izawa et al., 1963; Vernon and Garcia, 1967; Ke et al., 1968; Fujimori, 1969). These complexes shift their absorption spectra in aqueous solution to the blue region with a main peak at 370 nm.

Since hydrophobic interactions are recognised to be the important forces in protein structure and binding of the pigment–protein complexes (Siegelmann, 1981) a new separation technique, hydrophobic interaction chromatograhy (Siegelmann and Kycia, 1982), seems to be most promising for the separation and purification of pigment–protein complexes.

More detailed methods for the isolation, separation, and identification of carotenoproteins are reported for experiments using invertebrates as the starting material. Lobster skin was homogenised, the extract fractionally precipitated with ammonium sulphate, chromatographed on DEAE–cellulose, further purified by gel filtration, concentrated by dialysis, and further fractionated by electrophoresis (Cheesman et al., 1966; Quarmby et al., 1977). Using this procedure, carotenoprotein complexes and the distinct components of the apoprotein subunit have been separated.

Another elaborate method is described for the separation of several carotenoproteins from the starfish Asterias rubens (Elgsaeter et al., 1978). The method follows 10 different steps involving extraction, concentration by ultrafiltration, ammonium sulphate precipitation, chromatography on Sephadex, and nondenaturing preparative polyacrylamide-gel electrophoresis. The main carotenoprotein of the starfish, asteriarubin, isolated by the above method, has an absorption maximum at 554 nm and shifts toward 485 nm when the carotenoprotein is denatured and the carotenoid is subsequently measured in benzene. Isolation methods and characterisation of carotenoproteins from invertebrates have been discussed by Zagalsky (1976).

III. FLAVOPROTEINS

The most favoured pigment group for blue-light receptors are currently the flavoproteins. Like the carotenoids, the flavins are ubiquitous in living organisms and exist in many forms. The problem is to isolate a possible flavin photoreceptor, which is probably present only in small concentrations, from the high amount of bulk flavins.

The possibility that riboflavin may be the photoreceptor for phototropism was first discussed by Galston in 1949. Flavoproteins fit well for the peak around 370 nm of the common action spectrum of blue-light effects, although it should be more dominant versus the longer wavelength peaks. The triple peak of the action spectra around the 450-nm region is also not typical for absorption spectra of flavoproteins.

Most of the known primary reactions after photoreception of blue light are redox reactions. The majority of the physiological flavin reactions are also documented as being redox reactions. This is further evidence that flavins are more likely candidates for photoreceptors than carotenoids, which are so far not known to cause any redox reactions. The three most widespread flavins of biological importance are riboflavin (Rb), flavin mononucleotide (FMN), and flavin adenine dinucleotide (FAD), although more rarely occurring biologically active flavins are present in biological tissues, especially in plants. Flavins can occur in the free, nonbound state (Rb) or as complexes with proteins or other cell components. A retrospective on flavin compounds was given by Galston in 1977.

A. Isolation and identification of chromophores

Depending on the type of bonding the isolation procedures of the chromophore are different. The extraction procedure which is used depends on whether determination of total flavin content or the estimation of individual flavins is intended. When extracting from animal and plant tissue, the samples are treated with hot distilled water as described by Yagi (1962, 1971). For determination of total amount of flavins, NaOH is added to the extract which is then subjected to photolysis, forming lumiflavin, followed by extraction into $CHCl_3$ under acidic conditions. For determination and calculation of the individual flavins, the extract is separated by paper chromatography and, after development of the chromatogram, again photolysed to lumiflavin. For calculation of flavin concentrations, equations are given by Yagi (1962, 1971).

An elaborate extraction procedure which is more concerned with the problems in plants is described by Koziol (1971). Depending on whether total or individual flavin determinations are required, acid extraction or extraction with buffers or hot water are described. If high quantities of cellulose are present, concentrated LiCl solution can be used (Lempka and Andrezejewski, 1966).

If the chromophore is tightly bound to a protein, trypsin treatment can be used after acid or buffer extraction. If the tissue has a high starch content, amylase treatment prior to the trypsin treatment must be applied (Koziol, 1971; Yagi, 1962). To determine individual flavins in biological liquids or tissues rich in protein, deproteinisation can be carried out by treatment with 20% trichloroacetic acid at $0\,^{\circ}C$ to prevent hydrolysis of FAD, or by adding excess acetone (3–5 volumes) to the solution and cooling in a refrigerator. Proteins and other interfering substances are separated by centrifugation. The acetone is afterwards evaporated at temperatures below $0\,^{\circ}C$; trichloroacetic acid is neutralised by K_2HPO_4. If quantitative determination of the flavins by fluorescence measurements is intended, quenching compounds and other interfering substances must be removed (Koziol, 1971). The relatively high resistance of flavins toward oxidants enables interfering compounds to be destroyed by oxidation with excess $KMnO_4$ and subsequent decolouring of the solution by H_2O_2 (Koziol, 1971).

For separation of riboflavin and the flavin nucleotides, their different solubility in benzyl alcohol can be used (Burch, 1957). Enrichment of flavins can be achieved by a method described by Yagi (1962). Flavins are first extracted from aqueous solutions with phenol or its derivatives. Addition of a small amount of water and excess diethyl ether produces a multifold concentration in the water phase.

As mentioned earlier, flavins can be separated by paper chromatography and then be identified by their R_f values. Different solvent systems are described for separation (Koziol, 1971; Yagi, 1962, 1971). Spots can be visualised by UV light of 366 nm which shows different colours of fluorescence emission for each flavin derivative. Further characterisation of the spots can be performed by fluorescence and absorption spectroscopy. For absorption data under different conditions, see Whitby (1953), Roth and McCormick (1967), McCormick et al. (1967), Massey and Ganther (1965), Dudley et al. (1964), Koziol (1966, 1969, 1971), and Weber (1966). Fluorescence emission spectra and data are reported by Koziol (1971) and Cerletti and Giordano (1971).

A method for preparing pure FAD by cellulose column chromatography is described by Whitby (1953). Better separation of all types of flavins can be obtained by paper electrophoresis (Cerletti and Giordano, 1971) using sodium acetate buffer, pH 5.1, a current of 50 mA and a voltage of 15 V/cm or the procedure described by Yagi (1962). For further characterisation the above mentioned methods should be applied. Another method for the isolation, concentration, and partial separation of flavins is resin chromatography on phenol – formaldehyde type resins followed by paper chromatography (Koziolowa and Koziol, 1968). The great advantage of the method is the fact that, after resin separation, no contamination by other pigments or substances can be found. In addition to these classical methods for separation of flavins, another report (Pietta and Calatroni, 1982) describes the direct separation and quantitative determination of flavins by high performance liquid chromatography (HPLC) on a reversed phase (RP) column. Other authors use RP columns as well as normal columns for the separation and determination of nucleotides and related compounds like the flavin nucleotides (Hartwick *et al.*, 1979a,b; Schwenn and Jender, 1980; Light *et al.*, 1980; Pietta and Calatroni, 1982). The great advantage of this method is the quick and direct way of quantitative determination of flavins, also taking into account the hydrolysis of FAD and FMN (Pietta and Calatroni, 1982).

Studies on Raman spectroscopy, bio- and chemiluminescence and photochemistry of flavins were reviewed by Müller (1981). A very interesting flavin, FX, was isolated by Zenk (1967a,b) from *Avena* coleoptiles. It was later identified by Hertel *et al.* (1980) to be a naturally occurring riboflavin derivative.

B. Isolation and characterisation of holochromes

A general procedure for the isolation and purification of flavin – protein complexes is described by Brain *et al.* (1977) and one with slight variations by Song *et al.* (1980a,b). Depending on the starting material, the cells are homogenised mechanically in an *N*-morpholinopropanesulphonic acid (MOPS) buffer (ph 7.4) containing sucrose, SH reagent, EDTA, and $MgCl_2$. The crude homogenate is filtered through cheesecloth and the filtrate subjected to differential centrifugation. The $21,000 \times g$ and $50,000 \times g$ pellets show identical spectra to those of flavins with maxima at 422, 449, and 481 nm, which is consistent with the data of Briggs (1974) and Galston *et al.* (1977). Fluorescence emission and excitation spectra of these fractions were also attributable to flavins (Song *et al.*, 1980a,b). It could also be shown that the $21,000 \times g$ and $50,000 \times g$ pellets were heterogenous, indicating the presence of more than one flavin. The presence of a *b*-type cytochrome in both fractions and the flavin-mediated reduction of this complex could also be shown (Brain *et al.*, 1977; Song *et al.*, 1980a,b, Munoz and Butler, 1975). Further characterization of these flavoproteins was performed by assaying for several flavoenzymes (as can be done with all enzymatically active flavoproteins) using dichlorophenol–indophenol as electron acceptor. In this case, it was found that NADH-dehydrogenase (MW 175,000) was the major flavoprotein compound of the $21,000 \times g$ and $50,000 \times g$ pellet.

The activity on the SDS-gel was shown by Tetrazolium Blue assay (Song *et al.,* 1980a,b). SDS-gel electrophoresis was performed as described by Fairbanks *et al.* (1971). Using this method, another flavoprotein with a molecular weight of $15,000 \pm 2.000$ could be isolated which is not NADH-dehydrogenase and may possibly be a phototropic-receptor flavoprotein. With SDS-polyacrylamide-gel electrophoresis, the chromophore–protein complex can be isolated only if the chromophore is covalently bound to the protein; this seems to be the case with this holochrome. This correlates with the fact that other chromophores which act as photoreceptors are also covalently linked to the protein (Song, 1977). Data and interpretation of absorption spectra of flavoproteins are reported by Massey and Ganther (1965).

A more direct approach to the photoreceptor identification was made by replacing natural flavins by the flavin analogue roseoflavin after flavin-specific permease treatment of *Phycomyces* (Otto *et al.,* 1981). Although the action spectrum for the phototropic response of the sporangiophore could be extended to 529 nm, its effectiveness was only 0.1% of the natural photoreceptor. Spectroscopic characterisation of roseoflavin (Song *et al.,* 1980b) demonstrated that it is a kinetically less effective blue-light photoreceptor, thus explaining to some extent its low effectiveness in *Phycomyces.*

IV. HAEMOPROTEINS

The extension of some action spectra of blue-light effects into the green region has resulted in speculation about the participation of haemoproteins as photoreceptors (Schneider and Bogorad, 1978; Senger *et al.,* 1980). No direct evidence has been presented for a primary role of haemoproteins in blue-light perception. However, it has become quite clear that haemoproteins, like cytochromes, play an important role in the further processing of absorbed blue-light energy. In most cases of light-induced absorbance changes (LIAC; see Section VIII), the light absorption by flavin results in the reduction of cytochromes causing in turn the absorbance changes. The discussion of haemoproteins as photoreceptors, their role as a primary energy acceptor and their frequent occurrence as flavin–cytochrome complexes warrants a brief description of chromophore isolation and characterisation.

A. Isolation and identification of chromophores

The predominant haemoproteins in plants are cytochromes, peroxidases, and catalases. In cytochromes of the a, b, and d types, the chromophore is noncovalently bound to the protein (Goodwin and Mercer, 1983). In these cases the chromophores are easily cleaved from the protein by acidified organic solvents (Fischer, 1955; Labbe and Nishida, 1957), or by nonacidic solvent systems (York et al., 1967). In the case of c-type cytochromes, or other proteins with covalently bound chromophore, the silver sulphate method of Paul (1950, 1951) can be used; silver nitrate (Morrison and Stotz, 1955) or silver acetate (Margoliash et al., 1959) may be applied instead. Barrett and Kamen (1961) obtained better and more reproducible results by using mercury instead of silver salts.

All known methods, including spectroscopic and other data, are compiled by Lemberg (1961), Smith and Fuhrhop (1975), and DiNello and Chang (1978). For ^{1}H-NMR data, see Caughey et al. (1975); mass spectrometry data are given by Budzikiewicz (1978).

B. Isolation and characterisation of holochromes

The isolation of cytochromes from the electron-transport chains of respiration and photosynthesis is not a topic of this chapter but some general principles are of practical use for related experiments because, in connection with blue-light phenomena, much attention has been paid to the isolation of cytochromes and cytochrome–flavin complexes concerned with primary processes. Isolation methods have been compiled for cytochrome c from higher plants and yeast (Hagihara et al., 1956, 1959; Okunuki, 1966).

A system in which a haemoprotein is involved in a light-induced dark-reversible absorbance change was described by Schneider and Bogorad (1978). Similarities with a reduced peroxidase complex (Yokota and Yamasaki, 1965), and a peroxidase compound isolated by Lundegardh (1958) are discussed as well as similarities to spinach nitrate-reductase cytochrome (Vega and Kamin, 1977). The light sensitivity of CO-haemoproteins has been reported by Yamasaki et al. (1970). A membrane fraction containing b-type cytochrome which could not be attributed to mitochondria, proplastids, or endoplasmic reticulum was isolated from corn coleoptiles by Jesaitis et al. (1977) and Briggs (1974). It was suggested that the cytochrome was involved in energy transduction and that it acted possibly as a blue–UV photoreceptor for phototropism by itself. A similar complex was isolated from Neurospora by Munoz and Butler (1975); the fact that a Cyt-b-deficient mutant of Neurospora is far less light sensitive may support the suggestion that a b-type cytochrome is directly involved in photoreception (Brain et al., 1977; Brain and Briggs, 1977).

Membrane particles isolated from corn coleoptiles (Widell *et al.*, 1980; Britz *et al.*, 1979) contain a *b*-type cytochrome which can be reduced by red light in the presence of methylene blue. It is postulated that in this *in vitro* system, methylene blue replaces the flavin, photoreducing the cytochrome *in vivo*. Isolation of this low-potential b-type cytochrome was further improved by Widell (1980). By treatment with different detergents, a partial solubilisation of this *b*-type cytochrome responsible for the LIAC could be achieved, but it could not be totally separated from mitochondrial cytochromes. Widell and Larsson (1981) described a method for obtaining preparations of presumptive plasma membranes which were rich in blue-light reducible *b*-type cytochrome and had very low contamination with other cytochromes. Membranes were separated by partition between an upper aqueous polyethylene glycol-rich phase and a lower aqueous dextran-rich phase. As much as 80–90% of the cytochromes in the upper phase were active in LIAC, whereas only 10–15% activity was found in the lower phase. Most of the mitochondrial membrane proteins (cytochrome *c* oxidase) were recovered in the lower phase.

V. RHODOPSINS

Rhodopsins are typical blue-light-absorbing pigments (Traulich *et al.*, 1983; Schimz *et al.*, 1982), but unlike flavins, carotenoids, and cytochromes, they are not ubiquitous. Bacteriorhodopsin, which was discovered by Oesterhelt and Stoeckenius in 1971, undergoes a photoreaction cycle in order to translocate protons across the membrane and generate energy (Oesterhelt, 1976). It has been shown that the photoreaction cycle is not essential for the photosensory function of bacteriorhodopsin although it may contribute toward it (Hildebrand and Schimz, 1983; Schimz *et al.*, 1982). In a bacteriorhodopsin-deficient mutant of *Halobacterium halobium* (L-33), increased amounts of halorhodopsin could be found (Lanyi and Weber, 1980; Lanyi, 1982). In this organism, this represents the photosensory pigment as well as the light-energy-converting retinal pigment (Wagner *et al.*, 1980).

Thus apart from their role as an energy-generating system for the proton pumps of Halobacteria, rhodopsins seem to regulate the photosensory behaviour of *Halobacterium* (Hildebrand and Dencher, 1975; Nultsch and Häder, 1978; Dencher and Hildebrand, 1979; Spudich and Stoeckenius, 1979; Wagner *et al.*, 1980; Baryshev *et al.*, 1981). A third rhodopsin, termed slow rhodopsin, was discovered and is regarded as a photoreceptor for phototaxis (Roberto *et al.*, 1982). Nevertheless, because of its nonubiquity, rhodopsin cannot be considered a general blue-light receptor.

VI. ARTIFICIAL PHOTORECEPTORS

Since the pioneering experiments with mycelia of *Fusarium* (Lang-Feulner and Rau, 1975) it has been known that some artificially added pigments can alter action spectra of typical blue-light effects. By application of suitable pigments, the light-perception response can be shifted into regions of the spectrum where no interference with other pigments can be expected. It should be possible to reveal the involvement of screening or light harvesting of other pigments by comparison of the quantum efficiencies of natural and artificial photoreceptors.

Only photodynamically active dyes, such as methylene blue, toluidine blue, or neutral red can act as artificial photoreceptors. Photodynamically inactive dyes such as dichlorophenol – indophenol or malachite green do not mediate any photoreaction (Lang-Feulner and Rau, 1975). All *in vitro* reactions which have been studied have also shown that the photoreactions with artificial acceptor pigments were oxygen dependent. Whereas the light induction of carotenogenesis in *Fusarium* could be replaced by hydrogen peroxide (Theimer and Rau, 1970), this was not possible for LIACs in preparation from corn coleoptiles (Widell *et al.,* 1980).

A more direct approach to identify the photoreceptor is the addition of endogenous flavoproteins or exogenous flavins to cell-free preparations and to determine subsequent changes in the LIACs (Goldsmith *et al.,* 1980; Manabe and Poff, 1978; Munoz and Butler, 1975; Ninnemann *et al.,* 1977; Poff and Butler, 1975; Schmidt *et al.,* 1977). The flavin analogue roseoflavin has also been applied to phycomyces by specific permease treatment after natural flavin had been removed (Otto *et al.,* 1981), but the reaction recovered only 10^{-3} of the natural effect. Techniques and results of artificial membrane – flavin systems have recently been summarised by Schmidt (1981).

VII. ACTION SPECTROSCOPY

The predominant procedure for identification of blue light effects is the determination of action spectra. A general problem in interpreting action spectra is the determination of how much screening or light harvesting effects of other pigments might be involved. The theory and techniques involved in action spectroscopy are described in Chapter 5 of this volume and by Hartmann (1982). Specific problems concerning action spectra for blue light effects have been pointed out by De Fabo (1980).

VIII. LIGHT-INDUCED ABSORBANCE CHANGE (LIAC)

There is no specific assay to identify the blue light receptor(s). However, in many cases the action spectra of dark, reversible light-induced absorbance changes (LIACs) *in vivo* resemble those of typical blue-light effects (Munoz and Butler, 1975). There is ample evidence that most LIACs represent a flavin-mediated reduction of *b*-type cytochromes, but there are LIACs reported *in vivo* and *in vitro* which do not agree with the involvement of *b*-type cytochromes (for literature, see Schmidt, 1981; Senger and Briggs, 1981). One should also bear in mind that dark, reversible LIACs have been reported to occur in carotenoids accompanying changes in potentials across the membranes of some photosynthetic bacteria (Itoh, 1982; Wraight *et al.*, 1978).

LIACs are used as a convenient assay for the nitrate-reductase system (Lipson and Pratap, 1981; Ninnemann, 1982) and as markers for plasma membrane fractions containing the blue-light receptor (Widell and Larsson, 1981). The identity of the photoreceptor of blue-light-mediated phototropism and LIAC was proposed for oat coleoptiles (Leong and Briggs, 1982).

LIACs are measured as absorption differences between irradiated and nonirradiated samples. The main peak is taken from difference spectra and varies in different materials between 423 and 428 nm (Munoz and Butler, 1975; Leong and Briggs, 1982; Widell, 1980; Ninnemann, 1982). The LIAC measurements are generally made with dual-wavelength spectrophotometers which record the absorption differences either immediately after turning off the actinic light (Munoz and Butler, 1975; Goldsmith *et al.*, 1980) or simultaneously by protecting the photomultiplier from the actinic light by certain filter combinations (Widell and Larsson, 1981). The wavelength of the reference beam passing through the nonirradiated sample is usually set at the absorption peaks of mitochondrial cytochromes. Actinic blue light is always provided in saturating intensities of various duration. In some cases, LIAC data have been corrected for irreversible light-induced absorbance changes, measured after a 10-min dark period following the exposure to saturating actinic light (Munoz and Butler, 1975). Units for LIAC photoactivity are proposed by Goldsmith *et al.* (1980). A new method of graphical estimations based on fluence-response data of LIACs allow quantum efficiency determinations (Lipson and Presti, 1980).

REFERENCES

Akoyunoglou, G., Anni, H. and Kalosaka, K. (1980). *In* "The Blue Light Syndrome" (H. Senger, ed.), pp. 473–484. Springer-Verlag, Berlin and New York.
Bara, M. and Galston, A. W. (1968). *Physiol. Plant.* **21**, 109–118.

Barrett, J. and Kamen, M. D. (1961). *Biochim. Biophys. Acta* **50,** 573–575.

Baryshev, U. A., Glagolev, A. N. and Skulachev, V. P. (1981). *Nature (London)* **292,** 338–340.

Bensasson, R. V. (1980). *In* "Photoreception and Sensory Transduction in Aneural Organisms" (F. Lenci and G. Colombetti, eds.), pp. 211–234. Plenum, New York.

Björn, L. O. (1967). Ph.D. Thesis, University of Lund, Sweden.

Björn, L. O. (1980). *In* "The Blue Light Syndrome" (H. Senger, ed.), pp. 455–464. Springer-Verlag, Berlin and New York.

Brain, R. D. and Briggs, W. R. (1977). *Res. Photobiol. [Proc. 7 Int. Congr.] 7th, 1976,* pp. 539–544.

Brain, R. D., Freeberg, J. A., Weiss, C. V. and Briggs, W. R. (1977). *Plant Physiol.* **59,** 948–952.

Braumann, T. and Grimme, L. H. (1981). *Biochim. Biophys. Acta* **637,** 8–17.

Briggs, W. R. (1974). *Year Book—Carnegie Inst. Washington* **74,** 807–809.

Brinkmann, G. and Senger, H. (1980). *In* "The Blue Light Syndrome" (H. Senger, ed.), pp. 526–540. Springer-Verlag, Berlin and New York.

Britton, G. and Goodwin, T. W. (1971). *In* "Methods in Enzymology" (D. B. McCormick and L. D. Wright, eds.), Vol. 18, Part C, pp. 654–701. Academic Press, New York.

Britz, S. J., Schrott, E., Widell, S. and Briggs, W. R. (1979). *Photochem. Photobiol.* **29,** 359–365.

Buchwald, M. and Jencks, W. P. (1968). *Biochemistry* **7,** 834–844.

Budzikiewicz, A. (1978). *In* "The Porphyrins" (D. Dolphin, ed.), Vol. 3, Part A, pp. 395–461. Academic Press, New York.

Bünning, E. (1937). *Planta* **26,** 719–736.

Burch, H. B. (1975). *In* "Methods in Enzymology" (S. P. Colowick and N. O. Kaplan, eds.), Vol. 3, pp. 960–967. Academic Press, New York.

Castle, E. S. (1935). *Cold Spring Harbor Symp. Quant. Biol.* **2,** 224–229.

Caughey, W. S., Smythe, G. A., O'Keeffe, D. H., Maskasky, J. E. and Smith, M. L. (1975). *J. Biol. Chem.* **250,** 7602–7622.

Cerletti, P. and Giordano, M. G. (1971). *In* "Methods in Enzymology (B. McCormick and L. D. Wright, eds.), Vol. 18, Part B, pp, 285–290. Academic Press, New York.

Cheesman, D. F. *et al.* (1966). *Proc. R. Soc. London, Ser. B* **16A,** 130–151.

Cheesman, D. F., Lee, W. L. and Zagalsky, P. F. (1967). *Biol. Rev. Cambridge Philos. Soc.* **42,** 131.

Clauss, H. (1979). *Protoplasma* **99,** 341–346.

Davies, B. H. (1976). *In* "Chemistry and Biochemistry of Plant Pigments" (T. W. Goodwin, ed.), 2nd ed., Vol. 2, pp. 38–165. Academic Press, London.

Davies, B. H., Hallet, C. J., London, R. A. and Rees, A. F. (1974). *Phytochemistry* **13,** 1209–1217.

De Fabo, E. (1980). *In* "The Blue Light Syndrome" (H. Senger, ed.), pp. 187–197. Springer-Verlag, Berlin and New York.

Dencher, N. A. and Hildebrand, E. (1979). *Z. Naturforsch., C: Biosci.* **34C,** 841–847.

DiNello, R. K. and Chang, C. K. (1978). *In* "The Porphyrins" (D. Dolphin, ed.), Vol. 1, Part A, pp. 289–339. Academic Press, New York.

Donkin, M. E. and Martin, E. S. (1981). *Z. Pflanzenphysiol.* **102,** 345–352.

Douce, R. and Joyard, J. (1982). *In* "Methods in Chloroplast Molecular Biology" (M. Edelman, R. B. Hallick, and N.-H. Chua, eds.) pp. 239–256. Elsevier/North-Holland and Biomedical Press, Amsterdam.

Drumm-Herrel, H. and Mohr, H. (1982). *Photochem. Photobiol.* **35,** 233–236.

Dudley, K. H., Ehrenberg, A., Hemmrich, P. and Müller, F. (1964). *Helv. Chim. Acta* **47,** 1354.

Edmunds, L. N. (1980). *In* "The Blue Light Syndrome" (H. Senger, ed.), pp. 584–596. Springer-Verlag, Berlin and New York.

Eger-Hummel, G. (1980). *In* "The Blue Light Syndrome" (H. Senger, ed.), pp. 555–562. Springer-Verlag, Berlin and New York.

Elgsaeter, A., Tauber, D. and Liaaen-Jensen, S. (1978). *Biochim. Biophys. Acta* **530**, 402–411.

Fairbanks, G., Steck, T. L. and Wallach, D. F. H. (1971). *Biochemistry* **10**, 2606–2617.

Fiksdahl, A., Mortensen, J. T. and Liaaen-Jensen, S. (1978). *J. Chromatogr.* **157**, 111–117.

Fischer, H. (1955). *Org. Synth., Collect. Vol.* **3**, 442–443.

Fong, F. and Schiff, J. A. (1978). *Plant Physiol.* **59**, 74–405.

Fujimori, E. (1969). *Biochim. Biophys. Acta* **180**, 360–367.

Furuya, M., Wada, M. and Kadota, A. (1980). *In* "The Blue Light Syndrome" (H. Senger, ed.), pp. 120–132. Springer-Verlag, Berlin and New York.

Galston, A. W. (1949). *Science* **109**, 485–486.

Galston, A. W. (1977). *Photochem. Photobiol.* **25**, 503–504.

Galston, A. W., Britz, S. J. and Briggs, W. R. (1977). *Yearbook—Carnegie Inst. Washington* **76**, 293–295.

Goldsmith, M. H. M., Caubergs, R. J. and Briggs, W. R. (1980). *Plant Physiol.* **66**, 1067–1073.

Goodwin, T. W. and Mercer, E. I. (1983). *In* "Introduction to Plant Biochemistry," 2nd ed., pp. 181–185. Pergamon, Oxford.

Gressel, J. (1979). *Photochem. Photobiol.* **30**, 749–754.

Hager, A. (1970). *Planta* **91**, 38–53.

Hager, A. and Meyer-Bertenrath, T. (1966). *Planta* **69**, 198–217.

Hager, A. and Stransky, H. (1970a). *Arch. Microbiol.* **71**, 132–163.

Hager, A. and Stransky, H. (1970b). *Arch. Mikrobiol.* **71**, 164–190.

Hager, A. and Stransky, H. (1970c). *Arch. Mikrobiol.* **72**, 68–83.

Hager, A. and Stransky, H. (1970d). *Arch. Mikrobiol.* **72**, 84–96.

Hager, A. and Stransky, H. (1970e). *Arch. Mikrobiol.* **73**, 77–89.

Hagibrahim, S. K., Tibbetts, P. I. C., Watts, C. D., Maxwell, J. R., Eglington, G.,Colin, H. and Guiochon, G. (1978). *Anal. Chem.* **50**, 549–553.

Hagihara, B., Horio, T., Yamashita, J., Nozaki, M. and Okunuki, K. (1956). *Nature (London)* **178**, 629–632.

Hagihara, B., Tagawa, K., Morikawa, I., Shin, M. and Okunuki, K. (1959). *J. Biochem. (Tokyo)* **46**, 321.

Hartmann, K. M. (1982). *In* "Biophysik" (W. Hoppe, W. Lohmann, H. Markl and H. Ziegler, eds.), 2nd ed., pp. 22–52. Springer-Verlag, Berlin and New York.

Hartwick, R. A., Assenza, S. P. and Brown, P. R. (1979a). *J.Chromatogr.* **186**, 647–658.

Hartwick, R. A., Krstulovic, A. M. and Brown, P. R. (1979b). *J. Chromatogr.* **186**, 659–676.

Hase, E. (1980). *In* "The Blue Light Syndrome" (H. Senger, ed.), pp. 512–525. Springer-Verlag, Berlin and New York.

Haxo, F. T., Kycia, H. J., Somers, F., Bennett, A. and Siegelmann, H. W. (1976). *Plant Physiol.* **57**, 297–303.

Hertel, R., Jesaitis, A. J., Dohrmann, U. and Briggs, W. R. (1980). *Planta* **147**, 317–319.

Hildebrand, E. (1977). *Biophys. Struct. Mech.* **3**, 69–77.

Hildebrand, E. and Dencher, N. (1975). *Nature (London)* **257**, 48.

Hildebrand, E. and Schimz, A. (1983). *Photochem. Photobiol.* **38**, 553–597.

Itoh, S. (1982). *Plant Cell Physiol.* **23**(4), 595–605.

Izwawa, S., Itoh, M., Ogawa, T. and Shibata, K. (1963). *In* "Studies on Microalgae and Photosynthetic Bacteria" (Jpn. Soc. Plant Physiol., ed.), pp. 413–422. Univ. of Tokyo Press, Tokyo.

Jeffrey, S. W. and Vesk, M. (1981). *In* "Photosynthesis VI—Photosynthesis and Productivity,

Photosynthesis and Environment" (G. Akoyunoglou, ed.), pp. 435–442. Balaban, Philadelphia, Pennsylvania.

Jesaitis, A. J., Heners, P. R., Hertel, R. and Briggs, W. R. (1977). *Plant Physiol.* **59**, 941–947.

Ji, T. H., Hess, J. L. and Benson, A. A. (1968a). *In* "Comparative Biochemistry and Biophysics of Photosynthesis" (K. Shibata, A. Takamiya, A. T. Jagendorf and R. C. Fuller, eds.), p. 36. Univ. of Tokyo Press, Tokyo.

Ji, T. H., Hess, J. L. and Benson, A. A. (1968b). *Biochim. Biophys. Acta* **150**, 676–685.

Kamiya, A., Ikegami, I. and Hase, E. (1981). *Plant Cell Physiol.* **22**, 1385–1396.

Ke, B. (1971). *In* "Methods in Enzymology" (A. San Pietro, ed.), Vol. 23, pp. 624–636. Academic Press, New York.

Ke, B., Green, M., Vernon, L. P. and Garcia, A. F. (1968). *Biochim. Biophys. Acta* **162**, 467–469.

Koka, P. and Song, P.-S. (1977). *Biochim. Biophys. Acta* **495**, 220–231.

Kowallik, W. (1982). *Annu. Rev. Plant Physiol.* **33**, 51–72.

Koziol, J. (1966). *Photochem. Photobiol.* **5**, 41–54.

Koziol, J. (1969). *Photochem. Photobiol.* **9**, 45–53.

Koziol, J. (1971). *In* "Methods in Enzymology" (D. B. McCormick and L. D. Wright, eds.), Vol. 18, Part B, pp. 253–285. Academic Press, New York.

Koziolowa, A. and Koziol, J. (1968). *J. Chromatogr,* **34**, 216–221.

Labbe, R. F. and Nishida, G. (1957). *Biochim. Biophys. Acta* **26**, 437.

Lang-Feulner, J. and Rau, W. (1975). *Photochem. Photobiol.* **21**, 179–183.

Lanyi, J. K. (1982). *In* "Methods in Enzymology" (L. Packer, ed.), Vol. 88, pp. 439–443. Academic Press, New York.

Lanyi, J. K. and Weber, H. J. (1980). *J. Biol. Chem.* **255**, 243–250.

Laudenbach, B. and Pirson, A. (1969). *Arch. Mikrobiol.* **67**, 226–242.

Lemberg, R. (1961). *Adv. Enzymol.* **23**, 265–321.

Lempka, A. and Andrezejewski, H. (1966). *Przem. Spozyw.* **10**, 41.

Leong, T. Y. and Briggs, K. R. (1982). *Plant Physiol.* **70**(3), 875–881.

Lichtenthaler, H. K., Buschmann, C. and Rahmsdorf, U. (1980). *In* "The Blue Light Syndrome" (H. Senger, ed.), pp. 485–494. Springer-Verlag, Berlin and New York.

Light, D. R., Walsh, C. and Marletta, M. A. (1980). *Anal. Biochem.* **109**, 87–93.

Lipson, E. D. (1980). *In* "The Blue Light Syndrome" (H. Senger, ed.), pp. 110–118. Springer-Verlag, Berlin and New York.

Lipson, E. D. and Pratap, P. R. (1981). *9th Annu. Meet. Photobiol. Am. Soc. Photobiol. Abstr.* MAM-C9. *Suppl. Photochem. Photobiol,* p. 66.

Lipson, E. D. and Presti, D. (1980). *Photochem. Photobiol.* **32**, 383–391.

Lundegardh, H. (1958). *Nature (London)* **181**, 28–30.

McCormick, D. B., Li, H.-C. and MacKenzie, R. E. (1967). *Spectrochim. Acta, Part A* **23**, 2353–2358.

Manabe, K. and Poff, K. L. (1978). *Plant Physiol.* **61**, 961–966.

Margoliash, E., Frohwirt, N. and Wiener, E. (1959). *Biochem. J.* **71**, 559–570.

Massey, V. and Ganther, H. (1965). *Biochemistry* **4**, 1161–1173.

Menke, W. (1940). *Naturwissenschaften* **28**, 31.

Morrison, M. and Stotz, E. (1955). *J. Biol. Chem.* **213**, 373–378.

Müller, F. (1981). *Photochem. Photobiol.* **34**, 753–759.

Munoz, V. and Butler, W. L. (1975). *Plant Physiol.* **55**, 421–426.

Ninnemann, H. (1982). *Photochem. Photobiol.* **35**, 391–398.

Ninnemann, H., Strasser, R. J. and Butler, W. L. (1977). *Photochem. Photobiol.* **26**, 41–47.

Nishimura, M. and Takamatsu, K. (1957). *Nature (London)* **180**, 699–700.

Nishimura, M. and Takamatsu, K. (1960). *Plant Cell Physiol.* **1**, 305–309.

Nultsch, W. (1980). *In* "The Blue Light Syndrome" (H. Senger, ed.), pp. 38–79. Springer-Verlag, Berlin and New York.

Nultsch, W. and Häder, M. (1978). *Ber. Dtsch. Bot. Ges.* **91**, 441–453.

Oesterhelt, D. (1976). *Angew. Chem., Int. Ed. Engl.* **15**, 17–24.

Oesterhelt, D. and Stoeckenius, W. (1971). *Nature (London), New Biol.* **233**, 149–152.

Ogawa, T. (1981). *Plant Sci. Lett.* **22**, 103–108.

Okunuki, K. (1966). *Compr. Biochem.* **14**, 232–308.

Otto, M. K., Jayaram, M., Hamilton, R. M. and Delbrück, M. (1981). *Proc. Natl. Acad. Sci. U.S.A.* **78**, 266–269.

Paul, K. G. (1950). *Acta Chem. Scand.* **4**, 239–244.

Paul, K. G. (1951). *Acta Chem. Scand.* **5**, 389–405.

Pietta, P. and Calatroni, A. (1982). *J. Chromatogr.* **229**, 445–449.

Poff, K. L. and Butler, W. R. (1975). *Plant Physiol.* **55**, 427–429.

Presti, D. and Delbrück, M. (1978). *Plant, Cell Environ.* **1**, 81–100.

Presti, D., Hsu, W.-J. and Delbrück, M. (1977). *Photochem. Photobiol.* **26**, 403–405.

Quarmby, R., Norden, D. A., Zagalicky, P. F., Ceccaldi, H. J. and Daumas, R. (1977). *Comp. Biochem. Physiol. B* **56B**, 55–61.

Rau, W. (1976). *Pure Appl. Chem.* **47**, 237–243.

Rau, W. (1980). *In* "The Blue Light Syndrome" (H. Senger, ed.), pp. 283–298. Springer-Verlag, Berlin and New York.

Richter, G., Reihl, W., Wietoska, B. and Beckmann, J. (1980). *In* "The Blue Light Syndrome" (H. Senger, ed.), pp. 465–472. Springer-Verlag, Berlin and New York.

Roberto, A., Bogomolni, J. L. and Spudich, J. L. (1982). *Proc. Natl. Acad. Sci. U.S.A.* **79**, 6250–6254.

Roth, J. A. and McCormick, D. B. (1967). *Photochem. Photobiol.* **6**, 657–664.

Ruyters, G. (1982). *Photochem. Photobiol.* **35**, 229–231.

Schiff, J. A. (1980). *In* "The Blue Light Syndrome" (H. Senger, ed.), pp. 495–511. Springer-Verlag, Berlin and New York.

Schimz, A., Sperling, W., Hildebrand, E. and Köhler-Hahn, D. (1982). *Photochem. Photobiol.* **36**, 193–196.

Schmidt, W. (1980). *In* "The Blue Light Syndrome" (H. Senger, ed.), pp. 212–220. Springer-Verlag, Berlin and New York.

Schmidt, W. (1981). *Symp. Soc. Exp. Biol.* **36**, 299–324.

Schmidt, W., Thompson, K. S. and Butler, W. L. (1977). *Photochem. Photobiol.* **26**, 407–411.

Schneider, Hj. A. W. and Bogorad, L. (1978). *Plant Physiol.* **62**, 577–581.

Schwenn, J. D. and Jender, H. G. (1980). *J. Chromatogr.* **193**, 285–290.

Senger, H. (1981). *Photosynth., Proc, Int. Congr. 5th, 1980* Vol. V, pp. 433–445.

Senger, H. (1982). *Photochem. Photobiol.* **35**, 911–920.

Senger, H. and Briggs, W. R. (1981). *Photochem. Photobiol. Rev.* **6**, 1–38.

Senger, H., Klein, O., Dörnemann, D. and Porra, R. J. (1980). *In* "The Blue Light Syndrome" (H. Senger, ed.), pp. 541–542. Springer-Verlag, Berlin and New York.

Sharkey, T. D. and Raschke, K. (1981). *Plant Physiol.* **68**, 1170–1174.

Shropshire, W., Jr. (1980). *In* "The Blue Light Syndrome" (H. Senger, ed.), pp. 172–186. Springer-Verlag, Berlin and New York.

Siegelmann, H. W. (1981). *In* "Photosynthesis III, Structure and Molecular Organisation of the Photosynthetic—Membrane" (G. Akoyunoglou, ed.), pp. 297–299. Balaban Int. Sci. Serv., Philadelphia, Pennsylvania.

Siegelmann, H. W. and Kycia, J. H. (1982). *Plant Physiol.* **70**, 887–897.

Smith, K. M. and Fuhrhop, J.-H. (1975). *In* "Porphyrins and Metalloporphyrins" (K. M. Smith, ed.), pp. 757–869. Elsevier, Amsterdam.

Song, P.-S. (1977). "C. A. Lee 60th Birthday Commemoration Issue," No. 20, pp. 10-25. J. Agric. Chem. Soc., Korea.

Song, P.-S. (1980a). In "Photoreception and Sensory Transduction in Aneural Organisms" (F. Lenci and G. Colombetti, eds.), pp. 189-208. Plenum, New York.

Song, P.-S. (1980b). In "The Blue Light Syndrome" (H. Senger, ed.), pp. 157-171. Springer-Verlag, Berlin and New York.

Song, P.-S., Koka, P., Prezelin, B. B. and Haxo, F. T. (1976). Biochemistry 15, 4422-4427.

Song, P.-S., Fugate, R. D. and Briggs, W. R. (1980a). Flavins Flavoproteins, Proc. Int. Symp., 6th, 1979 pp. 443-453.

Song, P.-S., Walker, E. B., Vierstra, R. D. and Poff, K. L. (1980b). Photochem. Photobiol. 32, 393-398.

Spudich, J. L. and Stoeckenius, W. (1979). Photobiochem. Photobiophys. 1, 43-53.

Stahl, E. (1967). "Dünnschichtchromatographie." Springer-Verlag, Berlin and New York.

Stewart, J. and Leuenberger, U. (1976). Alimenta 15, 33-36.

Taylor, R. F. and Davies, B. H. (1974). Biochem. J. 139, 751-760.

Taylor, R. F. and Davies, B. H. (1975). J. Chromatogr. 103, 327-340.

Theimer, R. R. and Rau, W. (1970). Biochim. Biophys. Acta 177, 180-181.

Traulich, B., Hildebrand, E., Schinz, A., Wagner, G. and Langi, J. K. (1983). Photochem. Photobiol. 37, 577-579.

Tsuchiya, T. and Ishiguri, Y. (1981). Plant Cell Physiol. 22(3), 525-532.

Vega, J. M. and Kamin, H. (1977). J. Biol. Chem. 252, 896-909.

Vernon, L. P. and Garcia, A. F. (1967). Biochim. Biophys. Acta 143, 144-153.

Vernon, L. P., Ke, B., Mollenhauer, H. H. and Shaw, E. R. (1969). Prog. Photosynth. Res. 1, 137.

Wagner, G., Geissler, G., Linhardt, R., Mollwo, A. and Vonhof, A. (1980). In "Plant Membrane Transport: Current Conceptual Issues" (R. M. Spanswick, W. J. Lucas, and J. Dainty, eds.), pp. 641-644. Elsevier, Amsterdam.

Weber, G. (1966). In "Flavins and Flavoproteins" (E. C. Slater, ed.), pp. 15-21. Elsevier, Amsterdam.

Whitby, L. G. (1953). Biochem. J. 54, 437-442.

Widell, S. (1980). Physiol. Plant. 48, 353-360.

Widell, S. and Larsson, C. (1981). Physiol. Plant. 51, 368-374.

Widell, S., Britz, S. J. and Briggs, W. R. (1980). Photochem. Photobiol. 32, 669-677.

Wraight, C. A., Cogdell, R. J. and Chance, B. (1978). In "The Photosynthetic Bacteria," (R. K. Clayton and W. R. Sistrom, eds.), pp. 471-512. Plenum, New York.

Yagi, K. (1962). Methods Biochem. Anal. 10, 319-356.

Yagi, K. (1971). In "Methods in Enzymology" (D. B. McCormick and L. D. Wright, eds.), Vol. 18, Part B, pp. 290-296. Academic Press, New York.

Yamasaki, H., Ohishi, S. and Yamasaki, I. (1970). Arch. Biochem. Biophys. 136, 41-46.

Yokota, K. and Yamasaki, I. (1965). Arch. Biochem. Biophys. 105, 301-312.

York, J. L., McCoy, S., Taylor, D. N., and Caughey, W. S. (1967). J. Biol. Chem. 242, 908-911.

Zagalsky, P. F. (1976). Pure Appl. Chem. 47, 103-120.

Zechmeister, L. (1962). "Cis-trans Isomeric Carotenoids, Vitamins A and Arylopolyenes." Springer-Verlag, Berlin and New York.

Zieger, E., Field, C. and Mooney, H. A. (1981). In "Plants and the Daylight Spectrum" (H. Smith, ed.), pp. 391-407. Academic Press, New York.

Zenk, M. H. (1967a). Z. Pflanzenphysiol. 56, 57-69.

Zenk, M. H. (1967b). Z. Pflanzenphysiol. 56, 122-140.

Zurzycki, J. (1980). In "The Blue Light Syndrome" (H. Senger, ed.), pp. 50-68. Springer-Verlag, Berlin and New York.

Appendix: Useful Addresses

Fibre Optics

Dolan-Jenner Industries, Inc., P.O. Box 1020, Blueberry Hill Industrial Park, Woburn, MA 01801, USA. 617/935-7544.

Ealing Beck Ltd., Greycaine Rd., Watford, Hertfordshire WD2 4PW, England. Telex: 93-5726.

Ealing Corp., 22 Pleasant St., South Natick, MA 01760, USA. 617/655-7000.

Fibronics Ltd., M.T.M. Industrial Park, Haifa 31905, Israel. Telex: 46-744.

Focom Systems Ltd., Millshaw Industrial Estate, Leeds 11, England. Telex: 55186 FOCUM G.

Fort, 16 Rue Bertin Poiree, 75001 Paris, France. Telex: FORT 24-0316 F.

Hytran Products, Glascoed Rd., St. Asaph, Clwyd LL17 OLL, Wales. Telex: 61291.

KDK Fiberoptics Corp., 10 Bunker Hill Pkwy., West Boylston, MA 01583, USA. 617/835-3200.

Filters

Balzers AG, FL-9496 Balzers, Liechtenstein.

Corion Corp., 73 Jeffrey Ave., Holliston, MA 01746, USA. 617/429-5065.

Corning Glass Works, Mail Station 5124, Corning, N.Y. 14831, USA. 607/974-9000.

Ealing Corp., 22 Pleasant St., South Natick, MA 01760, USA. 617/655-7000.

Eastman Kodak Co., 343 State St., Rochester, N.Y. 14650, USA.

Kligle Brothers (Roscolene), 3232-48th Ave., Long Island City, New York 11101, USA.

Oriel Corp., 15 Market St., P.O. Box 1395, Stamford, CT 06904, USA. 203/357-1600.

Röhm GmbH Chemische Fabrik, Postfach 4166, Kirschenallee, D-6100 Darmstadt, West Germany. 6151-8061.

Schott Glaswerke, Hattenbergstr. 10, Postfach 2480, D-6500 Mainz, West Germany.

Maxlight Fiber Optics, 3035 North 33rd Dr., Phoenix, AZ 85017, USA. 602/269-8387.

Optronics Ltd., Cambridge Science Park, Milton Rd., Cambridge CB4 4BH, England. 223-64364.

Oriel Corp., 15 Market St., P.O. Box 1395, Stamford, CT 06904, USA. 203/357-1600.

Sumita Optical Glass Mfg. Co., Ltd., 3-15-10, Uchikanda, Chiyoda-Ku, Tokyo 101, Japan. 3-252-8261.

Volpi AG, Bernstrasse 129, Postfach, CH-8902 Urdorf, Switzerland. Telex: 56591.

Light Detectors

Bentham Instruments Ltd., 2 Boulton Rd., Reading, Berkshire RG2 ONH, England. Telex: 84-8686.

Biospherical Instruments, Inc. 4901 Morena Blvd., Ste. 1003, San Diego, CA 92117, USA. 714/270-1315.

Cathodeon Ltd., Nuffield Rd., Cambridge CB4 1TF, England. Telex: 81-685.

EG & G, Inc., 35 Congress St., Salem, MA 01970, USA. 617/745-3200.

EG & G Gamma Scientific, Inc., 3777 Ruffin Rd., San Diego, CA 92123, USA. 714/279-8034.

Eppley Laboratory, Inc., 12 Sheffield Ave., P.O. Box 419, Newport, RI 12840, USA. 401/847-1020.

Glen Creston Instruments Ltd., 16 Dalston Gardens, Stanmore, Middlesex HA7 1DA, England. Telex: 92-5791.

Hilger-Watts Ltd., 98 St. Pancras Way, Camden Rd., London NW1, England.

Kipp & Zonen Vertriebs GmbH, Wiesenau 5, 6242 Kronberg/Taunus, West Germany.

Li-Cor, Inc., 4421 Superior St., P.O. Box 4425, Lincoln, NE 68504, USA. 402/467-3576.

Macam Photometrics Ltd., 10 Kelvin Square, Livingston EH54 5DG, Scotland. 506-37391.

Optronic Laboratories, Inc., 730 Central Florida Pkwy., Orlando, FL 32809, USA. 305/857-9000.

Oriel Corp., 15 Market St., P.O. Box 1395, Stamford, CT 06904, USA. 203/357-1600.

Tektronix, Inc., P.O. Box 1700, Beaverton, OR 97075, USA. 503/627-7111.

Light Sources

Canrad-Hanovia, Inc., 100 Chestnut St., Newark, N.J. 07105, USA. 201/589-4300.

Cathodeon Ltd., Nuffield Rd., Cambridge CB4 1TF, England. Telex: 81-685.

Ealing Beck Ltd., Greycaine Rd., Watford, Hertfordshire WD2 4PW, England. Telex: 93-5726.

Ealing Corp., 22 Pleasant St., South Natick, MA 01760, USA. 617/655-7000.

EG & G, Inc. Electro-Optics Div., 35 Congress St., Salem, MA 01970, USA. 617/745-3200.

GTE Sylvania, 100 Endicott St., Danvers, MA 01923, USA. 617/777-1900.

ICL Technology, Inc., 399 Java Dr., Sunnyvale, CA 94086, USA. 408/745-7900.

Kratos Analytical Instruments, 170 Williams Dr., Ramsey, N.J. 07446, USA. 201/934-9000.

Leitz Wetzlar GmbH, Postfach 2020, 6330 Wetzlar, W. Germany.

Oriel Corp., 15 Market St., P.O. Box 1395, Stamford, CT 06904, USA. 203/357-1600.

Osram GmbH, Hellabrunnerstr. 1, D-8000 München, W. Germany.

N. V. Philips Gloeilampenfabrieken, Eindhoven, Netherlands. Telex: 51-121 PHTC NL.

Rofin Ltd., Winslade House, Egham Hill, Egham, Surrey TW20 OAZ, England. Telex: 93-4534.

Volpi AG, Bernstr. 129, Postfach, CH-8902, Urdorf, Switzerland. Telex: 56591 volpi ch.

Monochromators

Acton Research Corp., P.O. Box 215-OD, Acton, MA 01720, USA. 617/263-3584.

Anasearch Research Laboratories Ltd., Pearl House, Bartholomew St., Newbury, Berkshire RG14 5LL, England. Telex: 84-9266.

Bentham Instruments Ltd., 2 Boulton Rd., Reading, Berkshire RG2 ONH, England. Telex: 84-8686.

Edinburgh Instruments Ltd., Riccarton Currie, Edinburgh EH14 4AP, Scotland. Telex: 72-553 Edinst. G.

EG & G Gamma Scientific, Inc., 3777 Ruffin Rd., San Diego, CA 92123, USA. 714/279-8034.

Instruments SA, Inc., 173 Essex Ave., Metuchen, NJ 08840, USA. 201/494-8660.

Jarrell-Ash, 590 Lincoln St., Waltham, MA 02254, USA. 617/890-4300.

Kratos Analytical Instruments, 170 Williams Dr., Ramsey, NJ 07446, USA. 201/934-9000.

Macam Photometrics Ltd., 10 Kelvin Square, Livingston EH54 5DG, Scotland. 506-37391.

Oriel Corp., 15 Market St., P.O. Box 1395, Stamford, CT 06904, USA. 203/357-1600.

PTR Optics Corp., 145 Newton St., Waltham, MA 02154, USA. 617/891-6000.

Schoeffel/McPherson Instruments, 530 Main St., Acton, MA 01720, USA. 617/263-7733.

Optical Accessories

Bentham Instruments Ltd., 2 Boulton Rd., Reading, Berkshire RG2 ONH, England. Telex: 84-8686.

Corion Corp., 73 Jeffrey Ave., Holliston, MA 01746, USA. 617/429-5065.

Cryophysics GmbH, Butzbacher Str. 6, D-6100 Darmstadt, West Germany.

Ealing Corp., 22 Pleasant St., South Natick, MA 01760, USA. 617/655-7000.

Labsphere, North Rd., P.O. Box 70, North Sutton, NH 03260, USA. 603/927-4266.

Karl Lambrecht Corp., 4204 North Lincoln Ave., Chicago, IL 60618, USA. 312/472-5442.

Li-Cor, Inc., 4421 Superior St., P.O. Box 4425, Lincoln, NE 68504, USA. 402/467-3576.

Macam Photometrics Ltd., 10 Kelvin Square, Livingston EH54 5DG, Scotland. 506-37391.

Melles Griot BV, Edisonstraat 98, Postbus 272, 6900 AG Zevenaar, Netherlands. Telex: 45-940.

Oriel Corp., 15 Market St., P.O. Box 1395, Stamford, CT 06904, USA. 203/357-1600.

PAR GmbH, Waldstr. 2, D-8034 Unterpfaffenhofen, West Germany.

P.T.I. Co. Ltd., Coombe Rd., Hill Brow, L1SS, Hants, England. Telex: 86-172 ACEHB.

United Detector Technology, 3939 Landmark St., Culver City, CA 90230, USA. 213/204-2250.

Photomultipliers

Centronic, Inc., 1101 Bristol Rd., Mountainside, NJ 07092, USA. 201/233-7200.

Centronic, Ltd., King Henry's Dr., Croydon CR9 OBG, England. Telex: 89-6474 Centro G.

Hamamatsu Photonics K.K., 1126 Ichino-cho, Hamamatsu City, Japan. 534/34-3311.

Oriel Corp., 15 Market St., P.O. Box 1395, Stamford, CT 06904, USA. 203/357-1600.

RCA Corp., New Holland Ave., Lancaster, PA 17604. 717/397-7661.

Thorn EMI Electron Tubes Ltd., Bury St., Ruislip, Middlesex HA4 7TA, England. Telex: 93-5261.

Phytochrome Immunology and Purification

Antibodies Incorporated (antibodies, antisera)
P.O. Box 442
Davis, California 95616, USA
916-758-4400

Bellco Glass, Inc. (glassware, tissue culture supplies)
P.O. Box B
340 Edrudo Road
Vineland, New Jersey 08360, USA
609-691-1075

Bio-Rad Laboratories (chromatography, electrophoresis, immunochemistry)
2200 Wright Avenue
Richmond, California 94804, USA
415-234-4130

Cappel Laboratories, Inc. (antibodies, antisera)
Thud Ridge Farm
Cochranville, Pennsylvania 19330, USA
215-593-6914

DIFCO Laboratories (tissue culture reagents, supplies for immunology)
P.O. Box 1058A
Detroit, Michigan 48232, USA
313-961-0800

Flow Laboratories (sera, medium for monoclonal antibodies)
7655 Old Springhouse Road
McLean, Virginia 22102, USA
703-893-5925

Gelman Sciences, Inc. (filters, petri dishes, etc.)
600 South Wagner Road
Ann Arbor, Michigan 48106, USA
313-665-6511

Grand Island Biological Company (GIBCO) (sera, culture medium)
3175 Staley Road
Grand Island, New York 14072, USA
716-773-0700

Miles Laboratories, Inc. (biochemicals, immunochemicals)
Research Products Division
P.O. Box 2000
Elkhart, Indiana 46515, USA
219-264-8804

TAGO, Inc. (antibodies, antisera)
Immunodiagnostic Reagents
P.O. Box 4463
One Edwards Court
Burlingame, California 94010, USA
415-342-8991

Spectrophotometers

American Instrument Co. (Aminco), 8030 Georgia Ave., Silver Spring, MD 20910, USA.
301/589-1727.

Beckman Instruments, Inc., P.O. Box C-19600, Irvine, CA 92713, USA.

Perkin-Elmer Corp., Main Avenue, Norwalk, CT 06856, USA.

Shimadzu Scientific Instruments, 9147H Red Branch Rd., Columbia, MD 21045, USA.

Varian Associates (Cary), 611 Hansen Way, Palo Alto, CA 94303, USA.   415/493-4000.

Index

A

Absorbed energy fluence, 83
Absorbed photon fluence, 83
Absorption spectra, 20, 120, 122, 133, 143, 279, 280, 281
Action spectroscopy, 3, 6, 109, 110, 119, 124, 125, 127, 290
  continuous irradiation conditions, 121–127
  induction conditions, 115–121
  optical problems, 110–115
Action spectrum, 14, 15, 18, 20, 29, 30, 33, 34, 63, 114, 120, 121, 122, 126, 127, 279, 280, 281, 287, 290
Avogadro's number, 82, 168
Azimuth angle, 88, 105

B

Bandpass, 102
Bandpass filters, 57
Bandwidth, see half-power bandwidth
Beam splitters, 74, 75
Beer's law, see Beer–Lambert–Bouguer law
Beer–Lambert–Bouguer law, 142, 148
Blocking range, 55
Blue light photoreceptor
  absorption spectra, 279, 280, 281
  action spectra, 279, 280, 281, 287, 290
  action spectroscopy, 290
  artificial photoreceptors, 290
  carotenoids, 280, 281, 282, 289, 291
  carotenoproteins, 280–284
  chlorophyll, 279
  chromophore isolation and identification, 281, 282, 284–286, 288
  flavoproteins, 284–287, 290
  haemoproteins, 280, 287–289
  holochrome isolation and identification, 282, 284, 286, 288–289
  light-induced absorbance changes, 290, 291
  photochromicity, 280
  photosynthesis, 280, 288
  phototropism, 279, 287
  rhodopsins, 280, 289
Blue light sources, 61, 62, 63
Blue/UV-A photoreceptor, criteria for involvement, 13, 34–39, 109
Blue/UV-A photoreceptor and phytochrome
  concomitant action, 34–36
  facultative coaction, 36–39
  obligatory coaction, 36–37, 39
Breakdowns and their prevention, 69
Bunsen–Roscoe reciprocity law, 110

C

Calibration, 87–90, 101, 149–150
Canopy-cover, 5
Canopy-shade, 6, 161
Carotenoids, 3, 19, 109, 280–282, 289, 291
Central wavelength, 54
Chloronemata, 35
Chlorophyll, 3, 17, 23, 113, 125, 126, 127, 131, 159, 160, 279
Chloroplast, 25, 26, 113, 131
Chloroplast movement, 25
Cinemoid filters, 55
Cold mirror, 71, 72
Continuous irradiation, 121, 125, 127
Continuous light, 29, 30, 31, 32, 33, 124, 125
Controlled environment chamber, 10, 64–70
Conversion factors for light units, 86
Cosine correction, 83, 84, 90–93
Criteria for photoreceptor involvement, 13–42
Cross-section
  absolute, 109
  absorption, 110
  apparent, 109
  conversion, 109, 119
  excitation, see Conversion cross-section
Cryptochrome, see Blue/UV-A photoreceptor

Phonon, 83
Photoacoustic spectroscopy, 156
Photochromic, 7, 35, 110, 113, 118, 120, 124, 133, 138, 139, 280
Photoconversion cross-section, 114, *see also* Phytochrome photoconversion cross-section
Photographic flash guns, 76
Photogravitropism, 110
Photomodulation, 34
Photomorphogenesis, 1–10
Photomultiplier, 98–99, 103, 105, 132, 134
Photon, 83
  flow rate, 82, 83
  fluence, 82, 83
  fluence rate, 83, 84
  flux, 83, 84
Photons absorbed, 83
Photons applied, 83, 84
Photoperiodism, 1, 4, 7, 24
Photoresponse
  induction, 21
  reversion, 21
Photosynthesis, 40, 41, 127, 280, 288
Photosynthetically active radiation, 65, 67, 68, 69, 127
Phototropism, 1, 34, 63, 118, 279, 287
Phytochrome
  criteria for involvement, 20–34
  cycling rate, 27, 28, 160
  dark reactions, 116, 125, 159
  dark reversion, 110, 114
  destruction, 16, 114, 123
  extraction, *see* Phytochrome purification
  immunochemical characteristics, 7, 115
  immunochemistry, *see* Phytochrome immunochemistry
  induction response, 20, 21, 110
  isoelectric point, 7
  light reactions, 159, 160, 168, 169, 170, 171, 172
  membrane association, *see* Phytochrome in membranes
  model compounds, *see* Phytochrome chromophore model compounds
  monochromatic light, 163
  photoconversion, 115, 116, 118, 123, 125, 155, 160
  photoconversion cross-section, 116, 117, 119, 155, 162, 169, 170
  photoconversion rate, 167, 168, 170
  photoequilibrium, 16, 17, 18, 19, 36, 160, 161, 162, 163, 169, 172
  photoreversibility, 3, 21, 115, 127
  photosteady state, 29, 162
  phototransformation, 16, 17, 166
  physiologically active, 16
  polychromatic light, 164
  purification, *see* Phytochrome purification
  rate constants, 172
  red:far-red ratio, 6, 164, 165, 166, 167
  relative quantum efficiency, 162
  steady state, 17
  structure and properties, 7, 227–254
  synthesis, 16, 115, 123
  transformation kinetics, 7
Phytochrome chromophore model compounds
  chromophore cleavage reactions, 246, 247
  chromophore degradation reactions, 247–254
  general considerations, 228
  nomenclature, 228, 229
  phycobiliproteins and chromopeptides, 241–246
  semisynthetic bile pigments, 234–241
  synthetic bile pigments, 229–234
  usefulness, 227, 228
Phytochrome immunochemistry
  antibody production, 213
  antibody specificity, 214–217
  cell culture, 208, 209
  ELISA, 204, 205, 211, 218–220, 224
  immunoblotting, 222–224
  immunochemical applications, 217–225
  immunocytochemistry, 220–221
  immunoelectrophoresis, 218
  immunoglobulin purification, 204–205, 213–214
  injection, 203, 204, 208
  monoclonal antibody preparation, 205–214
  Ouchterlony double immunodiffusion, 217, 218
  phytochrome preparation, 202–203
  phytochrome quantitation, 218–220
  polyclonal antibody preparation, 203–205
  screening, 211–213
  serum preparation, 203–204
Phytochrome in membranes
  cell choice, 261
  cell preparation, 262–263
  cellular level, 261–265
  desorbing treatment, 260